Functional Mode

Studies in Cybernetics

A series of texts and monographs covering all aspects of cybernetics

Edited by F.H. George, Brunel University, UK

Volume 1
INTRODUCTORY READINGS IN EXPERT SYSTEMS
Edited by D. Michie

Volume 2
A CYBERNETIC APPROACH TO COLOUR PERCEPTION
N.C. Paritsis and D.J. Stewart

Volume 3
MANAGEMENT PRINCIPLES AND PRACTICE
A Cybernetic Analysis
R.H.D. Strank

Volume 4
THE CYBERNETICS OF ART
Reason and the Rainbow
M.J. Rosenberg

Volume 5
CREATIVITY AS AN EXACT SCIENCE
The Theory of the Solution of Inventive Problems
G.S. Altshuller

Volume 6
GOAL-DIRECTED BEHAVIOUR
M. Weir

Volume 7
THE SCIENCE OF HISTORY
A Cybernetic Approach
W.J. Chandler

Volume 8
PURPOSIVE BEHAVIOUR AND TELEOLOGICAL EXPLANATIONS
Edited by F.H. George and L. Johnson

Volume 9
ARTIFICIAL INTELLIGENCE
Its Philosophy and Neural Context
F.H. George

Volume 10
MAN, DECISIONS, SOCIETY
The Theory of Actor-System Dynamics for Social Scientists
T.R. Burns, T. Baumgartner and P. DeVille

Volume 11
THE SHAPING OF SOCIO-ECONOMIC SYSTEMS
The Application of the Theory of Actor-System Dynamics to Conflict, Social Power, and Institutional Innovation in Economic Life
T. Baumgartner, T.R. Burns and P. DeVille

Volume 12
SUBJECTIVITY, INFORMATION, SYSTEMS
Introduction to a Theory of Relativistic Cybernetics
G.M. Jumarie

Volume 13
ART IN THE SCIENCE DOMINATED WORLD
Science, Logic and Art
E.L. Feinberg

Volume 14
APPLIED INFORMATION THEORY
I.M. Kogan

Volume 15
INTRODUCTION TO MULTI-PLAYER DIFFERENTIAL GAMES AND THEIR APPLICATIONS
E.M. Vaisbord and V.I. Zhukovskii

Volume 16
SYSTEMOLOGY AND LINGUISTIC ASPECTS OF CYBERNETICS
G.P. Melnikov

Volume 17
STRATEGY AND UNCERTAINTY
A Guide to Practical Systems Thinking
A. Clementson

Volume 18
SELF-ORGANIZING SYSTEMS
A.M. Andrew

Volume 19
THE PARADIGM OF SELF-ORGANIZATION
Current Trends in Self-Organization
G.J. Dalenoort

Volume 20
METHODS OF RECOGNITION
A.L. Gorelik and V.A. Skripkin

Volume 21
FUNCTIONAL MODELING OF SYSTEMS
E.N. Baylin

Other volume in preparation

Volume 22
SELF-STEERING AND COGNITION IN COMPLEX SYSTEMS
Toward a New Cybernetics
F. Heylighen, E. Rosseel and F. Demeyere

This book is part of a series. The publisher will accept continuation orders which may be cancelled at any time and which provide for automatic billing and shipping of each title in the series upon publication. Please write for details.

Functional Modeling of Systems

Edward N. Baylin

STM Systems Corp., Ottawa, Canada

GORDON AND BREACH SCIENCE PUBLISHERS

New York • Philadelphia • London • Paris • Montreux • Tokyo • Melbourne

Gordon and Breach Science Publishers

Post Office Box 786
Cooper Station
New York, New York 10276
United States of America

5301 Tacony Street, Slot 330
Philadelphia, Pennsylvania 19137
United States of America

Post Office Box 197
London WC2E 9PX
United Kingdom

58, rue Lhomond
75005 Paris
France

Post Office Box 161
1820 Montreux 2
Switzerland

3-14-9, Okubo
Shinjuku-ku, Tokyo 169
Japan

Private Bag 8
Camberwell, Victoria 3124
Australia

Library of Congress Cataloging-in-Publication Data

Baylin, Edward N., 1945-
Functional modeling of systems / Edward N. Baylin.
p. cm. -- (Studies in cybernetics ; v. 21)
Includes bibliographical references.
ISBN 2-88124-731-8
1. System design. 2. System analysis. I. Title. II. Series.
QA76.9.S88B38 1989
004.2'1--dc20 89-23672
CIP

 Printed in the United Kingdom by Billings & Sons Ltd.

Contents

Introduction to the Series

The subject of cybernetics is quickly growing and there now exists a vast amount of information on all aspects of this broad-based set of disciplines. The phrase "set of disciplines" is intended to imply that cybernetics and all the approaches to artificial (or machine) intelligence have a near identical viewpoint. Furthermore, systems analysis, systems theory and operational research often have a great deal in common with (and are in fact not always discernibly different from) what is meant by cybernetics, as far as this series is concerned: inevitably, computer science is bound to be involved also.

The fields of application are virtually unlimited and applications are discovered in the investigation or modelling of any complex system. The most obvious applications have been in the construction of artificially intelligent systems, the brain and nervous system, and socio-economic systems. This can be achieved through either simulation (copying as exactly as possible) or synthesis (achieving the same or better end result by any means whatsoever).

The range of applications today has become so broad it now includes such subjects as aesthetics, history and architecture. Modelling can be carried out by computer programs, special purpose models (analog, mathematical, statistical, etc.), and automata of various kinds, including neural nets and TOTES. All that is required of the system to be studied is that it be complex, dynamic and capable of "learning", and also have feedback, feedforward, or both.

This is an international series.

FRANK GEORGE

Dedicated with love to the late Samuel Baylin, inventor and my loving father, whose creative and sensitive spirit has been the inspiration of all my research.

Preface

WHERE THIS BOOK FITS

The analysis and design of different types of systems has been the subject of many books. Typically, with regard to computerized information systems, the following areas have been covered: system components, steps in system development, structuring of the development process, diagramming methods, types of applications, and cases. Distinctions are made between a system's:

- **functions** (operations/procedures/actions). Functions are represented using various forms of procedural diagrams. Also, the workings of *control* functions are often shown by so-called "state-transition" diagrams. The latter have come to be known as charts for showing "time-dependencies," *rather than* "functions," although they do directly and accurately mirror the workings of *functions.*
- **inputs/outputs and stores.** These are, for example, represented in computer systems by entity-relationship charts for data structures.

However—and although the first thrust of much of the literature was concentrated on system functions—existing texts give a merely superficial coverage of how to conceptualize a system's functional (operational) structure. This is true both of control functions and of the functions being controlled (the "baseline" ones). Because it underlies all organization of systems work, the conceptualization of the functional structural properties is not only an abstract matter of interest to the systems scientist, but is also of interest to the practitioner. Existing literature seems to complacently assume that the subject of system functions—and especially of baseline functions (the functions other than control ones)—was adequately handled long ago, as it was perhaps the earliest practical systems subject to be addressed in systems literature. This assumption is, unfortunately, quite fallacious.

An Approach to Fill the Gaps

It would appear that much of the reason for superficiality in covering the functional aspects of systems is due to a lack of good material on the subject, and, in particular, lack of a truly adequate method dealing with how in general a system's control structure looks. This book contains a new set of concepts with associated terminology and innovative diagramming techniques which attempt to bridge the gap between GST (General Systems Theory) and current methods of procedural diagramming (i.e., charting methods for showing system functions). It develops a number of modeling concepts and techniques useful in working with all types and levels of system—although examples in this edition of the book usually involve business organizations and information systems in general. Moreover, the same methodology may be used in the analysis (strategic) and design (tactical) levels of system definition.

What, in fact, is developed here is an **applied** GST approach. In GST, one might learn that both organizations and information systems are such that they can be "classified as systems that are open, purposeful, adaptive, closed-loop, man-made, man/machine, concrete, complex, social, and probabilistic."* Although this is useful background, just as are the current ideas on cybernetics, feedback, feedforward, etc., GST of this nature does not bridge that enormous gap between the computer-oriented literature on structured analysis and design and highly conceptual, universalized thinking. In this area, between the diagramming methods and associated low-level theory, on the one hand, and the non-pragmatic GST isomorphisms (structural characteristics applicable to all systems), on the other, a route exists to take us forward, which will demonstrate the weaknesses of the current popular approaches to structured analysis and design of information systems.

Relationship to Structured Development Literature

The methodologies here expand those of the various authors in the computer and information systems field, who are noted for their contributions to the science of structured analysis and design of system functions. In fact, one can view these methodologies as a new way of understanding the existing documented approaches. As new concepts and techniques are developed, they are often related to existing ones, sometimes along with a suggestion or two on how to improve existing methods without radical change.

Nevertheless, this book does not just re-explain and reiterate known methodologies. The approaches contained herein are essentially new, being **derived** in a far more top-down, rigorous, and conceptual way. The rigor and conceptuality does not mean that the new approaches are not applicable or useful. In fact, they are just as practical as, and perhaps more useful than, any of the existing ones.

"How would reading your materials be useful to me?" is a question that seems to arise whenever I have had the opportunity to discuss the

---

* Reprinted with permission—from Ahituv. N. and S. Neumann, Principles of Information Systems for Management, Dubuque, Iowa: W.C. Brown©, 1982, p. 107.

materials in my books. In particular, this question comes from persons who are computer systems analysts versed in the latest structured system methodologies. My immediate reaction to that question might be similar to that of anyone who ventures into a new paradigm or paradigm variation, as it seems obvious that any scientific developments are "useful," but that the use is seldom immediately tangible.

With respect to my own materials, they will eventually lead to a new generation of the so-called "structured techniques" for handling system functions, since they provide a new framework in which these techniques can grow. If this book is not seen as being practical by the experienced systems analyst, it would be for the same reasons that producing a logical model, to capture the essence of a system, would not be seen by the system's user as being useful in comparison with the more physical system model, to show implementation details. Every systems analyst knows the importance of the logical model. Yet, the current methods of functional modeling of systems which have come to be known as methods of structured analysis are more at the physical than at the logical level of explanation of the very subject of structured analysis. Is it possible that the users of structured systems development methods—i.e., the systems analysts themselves—are just as resistant to logical modeling of systems as those users of systems with whom the systems analysts have to communicate in their work?

It can only follow that this book would be of use to the information systems practitioner, as it resolves many of the difficulties encountered when using current structured analysis methods for functional modeling. The experienced systems specialist can benefit by the ordering of the system analysis and design experience into a consistent pattern, and through the utilization of the same methodology at all stages of system development.

To respond more specifically to the above question from a skeptical computer systems analyst, it should first be observed that the new functional modeling ideas are not in themselves a method of hierarchical (multi-level) functional decomposition, although they do involve a multi-dimensional, multi-level way of specifying how a system is controlled. The expert systems analyst is undoubtedly aware of how, for example, leveled data flow diagrams (or some type of hierarchical decomposition procedural diagramming method) may be combined with leveled state-transition diagrams in order to decompose both the basic system and its control structures. The frameworks provided by the ideas in this book in fact complement these techniques, as they can be used to organize the hierarchical functional decomposition work associated with the present structured techniques. Just as the data base structure in an information system provides a cradle in which the procedural (i.e., functional) detail can grow, so too will the materials in this book provide the basis for developing far more sophisticated control (i.e., information) systems. This is because they provide a dimension of understanding which actually *precedes* the hierarchical decomposition methods now used as the basis for structured techniques of system functional definition. As for the immediate, "push-button" practicality of what has been developed, perhaps none exists at present. Nevertheless, for a person with some conceptual orientation, an exposure to the ideas will immediately, in a subtle but real way, result in the formation of a better organized capacity to analyze systems. Parenthetically, it might be noted here that the ideas themselves are rather simple, which would appear to be the case with virtually all ideas which provide paradigms or paradigm variations.

A Different Approach to the Teaching of Systems

The subject of systems development in the information systems curriculum has generally been recognized, by both students and teachers, as one of the most important and also one of the most difficult areas of study. One of the problems with learning and teaching the subject is that systems science is really a discipline on its own, and is a subject which cross-cuts many disciplines. Even recognized as a discipline on its own, the existing body of literature on GST (General Systems Theory) is hardly developed enough to bridge the middle ground between vague generalities and the applied modeling of systems in various disciplines, especially in the "engineering" disciplines such as business organizational study and computers, or "business computer information systems." The subject of business (computer) information systems is currently taught using texts which generally have a similar structure and contents, along with a few other books for reference to cases or to particular techniques. In my opinion, while these materials perform an essential function in teaching systems in relation to the area of business and computers, they fail to cover a number of essential areas of the subject.

To me, the currently developed subject matter and "science" of systems needs a vast overhaul and re-orientation. If I had to summarize the current approach to teaching the subject, I would have to say that the subject is largely taught indirectly. This is because, in many respects no one has directly, scientifically addressed many of the basic principles. Once these principles are understood and more effectively applied, the subject suddenly begins to emerge as a coherent, generalizable science. Actually, by developing the science, I myself have for the first time really begun to understand the core of the subject, rather than skirting around this core with what are ostensibly the "practical" applications of various techniques and concepts. Once the effort is made to understand the principles of the subject, both teaching and learning systems analysis and design become much simpler.

Generally, systems as a discipline has been taught as a fairly soft science, such as marketing, management, or social science. Thus, the main way of conveying the ideas has involved providing the student with case studies and practical work experience on actual systems. In contrast, in the hard sciences, such as mathematics, the principles are taught before the applications. The so-called science of structured analysis and design is a first move in the direction of integrating concrete details and principles. Several books on this subject are presently available. However, their intuitive/inductive approach has allowed only a small move in the direction of being able to teach systems like a hard science. Since this text moves us considerably further in this direction, its place could be in the very early systems courses, rather than in the more specialized undergraduate and post-graduate courses.

For the student who has little experience in systems work, this book encourages the development of a conceptual ability, since it emphasizes its importance in systems work. It demonstrates how completely different concrete situations may be handled in a fairly similar way from a conceptual viewpoint. What a great shortcut this will prove to be in learning about the different business systems!

RELATED TEXTS

The related publications include the present book, along with two others. Although these different texts are interrelated, each is put together such that it can be read with understanding independently of the others. Overall comprehension of any one of these books will, however, improve as the others are read.

It was largely the working out the ideas connected with the present book that led to the development of the other two. These three texts are designed to form a significant extension to the platform upon which

At the time of publishing, further information about these books can be obtained, or the two further related books can be ordered, by contacting Baylin Systems, RR1, Box 55, Williamstown, Ontario, Canada K0C 2J0, phone (613) 931-2488. Baylin Systems also accepts correspondance at 1905 Mariposa St., Boulder, Colorado 80302, U.S.A. Finally, you may also phone (514) 483-4369, in Montreal, Canada.

rests the understanding of systems in all disciplines, with special application to business systems. The present book forms an original body of what might be termed "applied general systems theory," as it develops the middle ground between general concepts and overly pragmatic, specialized materials. The counterpart of this text is *Conceptual Prototyping of Business Systems.* The latter actually simplifies the teaching of business systems, since it uses a common template—developed to several levels of detail—for all of the example systems presented. Nothing like either of these books is now in existence. Forming an entirely new approach, they do far more than merely patch some weak spots, or fill in the cracks in the existing platform upon which systems understanding currently stands.

The third book, *Procedural Diagramming for System Development—from a More Scientific Viewpoint,* forms an approach to these subjects which is unlike any current approaches, as it explains procedural diagramming in a top-down mannner, starting with its functions, and relates all this to the idea of conceptual prototyping. In any case, only one other book in the subject area is known to me (as well as to the author of the other one).

The essential collective purpose of these three books is to further the understanding of certain fundamentals in the subject of systems, specifically, those related to functional modeling and their expression through procedural diagramming. The following will be obtained by the persons who read these books:

—a new conceptual framework which will allow for the development of future generations of structured techniques for system functional modeling;

—unique multi-level templates for logical description of system functions;

—uniform concepts for bridging different levels of detail and various stages of system development;

—a scientific approach for understanding procedural diagramming;

—a clearer way of comprehending systems in general, involving the building of a middle ground between general systems theory and practitioner-oriented structured techniques of system development—which results in reformulation and new types of materials in all the specific topic areas discussed.

Conceptual Prototyping of Business Systems

The business systems prototypes materials evolved in connection with exemplifying, verifying, and further developing the functional modeling ideas. These two bodies of materials complement each other, and together form a unique package that might be called the "conceptual prototyping method." This book provides examples of write-ups of business systems, developed using the functional modeling ideas. What in particular makes this book unique is the use of a conceptual prototype, i.e., a standard conceptual model, or a template, applicable to a wide variety of systems. In addition, this book presents unique models of the business organization and its information-control systems. Also, in order to explain the business systems, it provides new insights into classifying business computer information systems and application files, and into batch versus unary processing.

It may seem odd to some that a theoretical functional modeling method could actually have practical consequences. These are manifested in the easy readability and accuracy of the business system descriptions and their diagrams. Notably, identical data flow diagrams and action charts are used **to a number of levels of detail** for every system studied, thereby making system specification at the logical level a relatively simple, user-friendly matter of following a prototype. In addition, each process at each level of detail is presented in a standard format, and, what is unique, is presented such that the functional significances of inputs and outputs are evident. The latter makes the flow particularly easy to follow, as it distinguishes:

—objective outputs from other outputs for feedback or interface only;

—transactional inputs (the ones operated upon by the process, and/or which serve as the immediate cause of the process) from framework ones (i.e., control and instrumental inputs).

Overall, what is achieved for logical level system specification in the business system prototypes book might be paralleled to what has been achieved in computer programming in going from third to fourth-generation programming languages. No other approach known to myself uses a conceptual prototype (single template) for all its systems, not even to a single level of detail.

Procedural Diagramming for System Development

The subject of procedural diagramming for analysts and programmers has only recently come to the fore, as diagramming tools are among the chief communications tools of systems analysts.

Procedural Diagramming for System Development further probes the procedural diagramming principles given in the present book only as background. A new diagramming method, the structure-flow chart, incorporates many of the functional modeling concepts of the present book. Investigation of procedural diagramming **concepts** took place in conjunction with the development of this new charting method. As a result of this research, the following new perspectives were developed:

—useful ways in which diagrams, and especially procedural diagrams, can be classified;
—a scheme for categorizing diagramming symbols;
—a sophisticated scheme for evaluating the features of procedural diagrams, especially as related to their use in the development process for various types of systems.

Like the present text, the procedural diagramming book contains as background a clarification of the principles of system development. In fact, these ideas are explained more fully in *Procedural Diagramming for System Development*. Despite the popular notion that the analysis (logical) level of system development determines system requirements while the design phase looks at the means for achieving these, working with the various functional modeling topics makes it clearer than ever that analysis and design are essentially similar processes when seen from the general systems viewpoint. This has been clarified by analyzing ways in which different levels of detail are bridged during system development, and by properly defining the various meanings of the terms "logical" and "physical."

Essentially, what triggered further exploration of the system development process in the procedural diagramming book was the realization that, with the new functional modeling approach, the same methods for understanding the business organization as a system could be used to model the information systems at various levels, from applications areas to modules within computer programs or manual procedures. This is because these methods are extremely rigorous.

Because of the new conceptual perspectives developed in the procedural diagramming book, a range of existing diagramming methods—which are described, exemplified, and evaluated in that book—can be handled in a smaller amount of page space, yet with an increase of clarity. Overall, what has been developed there might be referred to as the first methodical attempt to explain procedural diagramming in terms of its functions. Another way of stating the latter is that the book approaches the subject of procedural diagramming, as well as that of system development, using deductive reasoning, as opposed to an approach based on intuition and inductive reasoning; that is, it approaches the subject from a higher and more general perspective.

CONTRIBUTION TO SCIENCE

The following explains the evolution of the functional modeling concepts in this text. Figure 0-1, a network diagram, is referenced throughout this discussion. The ten rectangles in this chart symbolize the steps taken in developing an original, integrated, generalized method of conceptualizing system structure.

Clarification of the Functional Cohesion Method

As the computer literature on structured analysis and design points out, the best path to designing system structure is usually the "functional" one, although other cohesion techniques are also useful in given circumstances. A primary step is to explain the functional cohesion method, i.e., how to identify subsystems along functional lines (see top block in Figure 0-1). In this book, the criteria upon which this method is based are detailed in a much more direct way than has previously been achieved in the computer literature. The latter has at times defined functional cohesion as the subsystem identification method which the other cohesion methods are **not**; e.g., it has been defined as "whatever is *not* sequential, communicational, procedural, temporal, logical, or coincidental*," i.e., as whatever is not one of the other cohesion methods. In addition, the computer literature's explanations of cohesion methods rely on the concept of "functional element," a concept hardly understood beyond a fairly intuitive level. Therefore, in conjunction with the development of a rigorous explanation of functional cohesion, it is also necessary to redevelop the explanation of a function (and thereby of a functional element), as is represented by a separate block in Figure 0-1.

---

* Yourdon E. and L. Constantine, Structured Design: Fundamentals of a Discipline of Computer Program and System Design. Englewood Cliffs, N.J.: Prentice-Hall©, 1979, p. 127.

New Heuristics for Identifying Functions

To say: "a partition may be considered **functional** when interfaces among the pieces are minimized"* is to define only the effect of the functional notion. In fact, the idea of a function needs to be clearly defined before it is even possible to speak of functional elements.

A very fundamental realization in the method developed here is that functional and subsystem analysis can be separate matters, even when subsystems are identified along functional lines. (Other purposeful techniques of subsystem identification correspond to the non-functional types of approaches to modular cohesion, some of which are already expounded in detail in the computer-oriented literature.) The two matters can be separated, because a function, as defined in this book, is a grouping of operations made cohesive not merely by being necessary for achieving an objective, but by being **directly** necessary. These points of departure represent very basic matters, which, if improperly understood, can derail the process of functional analysis, and thereby block the route towards the design of subsystems along functional lines.

A function, as currently understood, is approximately an activity fulfilling a system "requirement." The present book changes this to mean an activity, or set of operations, **directly** fulfilling a specific objective, or a mutually contingent set of specific objectives, which must be met within the system. A function consisting of all the operations which directly fulfill an objective within a given functional class in the system is referred to in this book as an "objective-defined" function. The factor of directness, and the rigorous exploration of a number of other matters connected with the fundamental concept of function, are very significant in the development of the remaining ideas shown in Figure 0-1.

A New Functional Classification Scheme

Besides referring to functions as "objective-defined," this book develops a classification scheme of these, based on deductive logic. This new scheme is comparable, and similar in many ways, to the currently popular one, attributed to Fayol, but is needed to clarify a number of matters. Within the context of the new functional classification scheme, the identification of functions as being objective-defined is a starting point.

---

* De Marco, T., Structured Analysis and System Specification, Englewood Cliffs, N.J.: Prentice-Hall©, 1978 and 1979, p. 42.

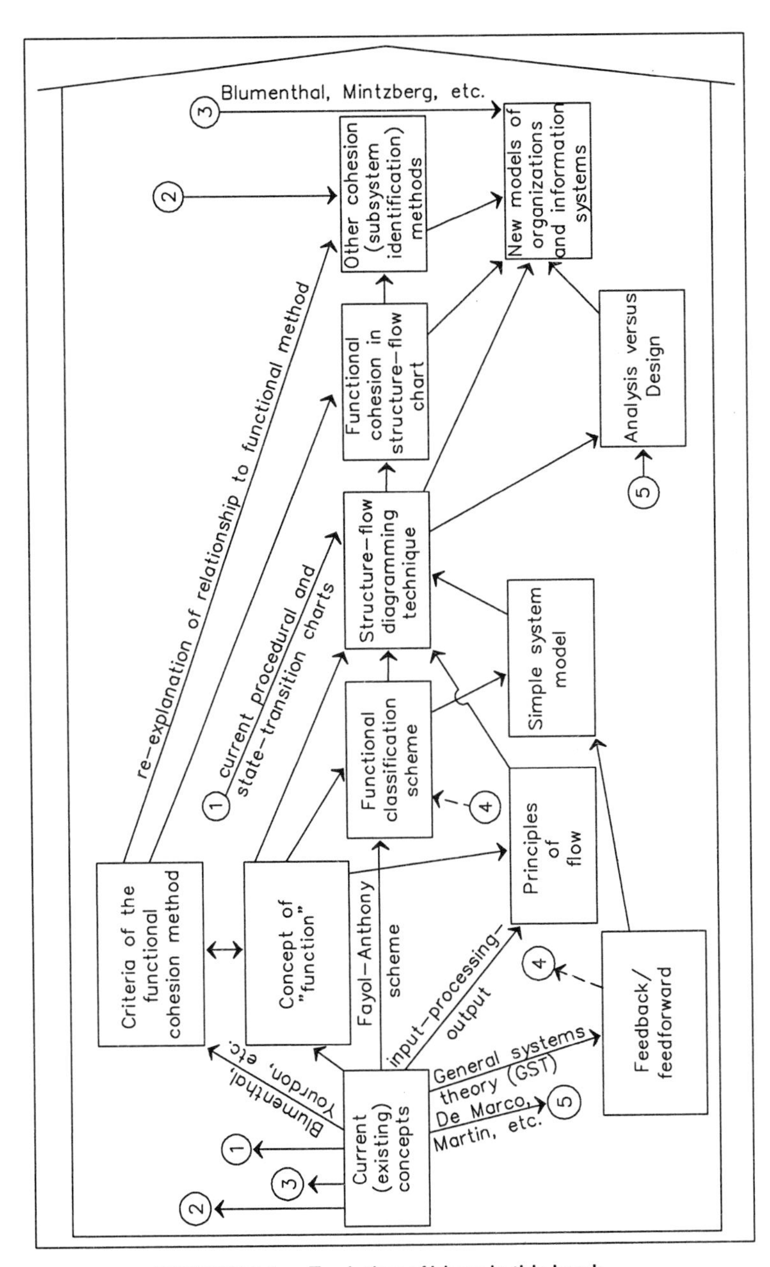

FIGURE 0-1: Evolution of ideas in this book.

Clarified Terminologies of Feedback, Feedforward, and Control

Feedback/feedforward is another fundamental subject which this book has had to review. In terms of specific diagrams, it must be known, for example, whether the current or the next run of the system is in a feedback loop. To include it, some terminology has to be clarified and developed/redeveloped.

A Simple Model of Systems

By joining the notions of feedback/feedforward with the functional classification scheme, a functional model of systems was derived. This model is the basis of the new diagramming technique, which uses a "wideshafted arrow" symbol to represent an objective-defined function.

The new model is also the basis of "conceptual prototyping," whereby one core model (conceptual prototype, or template) is developed as a basis for modeling/diagramming a wide variety of systems. The development of these template models is perhaps the most practical consequence of the ideas put forth in this text.

A New Method of Diagramming

The new diagramming technique using "wideshafted arrow" symbols is called the "structure-flow" chart. This new method incorporates the objective-defined function concept, the new functional classification scheme, the reinterpretations of feedback, feedforward, and control, the simple model of systems, etc.

This author believes that the structure-flow charting technique represents an important step forward in both theory and practical systems work, as it allows for a more multi-dimensional and in-depth description of systems, and especially because it is the first (known to this author) to incorporate conceptual prototyping. More specific firsts of this diagramming method include features such as:

—incorporation of functional classes/time-orientations into page layouts;

—heuristic restrictions on what system functional elements can and must be comprised by each operations symbol (the wideshafted arrow);

—sophisticated ability to demonstrate the functional significances of inputs and outputs flowing from operation to operation;

—the ability to serve effectively **both** as a flow and levels-of-control chart;

—three dimensional constructs to handle control levels, of which different kinds are made distinguishable.

More Understanding of all the Methods of Identifying Subsystems

With the structure-flow chart, it now becomes possible to explain the functional cohesion method in terms of a chart. The next step (see Figure 0-1) is to analyze the remaining, but less desirable, subsystem identification methods in a positive way, that is, from the perspective of a strong knowledge of the functional cohesion method. As a result, the rankings assigned to these cohesion methods can be understood by comparing them to the functional one.

The Idea of Input-Processing-Output Better Understood

Objective-defined functions contain "flow functions," i.e., input, processing, and output steps. The analysis of these requires more attention to how flow towards objectives actually works. For one thing, reservoirs (stocks) along the line of flow serve a variety of purposes in a number of ways. Secondly, flow in complex systems usually involves switching back and forth (intertwining) between and among various flow functions along an incrementally changing line. To illustrate: a little input is done, then some processing, then more input, then more processing, then some output, then more processing, and, finally the balance of output. An example of intertwining occurs within the flow of presentation of the various subjects identified in Figure 0-1; these subjects are interwoven in various ways to provide for a more interesting presentation. (Also, of course, there are a few interdependencies among the subjects noted in the figure which, for purposes of simplicity, are not included.)

Conceptual Prototyping

The whole arsenal of conceptual and diagramming materials developed in this text is needed in order to properly develop what may ultimately be seen as perfectly obvious conceptual prototypes applicable to all levels and types of systems. Yet, these templates could not have been developed, and cannot be well understood without an in-depth understanding of all the ideas in this book. Only then can one at last sensitively apply the new functional modeling method to, for example, the business organization and its information systems.

The applicability of conceptual prototype models to all levels and types of systems is no surprise, since the classification schemes for objective-defined and flow functions developed in this text are based on principles of "relativity of functional significance." These principles form an essential theme running throughout the materials.

An example of how different levels and types of subsystem are identified is provided by applying the functional cohesion method to the business organization. A model of the business is derived which is comparable to Blumenthal's model, since the latter also uses what is, in effect, the functional approach to subsystem identification within the business organization. In our model, within each functionally identified subsystem of the business—and in the business as a whole—the information-control subsystem is isolated by using one of the nonfunctional subsystem identification techniques, namely, the one known in the computer literature as "logical cohesion." Furthermore, within the information subsystem(s), the computerized information subsystem(s) is identified by yet another one of the techniques, called "procedural cohesion." Whatever the level or type of system, the same set of prototype models applies, and may be represented with the structure-flow diagramming technique. Moreover, whatever the type of business organization, the same conceptual prototype can be used.

Acknowledgements

I would like to thank the following individuals and organizations for their assistance in writing this book:

Individuals

- Brenda Shestowsky, for her various roles, as organizer, editor, indexing assistant, business assistant and advisor, for clarifying a number of key ideas, and for her article which appears in this book.
- Rosalynd Baylin, my mother, for being English editor and course assistant.
- Sylvie Gauthier, Jocelyne Caron, and Joanne Higgins-Laudi, for their CAD (computer-assisted drawing) work.
- Carol Lovelace (Folkins), for most of the desktop publishing.
- Gabor Lorenz, for editing and contributing ideas in the writing, the concepts, and the terminology.
- The other members of my employment project team, for the services they provided outside the direct context of this project. These individuals were:
  —Jeane Ennis, for project leadership.
  —Majeed Khoury, John Farkas, and Marie Vézina, for indexing.
  —Carol Lovelace (Folkins) and Marjaneh Pourmand, for completion of the CAD work.
  —Hicham Zahr, for wordprocessing in preparation for typesetting.
  —Nancy MacKenzie and Ed Hawco, for final proof reading and editing, and assistance with desktop publishing.
  —Louise Gauthier, for CAD assistance.
- the many students at the university whose courses served as essential development and testing grounds for the materials.

Organizations

- Government of Canada, Ministry of Employment and Immigration, for the *indirect* assistance to this book that was given through providing generous employment grants for the other two books related to this book.
- John Abbott College, and in particular my department (Computer Science), for it general support of my research.

E.N.Baylin

PART ONE—CONCEPTUAL FOUNDATIONS

CHAPTER ONE

INTRODUCTION

BACKGROUND SYSTEM DEVELOPMENT CONCEPTS*

This chapter is largely intended to lay the groundwork for discussing the new functional modeling, by showing where it might be used in the system development process. In addition, several terms and ideas needed in later discussion are introduced in this chapter.

The first section of this chapter discusses the following:

—logical versus physical perceptions of systems, and the use of the words "logical" and "physical";
—analysis versus design in system development;
—the different ways of adding detail in going from the logical to the physical levels in system development;
—prototyping in system development.

---

* A more complete discussion of the ideas in this section can be found in the related publication, authored by Ed Baylin, *Procedural Diagramming for System Development—From a More Scientific Viewpoint.*

The discussion of prototyping follows that of analysis versus design. A new term, "conceptual prototyping," is introduced here. This term expresses an idea central to the present book.

The second section of this chapter looks at how the functional modeling ideas have already been applied in a somewhat more practical sense to the description of a wide range of systems. This may help the reader to see the "bridge from theory to practice."

LEVELS OF SYSTEM PERCEPTION

Clarification of Related Terminology

Uses of the Terms "Logical" and "Physical"

Unfortunately, much confusion has surrounded the usage of these words during system development, the logical view often being interpreted as "what" the system has to do, and the physical view as "how" the system is to achieve its ends. Not only is the distinction between "what" and "how" often blurred, but many of the "whats" are in fact "hows," as will be seen in the next section (about the essence of the system definition process).

Following (see Table 1-1) are three different dichotomies or spectrums, which constitute the possible meanings of logical versus physical when these terms are used in the context of system development:

1) general/overall **versus** detailed/specific;
2) abstract/conceptual/symbolic **versus** tangible/concrete/actual;
3) flexible/unconstrained (except for that required) **versus** constrained beyond that required. That which is required is often referred to as "what" the system has to do, while that which is chosen to implement the required tasks is often called "how" the system is to do them. (Other terms include "objectives" for the "whats," and "means" instead of "hows.")

The first dichotomy probably constitutes the most common meaning. As a system is developed, its perception usually changes from being **general** and **abstract** and **flexible** (unconstrained except for that required) to being more **specific** and **concrete** and **selected** (beyond that required). That is, changes in the three dichotomies of meaning tend to coincide, although not in certain cases; e.g., that which is required may sometimes be at a very detailed level, while that which is at a general level may not be required, or that which is abstract may be detailed and not required, etc.

In this text, the terms may be used in any of the above three senses, or in any combination of them, depending upon the context. Thus, logical may have a number of different meanings, depending upon how the word is used and the context in which it is applied. (In fact, the word logical is very logical itself, since its meaning is so flexible.)

In sum, logical and physical have both specialized and varied meanings in relation to systems development, depending on the context. As will be seen, the words "analysis" and "design," and the word "structure" (e.g., structure chart versus structured system development), suffer from similar semantical problems.

Meanings Outside of the Frame of Reference Applicable in System Development — Table 1-1 also covers meanings of the words logical and physical situated outside the frame of reference of the system development process. It is best to steer away from using these meanings in order not to go off-track in discussing system development. These meanings are as follows:

—"rational/objective," which is linked to "conceptual," but is not the same idea;

CONTEXT WITHIN WHICH THE TERM IS USED: Discussing System Development		Other Contexts	
LOGICAL	PHYSICAL	LOGICAL	PHYSICAL
general, overall	detailed, specific		
conceptual, imagined	actual, concrete, actualized, materialized	objective, rational	
		informational	materially tangible, e.g., hard
		theoretical	practical
constrained as to method only by that which is required (hence the term "logical constraint")	constrained as to method by options selected at the discretion of the system developer (hence the term "physical constraint")		

TABLE 1-1: Meanings of the words "logical" and "physical." *(This table is also published in both of the related Ed Baylin publications, Procedural Diagramming for System Development—From a More Scientifc Viewpoint, and Conceptual Prototyping of Business Systems—A Templating Approach to Describing System Functions. Copyright* [illegible]

—"theoretical" (also linked to "conceptual"), for logical, and its opposite "practical" (linked to "concrete"), for physical;
—"material" (linked to "concrete," or "actualized," or "materialized"), as opposed to, say, "informational." This dichotomy, while useful in the discussion of **types** of systems—e.g., those systems which process data as opposed to those which move material cargo—is in a different dimension.

Uses of the Term "Structure" *

The word "structure" is one of the most overused in systems work, even more so than the word logical. Thus, it has often been used to refer to the following:

—*routineness/repetitiveness*, e.g., structured decision making;
—*formality*, e.g., structured system development;
—*standardization*, e.g., structured computer program flowcharting;
—*skeletal fabric*, e.g., system structure;
—*a particular form of chart*, i.e., the structure chart.

In the following sub-section, the term structure has the meaning *skeletal fabric* (of a system).

Logical Versus Physical Modeling (Perceiving) of Systems

Logical Modeling

"Logical modeling" of a system means perceiving a system in a general, flexible, and abstract way. Logical modeling renders system components, as defined by functional decomposition, and system control levels, into a relatively stable and consistent conceptual format.

The heuristics of logical models apply to any system or system aspect, even where flow patterns are highly volatile and decision making is of an ad-hoc, informal, spontaneous variety. As well, the heuristics of logical system modeling transcend the discussion of material structure; for instance, a given machine or worker in an organization may perform more than one conceptual function. Notwithstanding, it may be easier to apply logical system modeling concepts to concrete/formalized/pre-defined/regulated cases.

---

* The following discussion of structure contains elements from Ed Baylin, "Logical System Structure," Journal of Systems Management—see references list at end of book for details.

Benefits of Logical Over Physical Modeling of Systems

The logical modeling of a system provides a virtually constant way of viewing this system, in spite of physical changes. In contrast, in physical modeling, system details are subject to change over time, especially in dynamically evolving systems. As system boundaries, strategies, and overall objectives change, so does the structure of a system. In fact, depending upon the way system structure is perceived, even temporary adjustments in system operation could be considered as changing it.

The latter applies especially when the system is really nothing more than a one-shot affair, for instance, a project with a limited life-span. However, from a logical point of view, even a project (one-shot event) is amenable to structural analysis similar to that applicable to a system. In a physical sense, each group of operations in a project repeats itself but once, or at least a limited number of times; e.g., to build a bridge, the metal structure is mounted only once. In contrast, from a conceptual (logical) point of view, a system may be thought of as a project which repeats itself, e.g., a bridge-building system, which builds several bridges. Actually, in this way of looking at things, the only difference between a project and a system is the number of transactions handled.

The system structure may stay virtually constant, but this does not imply that there is only one way of perceiving it. For instance, a logical model of system structure may break down the system into either elements or subsystems, as will be discussed in Chapter Three.

BRIDGING THE LEVEL OF SYSTEM DETAIL

Logical modeling of systems includes the subject of conceptual prototyping. The notion of conceptual prototyping, and its provision of a different type of system definition within the system development process, is not explained until the next sub-section (about prototyping). In preparation for this discussion, the present section introduces analysis and design, which are explained in terms of the idea of "bridging the level of system detail." The groundwork for these materials has been established by the above discussion of the words "logical" versus "physical," whose meanings are basic to an understanding of the system definition process.

Ways Of Bridging Different Levels Of System Detail

Three different ways may be identified for bridging the gap between different levels of detail, although most texts on the subject cover only one or two of them. These are all discussed below, both in isolation, and in relation to one another. To facilitate this discussion, two illustrations are used, namely, Figures 1-1 and 1-2, each of which will be referenced, depending upon which is most useful for illustrating the point being made. The following paragraph is a brief introduction to Figure 1-1.

Figure 1-1a shows cities on each side of a river. A system is to be created to take tourists from city "A" to city "B" across the river. At present, a rather old-fashioned system for going from city to city is operational. In this existing system, the traveller is required either to build a raft to cross the river, or to travel around the source of the river, high in the mountains, by donkey cart.

Expanding/Contracting the Umbrella of System Elements — For present purposes, system elements include the following: operations (functions), inputs/outputs, flows, storage points, and environmental interfaces (environmental entities and system boundary lines). The most underlying way of adding detail is to "expand the umbrella of system elements," that is, to add elements which would not be obtained by explosion (see below), since they are not implied as being sub-elements within a more macro level of system elements.

Incorporating/Eliminating System Constraints — Early in the analysis task (see below discussion of organizing the process of bridging the level of system detail), the analyst decides that crossing the river is to be selected as the means for making the trip. This embodies an analyst-selected physical constraint into the model, and implies the following three sub-objectives:

1- Go from city "A" to the river.
2- Cross the river.
3- Go from the river to city "B."

At this point, the specific city and river entry and/or exit points are not yet established, since it is the task of design to decide specifically how each subsystem objective is to be achieved.

With respect to the sub-objective of crossing the river, three viable alternatives exist for this, as follows: 1- by motorboat, 2- by sailboat, and 3- by swimming. The choice of one of these means implies a specific arrival point at the other side of the river, since different dock facilities exist for swimmers, sailboats, and motorboats. Moreover, since there is an island in the middle of the river, an intermediate rest step is implied if the choice is made to swim the river. Once the choice is made, a further physical constraint is embodied into the model. Then still further physical details can be selected, e.g., what swimming stroke to use, what type of sailboat to use, etc. Thus, further levels of sub-steps/subsystems/sub-objectives are specified in a progressive fashion, until enough detail has been specified to enable implementation of the new system.

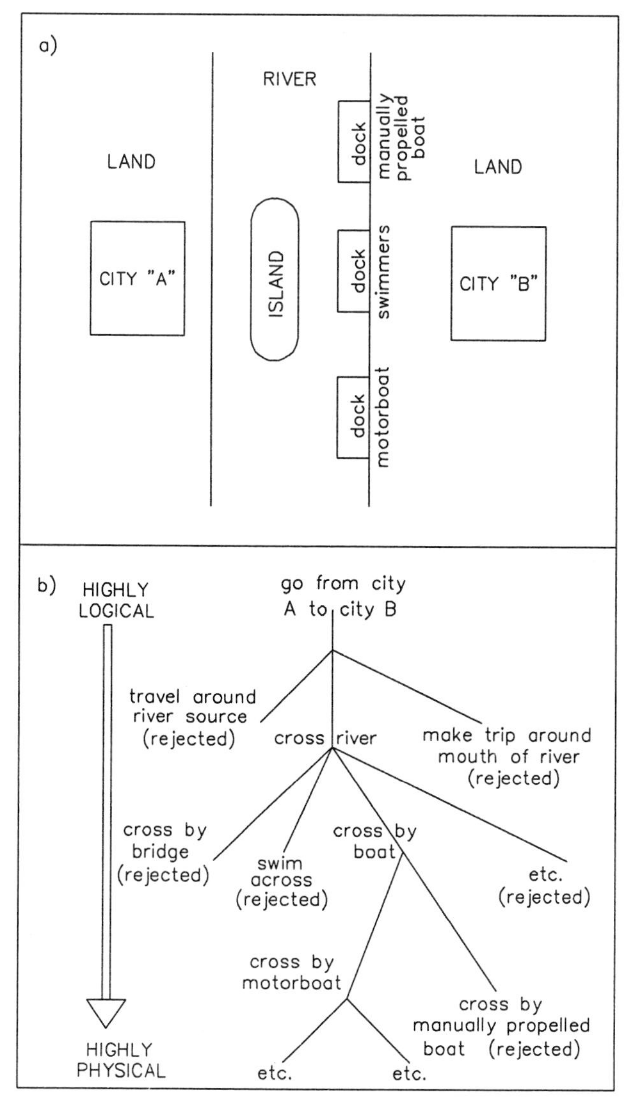

FIGURE 1-1: Example of a system and embodiment of physical constraints.
a) system to travel from city "A" to city "B";
b) decision tree of alternative means for city-to-city travel, and the progressive embodiment of physical constraints.

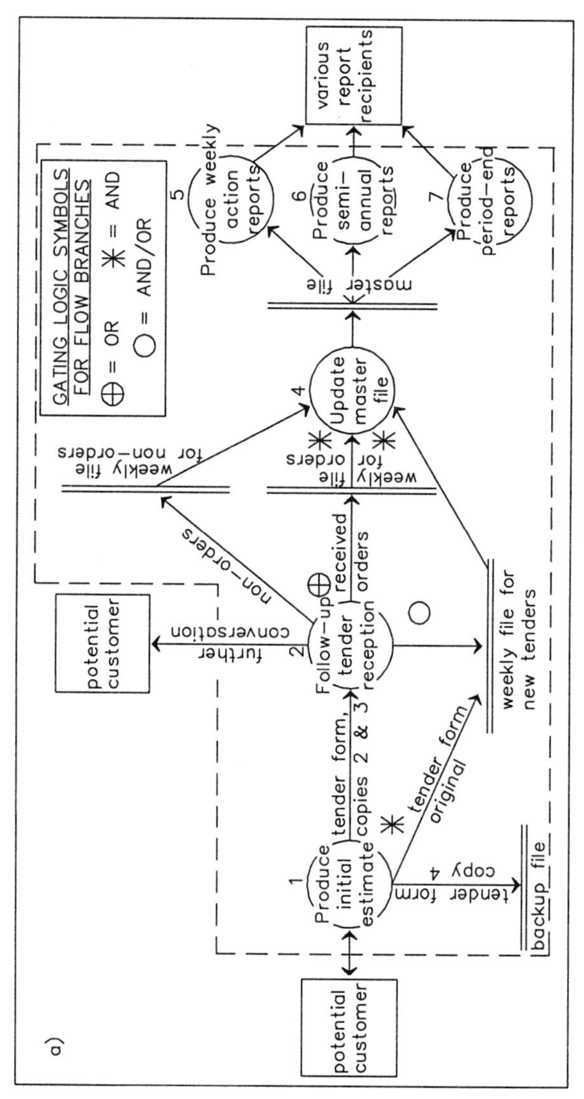

FIGURE 1-2: Data flow diagram (DFD) of the marketing information system.

a) ***left:*** *overview DFD;*

b) ***top facing page:*** *DFD exploding report production in marketing information system;*

c) ***bottom facing page :*** *DFD exploding the umbrella of functional elements in the marketing information system.*

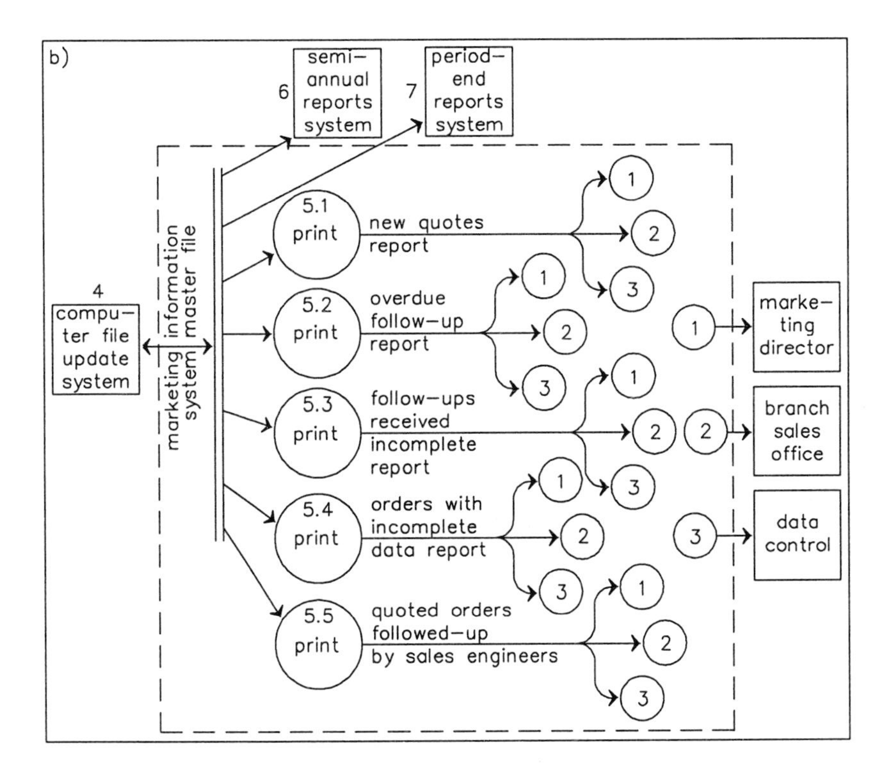
b)
6
semi-annual reports system
7
period-end reports system
4
compu-ter file update system
marketing information system master file
5.1 print
new quotes report
5.2 print
overdue follow-up report
5.3 print
follow-ups received incomplete report
5.4 print
orders with incomplete data report
5.5 print
quoted orders followed-up by sales engineers
1
2
3
marke-ting director
branch sales office
data control

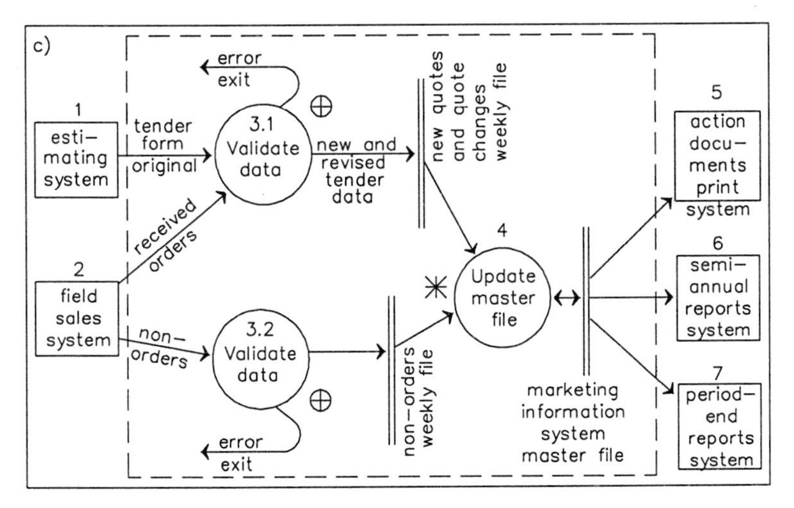
c)
1
esti-mating system
2
field sales system
tender form original
received orders
non-orders
3.1 Validate data
3.2 Validate data
error exit
new and revised tender data
new quotes and quote changes weekly file
non-orders weekly file
4
Update master file
marketing information system master file
5
action docu-ments print system
6
semi-annual reports system
7
period-end reports system

Decomposing/Concatenating System Elements — The previous principle is based on incorporating physical constraints into a more flexible picture, within the established umbrella of system elements. One further way exists of adding detail, namely, explosion (or decomposition), in which the many serial or parallel related sub-elements inherent in any single system element are specified. For instance, the "produce weekly actions reports" operation in Figure 1-2a may be exploded into five parallel sub-operations, each represented by its own separate operation, as in Figure 1-2b. In this case, each created sub-operation produces a different weekly action report, and explosion is not dependent upon chosen physical constraints. In spite of this, when sub-operations occur in serial to one another, explosion may only be possible after physical constraints have been chosen. This dependence may be illustrated in the example of going from city to city, since the means must be selected before the details of steps involved in going from city to city can be specified. For example, it must be known that the method to be used is to swim the river before the stop at the island in the middle of the river is put down as a step.

Organizing The Process Of Bridging Levels Of Detail

Analysis and design are activities performed in order to define the system to the system builders or maintainers. The essence of the distinction between analysis and design is, in the opinion of the author, that they deal with consecutive levels of detail; i.e., analysis deals with the logical view, while design deals with the physical view. This distinction does not necessarily mean that analysis always has to precede design in a chronological sense, nor does it imply many other ideas which are often assumed to be differences between analysis and design. It does, however, imply that analysis and design are essentially similar processes, although at different levels of detail (as will be discussed below).

Overview — The following summarizes the organization of the steps —to be described under the next series of sub-headers—potentially involved iVthe process of bridging levels of detail during system definition:

1) Perform the initial processing, at the least detailed level:
   a) Establish the umbrella of system elements to be considered, i.e., establish the mission and boundaries of the system.
   b) (Optional) Free the system from existing physical constraints.
   c) Set down new constraints within the system established by the definition of the umbrella.

d) Explode the system, as given in the umbrella definition, and in which new constraints have been embodied, into the first level of subsystems.

2) Repeat the following steps until a sufficient level of detail has been defined:
   a) Expand the umbrella of system elements, so as to include any necessary further elements
      > either within the subsystems identified in the previous explosion,
      > or forming a new subsystem.
   b) Set down new constraints within the subsystems established by the latest definitions of the umbrella of system elements.
   c) Explode the subsystems, as given in the latest umbrella definitions, and in which new constraints have been embodied, into the next level of subsystems.

Control Over the Umbrella of System Elements — The first step in analysis is achieved by identifying the most basic system elements to be considered. This may be referred to as "laying down an initial umbrella of system elements." Within this umbrella, various constraints may be considered (see next sub-header).

It is important for the analyst to avoid unnecessarily expanding the umbrella, i.e., to keep the umbrella in a reasonably contracted state in relation to each successive level of detail. Since certain indirectly related operations are less salient and important when the system is first studied, they should generally be excluded from the initial, highly logical picture of the system. They may be added at a later point of system definition. Similarly, many of the system inputs and outputs as well as external interfaces may at first be omitted from the umbrella of system elements. For example, no data validation operations need be implied by any of the bubbles (symbols for operations/functional elements) in Figure 1-2a. However, new bubbles may be added, as in Figure 1-2c, where data validation steps are inserted before the file update step. Here the added operations are control functions. Such functions are neither in the direct, nor in the principal, line of flow towards the system objectives; i.e., they are both indirectly related to and relatively unimportant in achieving the system objectives, albeit necessary for this purpose.

Freeing from Existing Physical Constraints — The analyst must begin by freeing himself/herself from the physical details of the current system, thereby opening the door to many other possible ways of going from city to city (see Figure 1-1), even ways involving technologies not yet in existence. For example, crossing the river by motorboat, flying by airplane, taking an ocean trip around the mouth of the river, etc. are all possibilities. One of these is to be chosen at some stage of system definition as the new means for going from city to city. Apparently, the first decision which has to be made is whether or not to make the trip by crossing the river.

The analyst does not consider the objective of going from city "A" to city "B" as being alterable, although the trip to city "B," as opposed to some other city, really represents a so-called "physical constraint" when going from city "A" to city "B" is viewed as a subsystem within a larger tourism system. However, from the local viewpoint of the city "A" to city "B" subsystem, this is seen as a logical constraint, since it cannot be altered by the system developer. By "logical" in this particular case is meant only that the system developer is pre-committed to the preset system objective.

Adding Detail Using a Top-Down Approach — The idea is to proceed in a step-by-step fashion, gradually expanding the umbrella of system elements to include more and more of the less important elements, and committing oneself to the fewest possible physical constraints at each successive level of detail, within the established umbrella of system elements at that level of detail. Of course, what is logical detail from the viewpoint of a more detailed level of system development may be perceived as physical detail from a more general level of the process. Thus, in Figure 1-1b the first choice is to go from city to city by crossing the river. Second, crossing the river by boat is selected. Finally, crossing the river by motorboat is chosen from among a number of alternatives applicable at that level of the decision tree shown in the figure. Now, if the analyst were to jump the middle part of the decision tree, from the most logical level (going from city to city) straight to the most physical level (crossing the river by motorboat), he or she would not be proceeding in a top-down fashion, since the choice of crossing the river in the first place must precede the choice of means of crossing the river.

PROTOTYPING

The preceding sub-section referred to the idea of prototyping. This method of system development is a key ingredient in diminishing the differences between analysis and design. What follows discusses prototyping, differentiating between conceptual (symbolic, in-the-mind) and actualized (materialized, working, operational) varieties.

Explanation of Prototyping

Conceptual Versus Actualized Prototyping

A new term, "conceptual prototyping," is introduced in this book. This term is fairly synonymous with the term "logical system structuring," insofar as the prototypes are conceptual (i.e., logical in the sense of symbolic/in-the-mind/in-the-perception) representations embodied in diagrams. However, this is not the current way in which the word prototype is used. Rather, the term prototype currently means "actual working model." In the computer systems trade, for example, prototyping has come to be associated with quickly created computer programs used as a starting point for modifications as the user progressively and iteratively defines what the system is to accomplish. In this book, the "actual working model" kind of prototype is referred to as "actualized." (The terms "materialized," "working," or "operational," can also be used.)

Further Clarification with an Example

The difference between a prototyping method of system conception and a non-prototyping one, as well as the difference between actualized and conceptual prototypes, is explained in the following story. An old man living in the mountains about 100 miles away from the nearest city wishes to buy a new car. Unfortunately, he cannot go into the city to visit car dealerships, since he is currently bedridden. Thus, he asks his son to go to the city to purchase the car. To help his son purchase a car which is possibly the right one, the father devotes considerable effort to specifying the requirements which must be fulfilled by the new car; e.g., it must be blue in colour, medium-sized, hatchback, etc. The son then goes to the city and visits various car dealerships before deciding which car might best fulfill his father's requirements.

In the actualized (materialized) prototyping method, the son would go to the city and drive a number of sample cars back for his father to see, on the assumption that this is a test car, on loan from the dealership. Upon inspecting and rejecting a given car, the father may obtain a more precise idea as to what other car he would prefer instead. Obviously, the prototyping method used here would be much more practical if the father were to live in the city, preferably near the various car dealerships, since driving a test car out to the country is a fairly large effort for the son, and may not even be permitted by any dealer. In other words, the technical and economic feasibility of actualized prototyping require that samples can easily be made. They also require that samples can be modified with ease.

In the conceptual (symbolic) prototyping (or "templating") method, the father would be able to study the various shapes, sizes, and characteristics of cars from diagrams produced jointly by the various car makers, or by an engineering professor of automotive technology. A particular prototype, or combination of prototypes, could then be selected. Such use of conceptual prototypes facilitates the specification of requirements **before** any actual product becomes available (as either the final product or as a test prototype). Such a method would be particularly useful if the father has had little contact with the outside world during the last twenty years, and is not familiar with modern cars, and is not geographically located for actualized prototyping to be feasible.

It can be concluded from the above example that a conceptual (symbolic) prototype (template) based on the essential type of system being considered, may be used prior to an actualized prototype in the definition of the same system. Both conceptual and actualized prototypes provide a starting reference against which modifications can be made.

Effects of Prototyping on the System Definition Process

Effects of Prototyping on Design

Actualized Prototyping at the Physical Level — Actualized prototyping applies only at the detailed level of system definition. In computer systems work, actualized prototyping is used in the later stages of analysis (which may also be called system design, or general design, or logical design, depending upon the author) and in design (also called detailed design or physical design). The effects of actualized prototyping on design have been studied extensively in connection with fourth-generation computer languages, where it has been noted that, with prototyping, design and programming of computer systems become essentially the same step. This greatly reduces the amount of time needed to go from the conception to the implementation stages of computer system development.

Conceptual Prototyping at the Physical Level — In fact, conceptual prototyping (templating) in what may be a very detailed sense is naturally practiced by any experienced systems analyst, since experience with one system helps to envisage how new systems will look. Thus, the template models of systems tend to become, so-to-say, "hardwired into the brain circuits" of the experienced and capable systems developer. Usually, such conceptual prototypes are very general in level. However, if the prototype models are detailed enough, they may be said to be at the "physical" level, i.e., where "physical" is opposed to the "general" meaning of "logical" in systems development, but not in the sense of "physical" opposed to the "symbolic" or "conceptual" meanings of "logical." (Thus, the "in-the-mind" versus "tangible" opposition of the two words "logical" and "physical" may not coincide with the "general" versus "specific" opposition of these same two words, although both types of uses of these words are applicable to the discussion of the system development process.)

The effect of conceptual prototyping at the physical level is to speed up the design process, since a conceptual prototype itself is a result of the completion of the initial part of the design process.

Effects of Conceptual Prototyping (Templating) on Analysis

Reduced Need to Eliminate Existing Physical Constraints — To a considerable extent, with current methods of analysis, it may be possible to obtain a logical system model devoid of unwarranted physical constraints without paying too much attention to the existing system. However, the analysis step which consists of studying the current system and then removing physical constraints is also needed, at least where the existing system is just being modified. In these circumstances, in parallel with *a priori* modeling, it is necessary to look at physical features.

In contrast, a conceptual prototyping method can be developed which can be used with some efficiency in working from the almost purely logical, either back to the physical details of an existing system, or directly towards the development of modifications to an existing system, or directly towards the physical details of a new system. This method can be used in parallel with the structured approach just described (i.e., the approach which begins by looking at the physical features of the existing system). However, what is being suggested here is not a general-purpose model applicable to all systems. Rather, conceptual (symbolic) prototyping is a set of concepts and diagramming techniques which has the degree of sophistication and adaptability to enable the developer to create a number of typical conceptual prototypes (templates) of systems, so that a fairly close fit can be found between any system and one or more system prototype models. All of this is based on what might be called a "core" prototype, e.g., the clothing model which provides the basic shape of all human bodies, not accounting for the occasional freak.

Therefore, just as a clothing manufacturer will develop a number of standard sizes and styles (shapes) for given clothing lines, a number of fairly standard system morphologies can be created. Minor changes and additions can be made to these, thereby tailoring the fit to individual cases. Moreover, this tailoring is made easier if the standard is kept fairly simple, so that new features can be added without having to remove too many of the existing features from the prototype. That is, keeping the standard at a more logical level allows different ways of adding physical detail without having first to free the prototype itself of **existing** physical detail.

Unstructuring Effect on Structured Analysis — Even in modifying part of an existing system, the ability to model through conceptual prototypes has a considerable effect on how the analysis process develops. In truth, are not the steps of analysis seen the various structured system development methods (see related publication on procedural diagramming) based on certain constraints assumed about analysis, namely, the general impracticality of basing the development of a logical model of system functioning on conceptual prototype models?

In contrast to the steps specified in connection with various traditional structured methods of system development, with the conceptual prototyping approach, a new sequence, seen in Figure 1-3, becomes more possible. Here, steps 2.1 and 2.2 execute in parallel, and together lead into step 2.3. Step 2.1 is the same as in the traditional structured methods; i.e., it studies the **not-to-be-modified** part of the existing system (referred to in the figure as the "current environment"). Meanwhile, step 2.2 obtains a model of the **to-be-modified** part of the current system which is as free as possible from physical constraints, even those imposed by the current workings of the **not-to-be-modified** part of the system. Step 2.2 bases itself on the conceptual prototyping approach, so that the **to-be-modified** part of the system does not have to be modeled as it is.

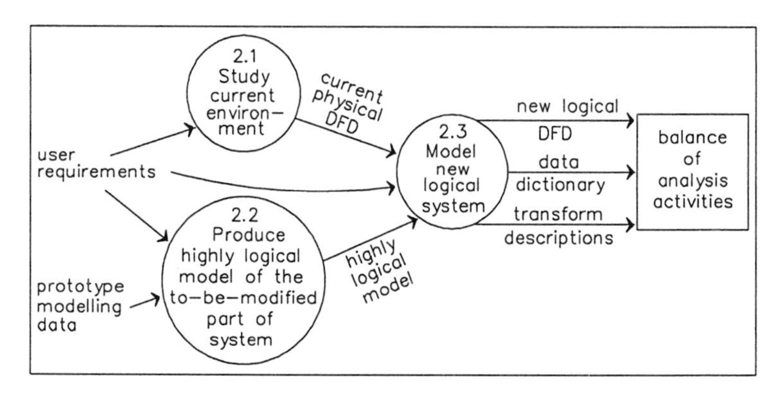

FIGURE 1-3: DFD showing the steps of the system definition process when conceptual prototyping is used.

This is different from the structured methods, in which step 2.2 would produce a model of the current **to-be-modified** part of the system, which, although fairly logical, is still reflective of the physical constraints embodied in the current system. Finally, step 2.3 in the conceptual prototyping approach shown in Figure 1-3 modifies the prototype model by adjusting it to reflect the new logical constraints. While the end product may be similar in both methods, the results of explosion of the analysis process are different, as different choices of constraints on the system development process itself are implied in each of these cases.

Summary of the Effects of Prototyping

Figure 1-4 demonstrates the roles of both conceptual (symbolic) and actualized prototyping in the system definition process.

The conceptual prototyping method enhances particularly the logical levels of system definition, such that analysis and design become rather intertwined, as freeing from previous physical constraints is largely eliminated from the definition process itself. As for the actualized (working) prototyping method, it essentially speeds up the design and construction of the system. As well, the earlier aspects of system definition need to be carried out less carefully with actualized prototyping, since the consequences of error are less costly.

Thus, in developing computer systems using fourth-generation languages (which have made actualized prototyping of computer systems cost-justified), most of the emphasis can be placed on the design phase, and less attention need be paid to analysis, on the one hand, and programming, on the other. In fact, the design and programming phases, in effect, become so intertwined with actualized prototyping that they may be considered a single phase of system development.

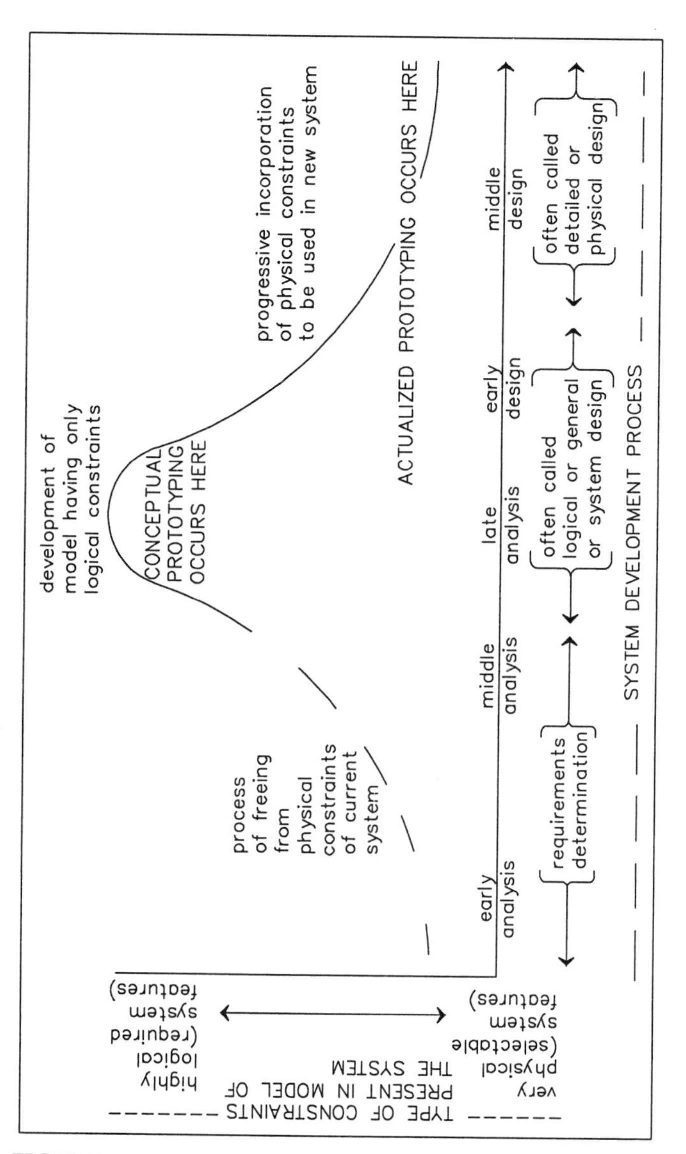

FIGURE 1-4:

Actualized and conceptual prototyping in system development. *The part of the curve drawn with a broken line represents the fact that freeing from physical constraints during early analysis becomes less needed when either type of prototyping is used. Thus, the process tends to go more quickly to middle analysis, and from there to middle design.*

A WIDELY APPLICABLE PROTOTYPE (TEMPLATE) FOR DESCRIBING SYSTEMS

In connection with developing the concepts in this book, a number of case examples of business information systems were evolved, and are now contained in the related Ed Baylin book, *Conceptual Prototyping of Business Systems—A Templating Approach to Describing System Functions*). The business system descriptions in the latter book are standardized in terms of:

- the format used in the write-ups of all examples;
- the conceptual prototype used as the basis of all of them, to several levels of detail. This prototype not only applies to simple business systems, but can also be used for many subsystems within each business system, and for most simple non-business systems as well.

The standard prototype is from the related publication is presented in the following, before the new charting method and concepts via which this standard template was derived, so that the reader can look ahead at a practical application of what is to be developed in the subsequent parts of the present text.

DEPICTION VIA DIAGRAMS

Figure 1-5 (spread over six pages later in this section) is an action chart representation of the standard prototype's various levels of detail.

NARRATIVE DESCRIPTION

Standard Coding Scheme Used for Identifying Operations

The first section of the code for each process is a series of one to three numbers, separated by dashes in the case of two or three numbers (e.g., 1-1-2, or 4-1). The number(s) here represent the class (and possibly subclass as well) of the operation, as follows:

- **1ST NUMBER**—indicates the major system stage (time-functional class), as follows:
  - 1- **Initiative current control**: those operations leading to the initiation of a run of the system baseline;

2- **Interruptive current control**: those operations leading to a change, cancellation, or rescheduling of either the start of the baseline operations (if they have not yet begun but have been prepared for start by the initiative operations), or of the baseline operations already in progress;
3- **Baseline**: those current operations which directly carry out the objectives of the system (as opposed to the control operations, 1, 2, and 4, which play an indirect role; in other words, steps 1, 2, and 4 serve step 3);
4- **Ensurance current control**: those operations which ensure the effective achievement by step 3 (the baseline) of the system objectives, by validating results of baseline execution and, in case problems are discovered, making corrections before the baseline can render delivery of its products to destinations served by the system as a whole.

- **2ND NUMBER**—(applicable only to major system stages 1, 2, and 4, i.e., the control operations)—indicates the decision-making level of the operation. For example, process 1-1 represents the decision to go ahead with a customer order, while operation 1-2 is the decision making connected with assigning the order for picking from the shelves at a particular instant of time. One level of decision making is more general than the other, and occurs before the other. When this distinction between superordinate and subordinate control levels is absent from the the description, a zero is used, e.g., 1-0.
- **3RD NUMBER**—(applicable only to major system stages 1, 2, and 4, i.e., control operations)—indicates the fact that the operation completes the task identified by the preceding two numbers, when completion is not immediately achieved by the first phase of operation execution, i.e., when scheduling delays occur in carrying out operations. For example, a customer order may not immediately be sent for picking because stock is backordered, or an accounts payable invoice may not be paid immediately because of a cash flow consideration. In this case, the first stage of processing is represented by, say, 1-1-1, while the decision-making flow related to releasing a held customer order, or a held invoice payment, is represented by 1-1-2. Had no delays been presented in the flow in this example, the entire process would have been represented by 1-1-0.

Further functional decomposition following the series of the first section of numbers separated by dashes is coded by numbers separated by periods, e.g., 1-1-2.**1**, 1-2.**3**, 2-1-1.**2.1**. A period with one further number, e.g., 2-1-1.**3**, represents a second level of decomposition of the operation represented by the series of numbers separated by dashes, while two further numbers separated by periods code a third level of detail, and so on.

Notwithstanding the coding in terms of numbers, alphabetical characters may sometimes be included as a suffix to any given number to represent parallel sub-operations. Thus, sub-operations 1**A**, 1**B**, and 1**C** are in parallel to one another. Following are a few examples of this:

1-2**A** vs. 1-2**B**, 1-1-2.2**a** vs. 1-1-2.2**b**, 1-1-1.2.1**a** vs. 1-1-1.2.1**b**,
1A-1.2**a** vs. 1A-1.2**b**, 1A-2B-2.3.4**a** vs. 1A-2B-2.3.4**b**

It should be noted that the paired A's and B's (or a's and b's) occur at the same detail level. The capital form of the letter is used in conjunction with the process numbers separated by dashes, while the small form of the letter is used in conjunction with the process number subqualifiers, i.e., the numbers separated by periods.

Customization Needed for Individual Applications of the Standard Prototype

Reduced Emphasis on Unimportant Operations — The importance of operations coded by the second and/or the third number in the series of numbers separated by dashes varies in each of the business systems which are prototyped. Therefore, while all categories of operations are noted in every use of this prototype, in certain cases, some of these categories may not be described at all, or may be described at the macro level only. For example, in an accounts payable system, process 1-2 is relatively unimportant, and so is not exploded further, while, in an accounts receivable system, process 1-2 is very important, and is therefore fully exploded.

As well, certain parallel processes vary in importance from example to example. Thus, in one case, operation 2B is important, while it is hardly described at all in other examples. Nevertheless, in keeping with a standard prototyping approach, an entry is made for 2B in all the examples of business systems. However, 2B is not decomposed (exploded) further when it is unimportant to a given system.

Input Systems and Output Systems — A tendency exists to customize the template to coincide with whether the system is of the "importing" or "exporting" type. Thus, although the example business systems are organized into groups according to the groupings of subsystems in the functional model of the business organization, a different type of grouping may be obtained, and was considered as the second best alternative for organizing the presentation of the prototype system examples. The input type of system has as its objective to import something into the organization, e.g., supplies in the case of the supply acquisition system, and money in the case of the accounts receivable system. The output system type has the objective of exporting something from the organization, e.g., the organization's products in the case of the various customer requirements systems, and money in the case of the accounts payable and payroll systems. For instance, the customization of the prototype for the two mentioned importing systems is rather similar.

Reapplicability of the Prototype to Systems at Different Levels of Detail

The student course registration system provides an interesting example of how the prototype may apply at different detail levels. Student registration is the initiative decision-making activity which precedes actual student admission to courses. The baseline operation of this system is the actual, material attendance of the student at courses. Student registration occurs at two levels. The first of these involves deciding whether to admit the student to register for courses at all, e.g., whether to accept a student for entry into a bachelor's programme. The second registration process occurs each semester. This is **course** registration, where the student selects particular courses in line with his or her programme. Together, these two levels of decision-making activity form the initiative operation of the school admission system as a whole.

A student **course** registration in itself is, however, a system at its own level of detail. It has a baseline operation at its own level, along with its own, local initiative, interruptive, and ensurance control operations to serve this baseline. For example, the initiative operation involves deciding whether to accept the student for course registration at a particular time, based on how particular days and hours during the registration period have been assigned in order to avoid bottlenecks in the course registration activity. At this time, it is assumed that the student has already been accepted into the programme, and so what is involved is simply a comparatively minute scheduling problem, e.g., whether to allow the student to enter the registration room at 10 a.m. on Thursday.

The following pages contain the action chart of the multi-level standard prototype:

Pages 24 to 25: Stage 1
Perform initiative time-orientation control operations

Pages 26 to 27: Stage 2
Perform interruptive time-orientation control operations

Page 28: Stage 3
Perform baseline operations

Pages 29 to 30: Stage 4
Perform ensurance time-orientation control operations.

(Note: The copyright on these diagrams has been retained by Ed Baylin, in connection with his related book, ***Conceptual Prototyping of Business System—A Templating Approach to Describing System Functions.***)

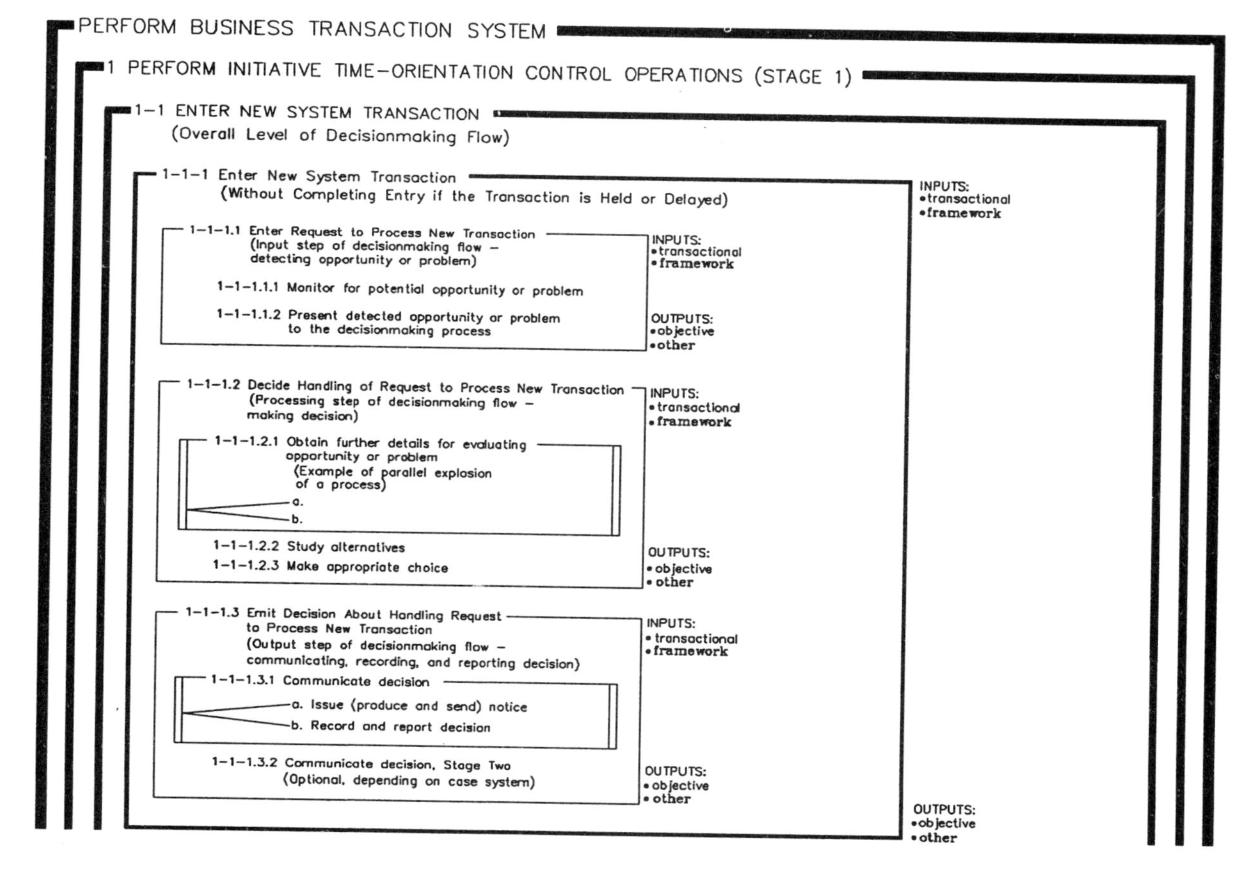
PERFORM BUSINESS TRANSACTION SYSTEM
1 PERFORM INITIATIVE TIME–ORIENTATION CONTROL OPERATIONS (STAGE 1)
1–1 ENTER NEW SYSTEM TRANSACTION
(Overall Level of Decisionmaking Flow)
1–1–1 Enter New System Transaction
(Without Completing Entry if the Transaction is Held or Delayed)
INPUTS:
•transactional
•framework
1–1–1.1 Enter Request to Process New Transaction
(Input step of decisionmaking flow – detecting opportunity or problem)
1–1–1.1.1 Monitor for potential opportunity or problem
1–1–1.1.2 Present detected opportunity or problem to the decisionmaking process
INPUTS:
•transactional
•framework
OUTPUTS:
•objective
•other
1–1–1.2 Decide Handling of Request to Process New Transaction
(Processing step of decisionmaking flow – making decision)
1–1–1.2.1 Obtain further details for evaluating opportunity or problem
(Example of parallel explosion of a process)
a.
b.
1–1–1.2.2 Study alternatives
1–1–1.2.3 Make appropriate choice
INPUTS:
•transactional
•framework
OUTPUTS:
•objective
•other
1–1–1.3 Emit Decision About Handling Request to Process New Transaction
(Output step of decisionmaking flow – communicating, recording, and reporting decision)
1–1–1.3.1 Communicate decision
a. Issue (produce and send) notice
b. Record and report decision
1–1–1.3.2 Communicate decision, Stage Two
(Optional, depending on case system)
INPUTS:
•transactional
•framework
OUTPUTS:
•objective
•other
OUTPUTS:
•objective
•other

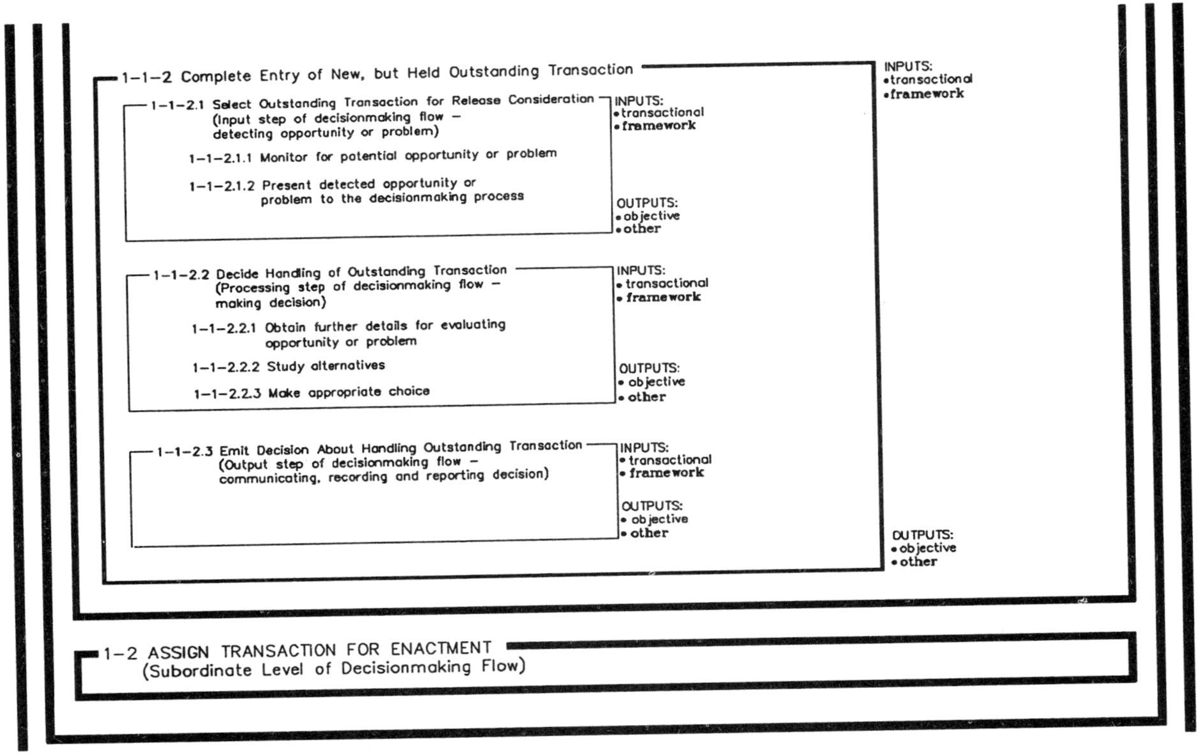

1–1–2 Complete Entry of New, but Held Outstanding Transaction

INPUTS:
• transactional
• framework

1–1–2.1 Select Outstanding Transaction for Release Consideration
(Input step of decisionmaking flow – detecting opportunity or problem)

1–1–2.1.1 Monitor for potential opportunity or problem

1–1–2.1.2 Present detected opportunity or problem to the decisionmaking process

INPUTS:
• transactional
• framework

OUTPUTS:
• objective
• other

1–1–2.2 Decide Handling of Outstanding Transaction
(Processing step of decisionmaking flow – making decision)

1–1–2.2.1 Obtain further details for evaluating opportunity or problem

1–1–2.2.2 Study alternatives

1–1–2.2.3 Make appropriate choice

INPUTS:
• transactional
• framework

OUTPUTS:
• objective
• other

1–1–2.3 Emit Decision About Handling Outstanding Transaction
(Output step of decisionmaking flow – communicating, recording and reporting decision)

INPUTS:
• transactional
• framework

OUTPUTS:
• objective
• other

OUTPUTS:
• objective
• other

1–2 ASSIGN TRANSACTION FOR ENACTMENT
(Subordinate Level of Decisionmaking Flow)

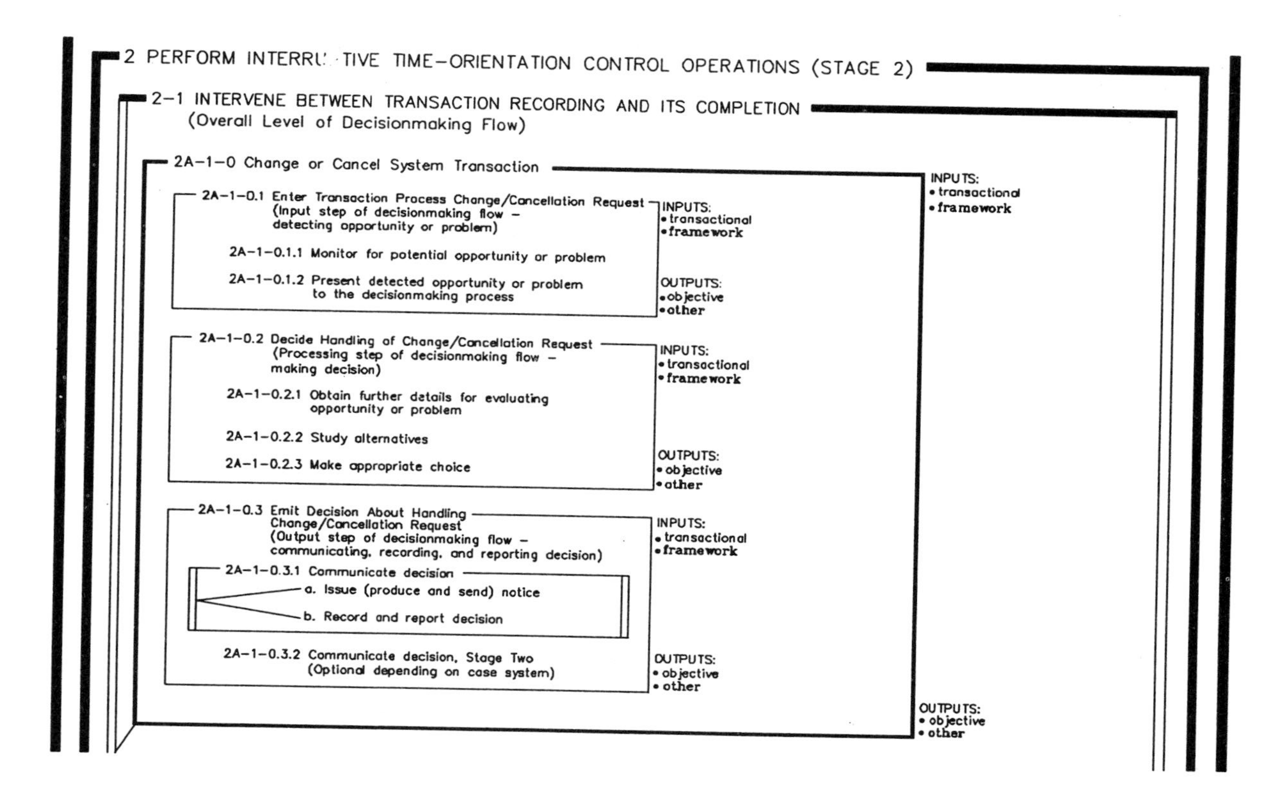
2 PERFORM INTERRU TIVE TIME-ORIENTATION CONTROL OPERATIONS (STAGE 2)
2-1 INTERVENE BETWEEN TRANSACTION RECORDING AND ITS COMPLETION
(Overall Level of Decisionmaking Flow)
2A-1-0 Change or Cancel System Transaction
INPUTS:
• transactional
• framework
2A-1-0.1 Enter Transaction Process Change/Cancellation Request
(Input step of decisionmaking flow – detecting opportunity or problem)
2A-1-0.1.1 Monitor for potential opportunity or problem
2A-1-0.1.2 Present detected opportunity or problem to the decisionmaking process
INPUTS:
• transactional
• framework
OUTPUTS:
• objective
• other
2A-1-0.2 Decide Handling of Change/Cancellation Request
(Processing step of decisionmaking flow – making decision)
2A-1-0.2.1 Obtain further details for evaluating opportunity or problem
2A-1-0.2.2 Study alternatives
2A-1-0.2.3 Make appropriate choice
INPUTS:
• transactional
• framework
OUTPUTS:
• objective
• other
2A-1-0.3 Emit Decision About Handling Change/Cancellation Request
(Output step of decisionmaking flow – communicating, recording, and reporting decision)
2A-1-0.3.1 Communicate decision
a. Issue (produce and send) notice
b. Record and report decision
2A-1-0.3.2 Communicate decision, Stage Two
(Optional depending on case system)
INPUTS:
• transactional
• framework
OUTPUTS:
• objective
• other
OUTPUTS:
• objective
• other

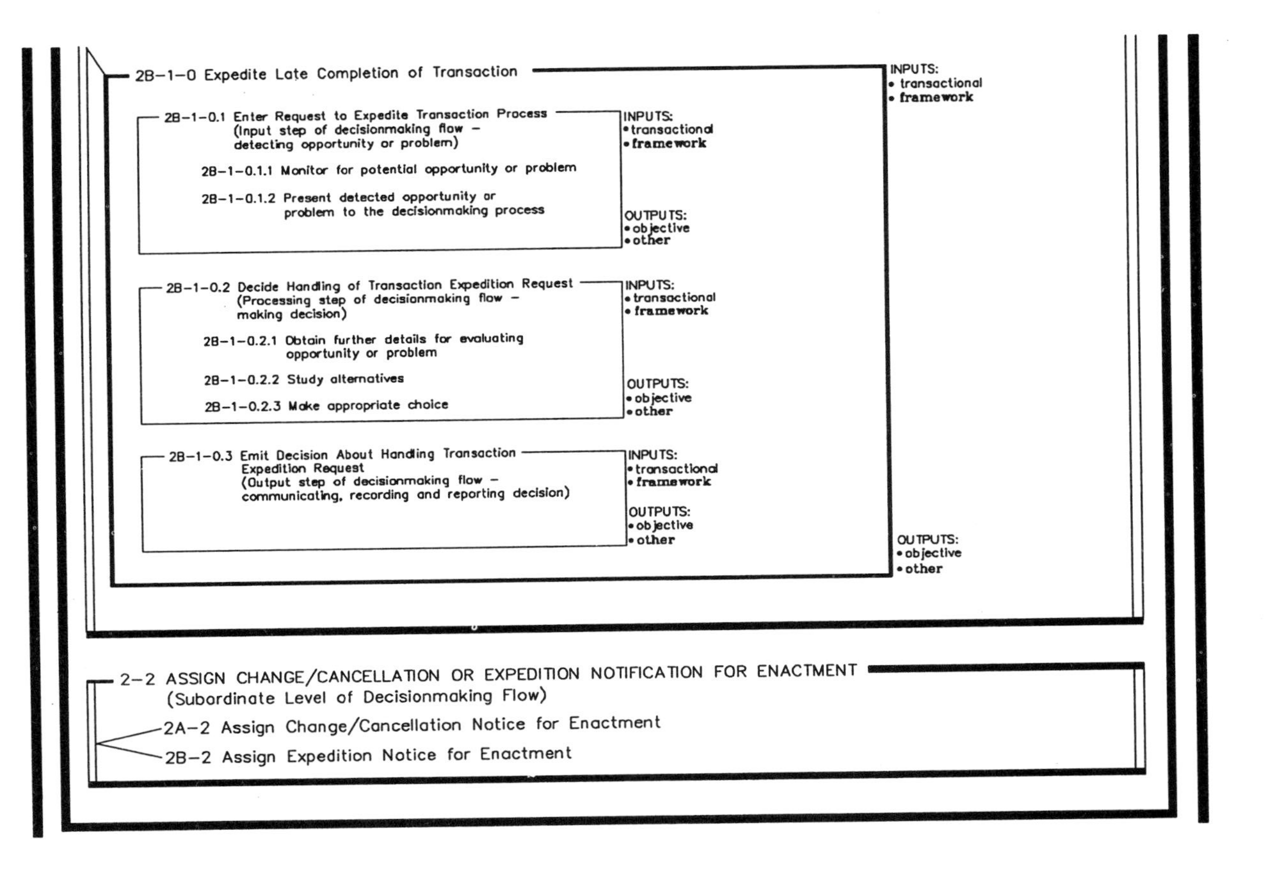
2B-1-0 Expedite Late Completion of Transaction
INPUTS:
• transactional
• framework
2B-1-0.1 Enter Request to Expedite Transaction Process
(Input step of decisionmaking flow – detecting opportunity or problem)
2B-1-0.1.1 Monitor for potential opportunity or problem
2B-1-0.1.2 Present detected opportunity or problem to the decisionmaking process
INPUTS:
• transactional
• framework
OUTPUTS:
• objective
• other
2B-1-0.2 Decide Handling of Transaction Expedition Request
(Processing step of decisionmaking flow – making decision)
2B-1-0.2.1 Obtain further details for evaluating opportunity or problem
2B-1-0.2.2 Study alternatives
2B-1-0.2.3 Make appropriate choice
INPUTS:
• transactional
• framework
OUTPUTS:
• objective
• other
2B-1-0.3 Emit Decision About Handling Transaction Expedition Request
(Output step of decisionmaking flow – communicating, recording and reporting decision)
INPUTS:
• transactional
• framework
OUTPUTS:
• objective
• other
OUTPUTS:
• objective
• other
2-2 ASSIGN CHANGE/CANCELLATION OR EXPEDITION NOTIFICATION FOR ENACTMENT
(Subordinate Level of Decisionmaking Flow)
2A-2 Assign Change/Cancellation Notice for Enactment
2B-2 Assign Expedition Notice for Enactment

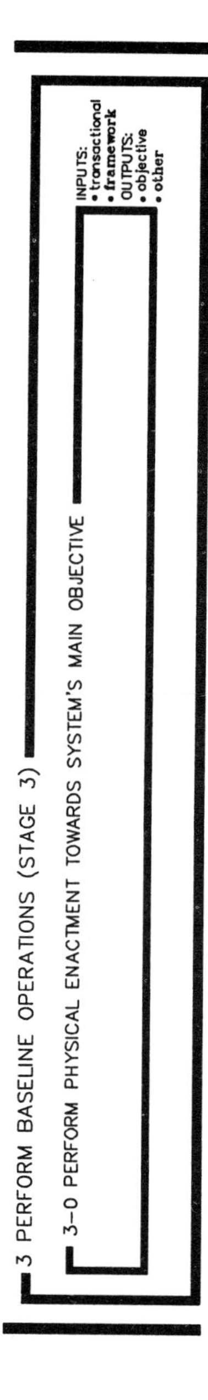
3 PERFORM BASELINE OPERATIONS (STAGE 3)
3–0 PERFORM PHYSICAL ENACTMENT TOWARDS SYSTEM'S MAIN OBJECTIVE
INPUTS:
• transactional
• framework
OUTPUTS:
• objective
• other

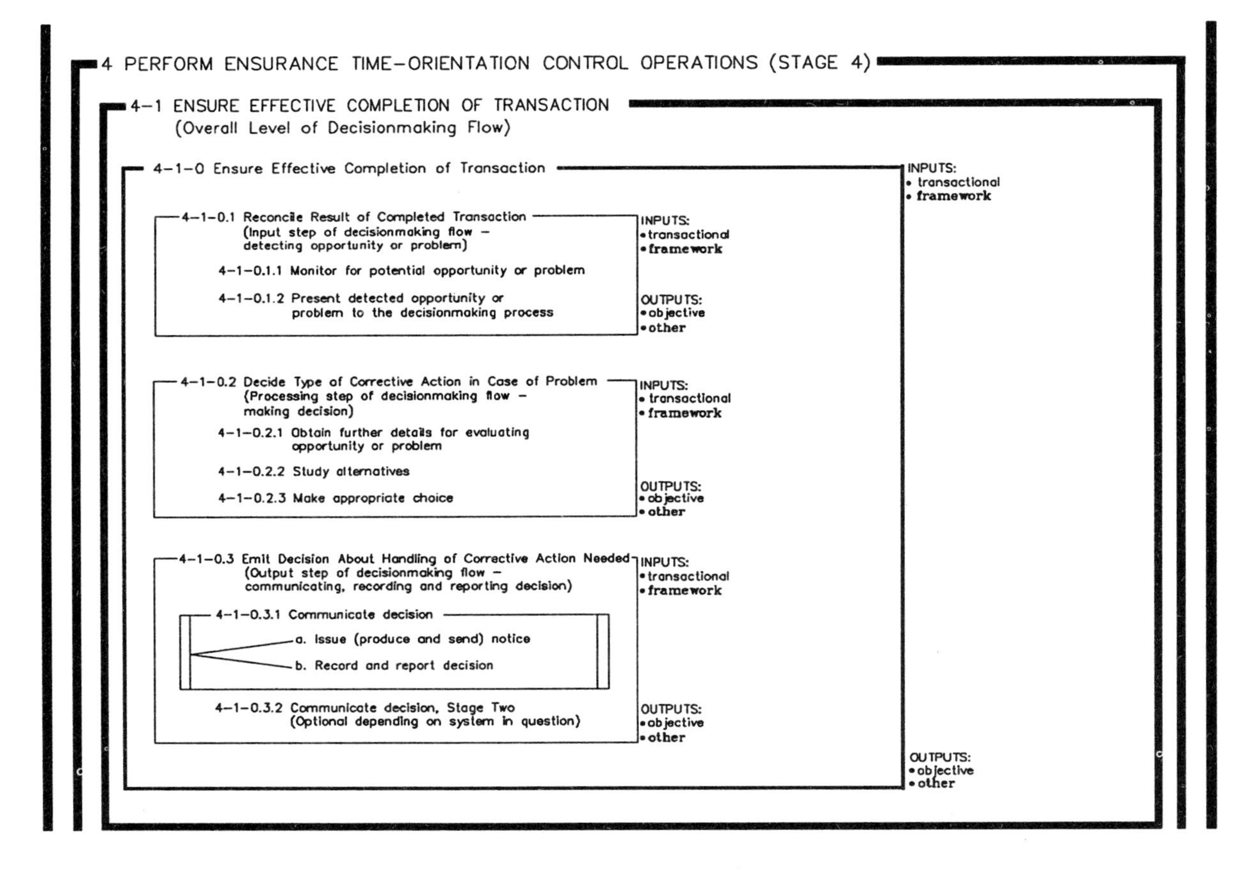
4 PERFORM ENSURANCE TIME-ORIENTATION CONTROL OPERATIONS (STAGE 4)
4-1 ENSURE EFFECTIVE COMPLETION OF TRANSACTION
(Overall Level of Decisionmaking Flow)
4-1-0 Ensure Effective Completion of Transaction
INPUTS:
• transactional
• framework
4-1-0.1 Reconcile Result of Completed Transaction
(Input step of decisionmaking flow – detecting opportunity or problem)
4-1-0.1.1 Monitor for potential opportunity or problem
4-1-0.1.2 Present detected opportunity or problem to the decisionmaking process
INPUTS:
• transactional
• framework
OUTPUTS:
• objective
• other
4-1-0.2 Decide Type of Corrective Action in Case of Problem
(Processing step of decisionmaking flow – making decision)
4-1-0.2.1 Obtain further details for evaluating opportunity or problem
4-1-0.2.2 Study alternatives
4-1-0.2.3 Make appropriate choice
INPUTS:
• transactional
• framework
OUTPUTS:
• objective
• other
4-1-0.3 Emit Decision About Handling of Corrective Action Needed
(Output step of decisionmaking flow – communicating, recording and reporting decision)
4-1-0.3.1 Communicate decision
a. Issue (produce and send) notice
b. Record and report decision
4-1-0.3.2 Communicate decision, Stage Two
(Optional depending on system in question)
INPUTS:
• transactional
• framework
OUTPUTS:
• objective
• other
OUTPUTS:
• objective
• other

4—2 ASSIGN ANY NEEDED CORRECTIVE ACTION FOR ENACTMENT
(Subordinate Level of Decisionmaking Flow)

CHAPTER TWO

FUNCTIONAL COHESION AND ORGANIZATIONAL MODELING*

This chapter discusses and illustrates the functional cohesion method, which is a concept needed in order to move further in this book. The method is illustrated in the following ways:

- by developing a model of human (specifically, business) organizations which is based upon functional cohesion;
- by also developing a model of human organization which is **not** essentially based on functional cohesion, thereby providing a way

* The following materials are based on two articles by Ed Baylin, both in the Cybernetics and Systems Journal: An International Journal. Washington, D.C.: Hemisphere Publishing Corp©. These articles are: 1) "Functional Modeling of the Business Organization," vol. 15, nos. 3—4, 1984; and 2) "Computer Systems Analyst's View of the Business Organization," vol. 16, 1985, pp. 305—323. It is indicated in the table and figure captions where copyright on a given table or figure is held by these journals.

of identifying subsystems which contrasts with the functional cohesion method;

- by including two supplements, which illustrate the following:
  - —the development by another author (Blumenthal) of an organizational model based on essentially the same ideas as the model using functional cohesion;
  - —the application of the functional cohesion method to the critique of the organizational structure of hospitals, based on an article by Brenda Shestowsky.

THE FUNCTIONAL COHESION CONCEPT

Functional cohesion means

1) keeping related things together, and
2) keeping unrelated things apart.

By achieving either one of these effects, the other effect is automatically achieved as well, since these are but two sides of the same coin. In other words, what is achieved within a system divided into subsystems (modules) by the functional cohesion method is the identification of a set of modules which are loosely coupled to each other (highly independent of one another) but internally highly coupled (each module is itself comprised of elements very related to one another).

An Example of Functional Cohesion — To give a simple example of such a set of modules, the reader might think of carving up a chicken with the cuts being made at the joints so that the pieces come apart easily. In other words, the cuts are made at the points of loose coupling. Such a carving pattern achieves the basics of functional cohesion, except for certain other factors to be considered (described below) which may not apply in this example. Of course, more than one way may exist of carving the chicken at points of loose coupling into pieces made up of tightly internally coupled clusters of parts.

WHY THE FUNCTIONAL COHESION METHOD IS WELL REGARDED

The functional cohesion method is recognized as the best one for identifying and defining most simple man-made systems, including computer information subsystems.

Ease of Comprehension and Communication of How the System Works — The functional cohesion method is the only decomposition method "by objectives," which makes the workings of the system easy to understand, communicate, and document. The mission component of each system's essence-identity is accomplished by achieving a number of intermediate objectives, up to and including fulfillment of the final system objective(s). Thus, with the functional method, a system is broken down in such a way that each subsystem corresponds to the sub-set of system operations that achieve one of the intermediate objectives, or the final (system) objective. Obviously, based on this objectives-oriented perception of the subsystem structure, the workings of the system become easy to see. This is the first benefit of the functional cohesion method.

Ease of Subsystem Maintenance — The second benefit of this method, which is the fact that maintenance of subsystems is easy, also largely results from its being the only one by objectives. Because this method produces the set of subsystem modules, within a parent system, which are the most optimally independent of one another, maintaining the subsystems is easy, as changes to one subsystem can be made without a proliferation of corresponding changes to other subsystems. In other words, the subsystems identified by the functional cohesion method are loosely coupled, i.e., have few interactions or interfaces. For example, in Figure 2-1a, three of the four subsystems, namely, B, C, and D, are loosely coupled to one another, as demonstrated by the two-way dashed arrows connecting them. As a result, changes may, for example, be made to the way C works without having to change B and D to accommodate new interfaces with C. Also, maintainability results from the fact that the functional cohesion method (at least as outlined below) avoids duplication (but not in ways which result in conflict between different subsystems).

CRITERIA OF THE FUNCTIONAL COHESION METHOD

How it is Explained in Current Literature

Keeping Highly Related Elements Together — One approach to obtaining a highly mutually independent but internally "cohesive" set of modules is to focus on the side of the coin which keeps highly related elements together. Thus, the functional cohesion method has been perceived as grouping all elements related to a single task into the same module. It has also been explained in a more circular fashion, by identifying types and degrees of cohesion, from least to most favourable, namely—coincidental, logical, temporal, procedural, communicational, and sequential—and then saying that functional cohesion is the method which the other methods fail to be.

Keeping Highly Unrelated Elements Apart — The other approach to achieving internally highly cohesive but mutually independent mod-

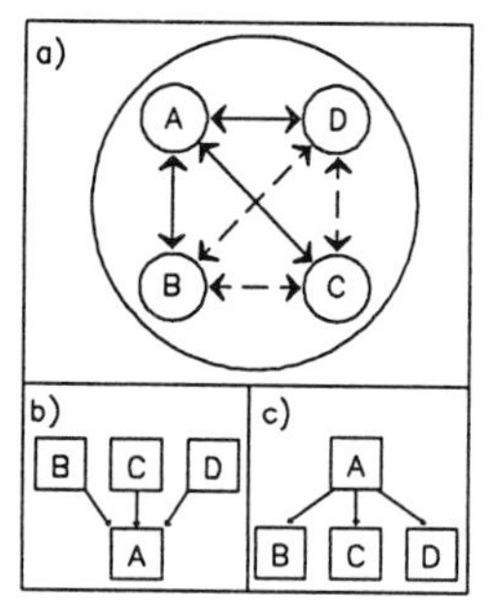

FIGURE 2-1: Functional cohesion method and avoidance of conflict.

a) avoiding subsystem conflict because of a system element which is highly coupled to other system elements. In this example, the reason for the high coupling is not specified. Even though system element A is highly coupled to the other system elements, A is separated from B, C, and D, by placing it into its own subsystem in order to avoid subsystem conflict. B, C, and D are loosely coupled to one another, and thus are separate subsystems;

b) avoiding subsystem conflict for a shared resource. To avoid subsystem conflict, the common resource system element, represented by A, is placed in its own subsystem. B, C, and D are loosely coupled to one another;

c) avoiding subsystem conflict due to a common controller. To avoid subsystem conflict, A, the supervisor (controller), is separated from B, C, and D. The latter are loosely coupled to one another.

ules in a system is to keep highly unrelated elements apart by studying the coupling between the various elements, and then eliminating the undesirable types of coupling. The one theory of coupling that is generally applied when this approach is used, has the following six degrees (categories) of coupling, from loosest to tightest: no coupling, data coupling, stamp coupling, control coupling, common coupling, and content coupling. These are defined in various publications on the subject of structured programming. Data coupling would roughly correspond to the pooled or sequential type of another, rather simpler, classification scheme for coupling. Thompson identifies four degrees of coupling, from loosest to tightest, as follows: no coupling, pooled (indirect) coupling, sequential coupling, and reciprocal coupling.* These two coupling schemes will be referred to in the following explanation.

This Author's Explanation

Principles of the Functional Cohesion Concept

The essence of the functional cohesion concept is as follows:
Keep together all sufficiently coupled elements having the same objective(s), unless duplication and/or conflict would result.

Listing and Explanation of the Functional Cohesion Principles

1) *All system elements which are closely related, either directly or indirectly, to the same intermediate/final system objective (or "mutually contingent" **set** of objectives) must be clustered together within the same subsystem, unless:*
   —some elements are not adequately coupled;
   —this would result in duplication of system elements;
   —this would result in subsystem conflict.
2) *All system elements clustered together within the same subsystem must be closely related, either directly or indirectly, to the same intermediate/final system objective (or mutually contingent set of objectives).*
3) *All system elements clustered together within the same subsystem must be sufficiently coupled.*

   This does not mean that every element must be **directly** coupled to every other element in the subsystem, but that the cluster of elements forming the subsystem must be sufficiently coupled by links between any one system element

---

* Thompson, J.D., Organizations in Action. New York: McGraw-Hill, 1967.

and the rest of the cluster. For example, A, B, and C may be placed in the same subsystem cluster, with both A and B being tightly coupled to C, although no direct coupling between A and B need exist.

Looking at the different degrees of coupling, sequential coupling is of an intermediate degree, and it is up to the person dividing the system elements to decide whether to keep merely sequentially coupled clusters together. The same may apply to data coupling (from the Thompson scheme of coupling), since this is a fairly normal type of coupling, and frequently corresponds to the sequential kind.

With respect to two elements coupled indirectly via a third element, if these form alternative sequential (uni-directional) links to (in the direction of) this third element, then the cluster of these three elements is insufficiently coupled. For example, where a business has two product lines, whose only significant link is that they both produce profit, the operations closely related to each product line should be placed in separate subsystems.

4) *In order to avoid duplication, when some system elements are closely related to more than one intermediate/final system objective (or mutually contingent set of objectives), the elements involved must not be placed in more than one subsystem.*

5) *In order to avoid subsystem conflict, when some system elements are closely related to more than one intermediate/final system objective (or mutually contingent set of objectives), the elements involved must not be placed in any one of the subsystems corresponding to those objectives, but, rather, into a separate, distinct subsystem of their own.*

This applies no matter how tightly coupled and how closely related to the same objective these system elements are, if conflict among the subsystems is thus engendered.

Functional cohesion attempts to achieve optimal independence of the different subsystems. This usually means that the subsystems are loosely coupled to one another. Sometimes, however, as in the case of subsystem A in Figure 2-1a (which is connected to the other subsystems by a solid line, indicating tight coupling), optimal independence does not mean loose coupling. In this case, separation of tightly coupled elements occurs when some of these need to be separated from each other in order to avoid conflict among the subsystems. This tight coupling could result from subsystem A containing the elements needed to control the activities of the other three subsystems, such as in the hierarchical arrangement of the subsystems in Figure 2-1c. Alternatively, subsystem A might, for instance, contain elements to maintain a machine used in enacting the activities of subsystems B, C, and D, as represented by the inverted hierarchy (the "fan-in" lines) in Figure 2-1b. In either case, the inclusion of the elements represented by subsystem A in any one of the other three subsystems would mean that the two subsystems left out would have to conflict with the subsystem containing these elements, in order to gain access to them. For example, if A's elements were contained in B, then C and D would have to invade B for access to A. To avoid such conflict among subsystems, the functional cohesion method requires that A be created as a separate, independent subsystem.

Avoiding Duplication as well as Subsystem Conflict — By considering principles 4 and 5 together, it can be realized that the avoidance of duplication does not in itself achieve functional cohesion when this avoidance leads to subsystem conflict, and vice versa. Consider two neighbours who share a common kitchen. If the kitchen is in a structure independent of the two neighbours' individual homes, duplication may be said to be avoided in a way which is more likely to avoid conflict than would be avoided by locating the kitchen in only one of the homes. This can be stated in terms more familiar to the computer programmer. Even though it effectively eliminates duplication, content coupling, where one module must traverse another module's territory in order to do its own operations, must be avoided, as it leads to conflict.

Table Presentation of the Functional Cohesion Principles — The principles of functional cohesion can also be stated in the form of a decision table, as in Table 2-1. In this, the condition entries are listed and numbered in the same order as the above principles. In the case of the first condition entry, however, a second condition must be included along with the primary one, to deal with possible implications with conditions 3, 4, and 5.

Applying the Functional Cohesion Method

List of Approximate Steps to Follow

1) Identify intermediate objectives of the parent system, and then group (cluster) together all the system elements closely related to each intermediate objective, or to the final objective.

CONDITION ENTRIES							
1) Are all elements which are related to the same objecive(s) clustered together?	N					Y	N
If not, is the separation only because of inadequate coupling or to avoid duplication or subsystem conflict?	N						Y
2) Are all elements in each cluster related to the same objective(s)?		N				Y	Y
3) Are all elements in each cluster adequately coupled?			N			Y	Y
4) Is duplication of elements absent?				N		Y	Y
5) Is subsystem conflict absent?					N	Y	Y
RESULTS							
Functional cohesion has not been achieved	X	X	X	X	X		
Functional cohesion has been achieved						X	X

TABLE 2-1: Decision table with rules for achieving functional cohesion. *This table may be used to determine whether subsystems within a system have been identified by the functional cohesion method. Note: All condition entries left blank could contain either Y or N without changing the result.*

2) Eliminate from each group (identified in #1) those elements not closely related to the objective(s) of the group (cluster). This step may be needed in case extraneous elements, which were for some reason highly coupled with objective-related elements, have been included in #1.
3) In the order given, separate (from each group left after #2) the following sub-groups of system elements into distinct subsystems of their own, thereby leaving the remaining part of each group as a subsystem identified by functional cohesion: (The separated sub-groups may also form subsystems identified by the functional cohesion method.)
   — those sub-groups of system elements which are insufficiently coupled within the group (cluster);
   — those sub-groups of system elements which are common to a number of different groups already identified (thus avoiding undesirable duplication);
   — those sub-groups of system elements which are tightly coupled to a number of different groups already identified (thus avoiding conflict between those different groups).

Examples of Functional Cohesion Being Applied — Figures 2-2 a, b, and c are hypothetical cases designed to give some practice in the application of the above heuristics. The first two figures show different ways in which the criteria can be met when applied to the same set of system elements, while the third figure shows an invalid grouping of system elements. Finally, Figure 2-3 is a structure-flow chart representing functions by wideshafted arrows. The circling of different arrows and arrow segments identifies subsystems along functional lines.

The Paradigm of Functional Cohesion at Different System Levels, and for Different System Types

The level at which functional elements are perceived to exist clarifies the functional cohesion method. To illustrate, Figure 2-4 shows the further analysis of one of the functionally identified subsystems into another level of subsystems identified along functional lines. Now, the question is, if the elements in subsystem D are all tightly coupled, how might it be possible to separate them further? In fact, a more micro perception of the elements exists at this next level of division into subsystems. This is analogous to thinking of protons and electrons, rather than atoms or molecules, as the elementary particles of matter. Within the more micro level, new patterns of high and loose coupling

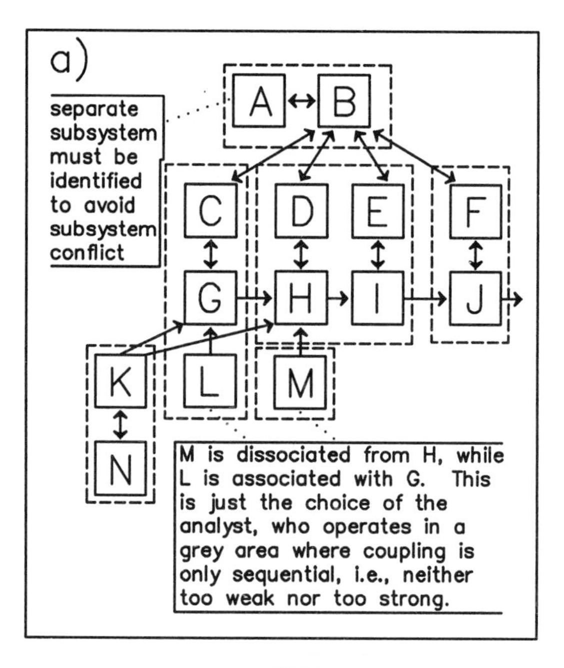

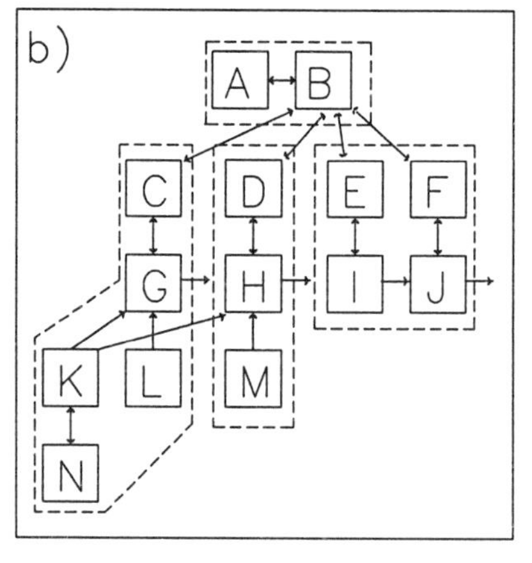

FIGURE 2-2:
Ways of grouping system elements into functional subsystems. *These show ways that the criteria of functional cohesion are or are not satisfied:*

a) ***(above)*** *Subsystems, obtained by enclosing functional elements (represented by squares) with dashed lines, are identified by the functional cohesion method. Where functional elements are sequentially coupled, the analyst has to decide whether or not to group them together, since sequential coupling is neither very strong nor very weak. This applies to the following pairs of elements: L and G, M and H, and H and I.*

b) ***(left)*** *This is another pattern of subsystem identification which satisfies the requirements of the functional cohesion method.*

c) ***(opposite page)***

of system elements are noticeable, so that the paradigm of functional cohesion may be applied once again—an instance of what is called the "relativity of functional conceptualization." It is this relativity which allows the application of the same paradigm to all levels and types of systems.

Comparison of the Different Explanations

Since the criteria of ensuring sufficient coupling, avoiding duplication, and avoiding subsystem conflict are included, the preceding criteria of functional cohesion go beyond just saying that a functionally cohesive module performs only one task (similar to step #1 in the above heuristic) and contains all the system elements closely related to the achievement of that task (similar to heuristic steps #1 and #2 together).

In terms of the above principles of functional cohesion, it would appear that all current definitions found in the literature of the functional cohesion method include references to the ideas stated more precisely in principles #1, #2, and perhaps #3, while reference to principles #4 and #5 may be entirely missing from current definitions.

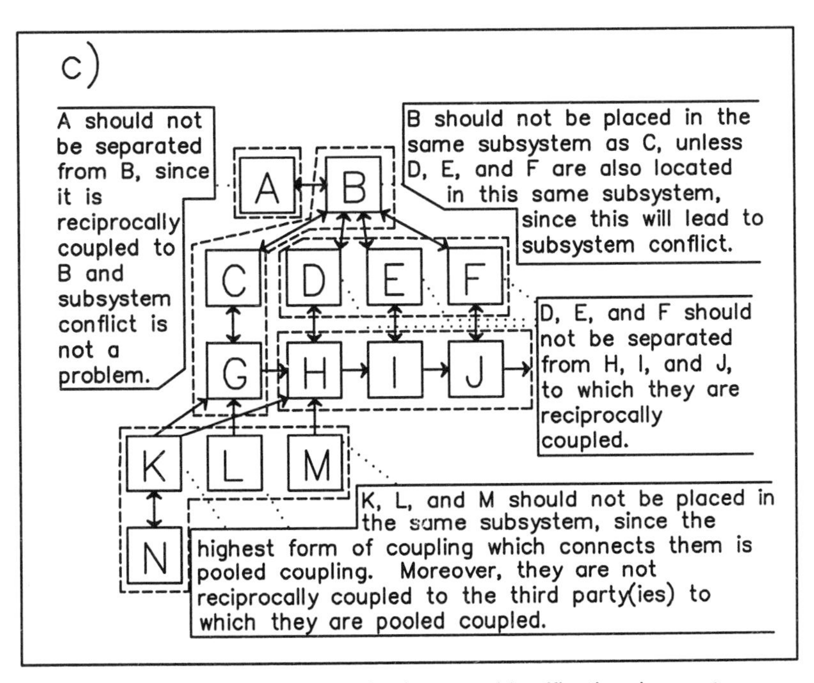

FIGURE 2-2c: *This pattern of subsystem identification does not entirely conform to the requirements of the functional cohesion method, since none of the above four subsystems meets all the requirements, for reasons indicated in each case.*

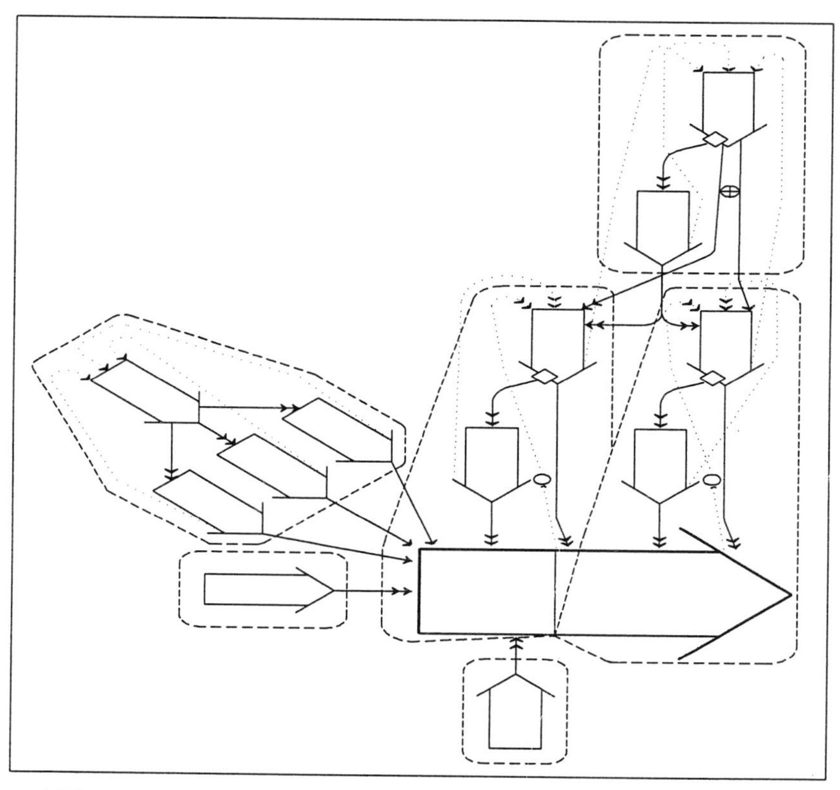

FIGURE 2-3: Functional cohesion illustrated by the structure-flow chart. *Functionally cohesive subsystems are encircled by dashed lines.*

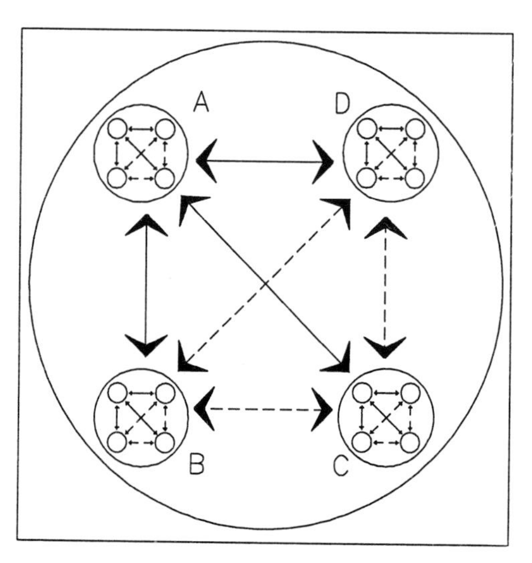

FIGURE 2-4: Functional cohesion nested within functional cohesion. *Based on a more detailed level of perception of system elements, the functional cohesion method may be applied to identify yet another level of subsystems.*

LINK TO FURTHER DISCUSSION

The application of the criteria of the functional cohesion method is a complex matter, since various problems may make it difficult and/or inappropriate to apply the method. A discussion of these matters is beyond the scope of this chapter, and requires materials from Chapter Ten as well as the various concepts of functions and functional elements to be provided in the next chapter. Meanwhile, a functional element should be taken to imply a sub-function, i.e., some part (or all) of a function.

FUNCTIONAL MODEL BREAKING DOWN THE ORGANIZATION INTO SUBSYSTEMS

This section of the chapter illustrates the use of the functional cohesion method to model the organization, while the next section illustrates the contrasting results obtained by using a number of different cohesion methods for this. Specifically, the second model, which may be dubbed "the computer applications analyst's view," is based on both the logical and procedural cohesion methods of subsystem identification, as well as on the functional method. In both views of the organization, the cohesion methods are applied over a much wider range of system levels than has hitherto been done by computer system specialists, who have tended to restrict the application of these methods to modular design of computer programs.

The terminology used in describing the two models presented in this chapter is actually based on a particular kind of organization, namely a manufacturing business. However, these terms can be altered to accommodate any type of industry.

The model developed in this section (see Figure 2-5) derives subsystems of the business organization using the functional cohesion method. In conjunction with reading this, the reader may want to review the two supplements to this chapter. Also, the reader may wish to refer forward via the index for some of the terms which have not yet been introduced.

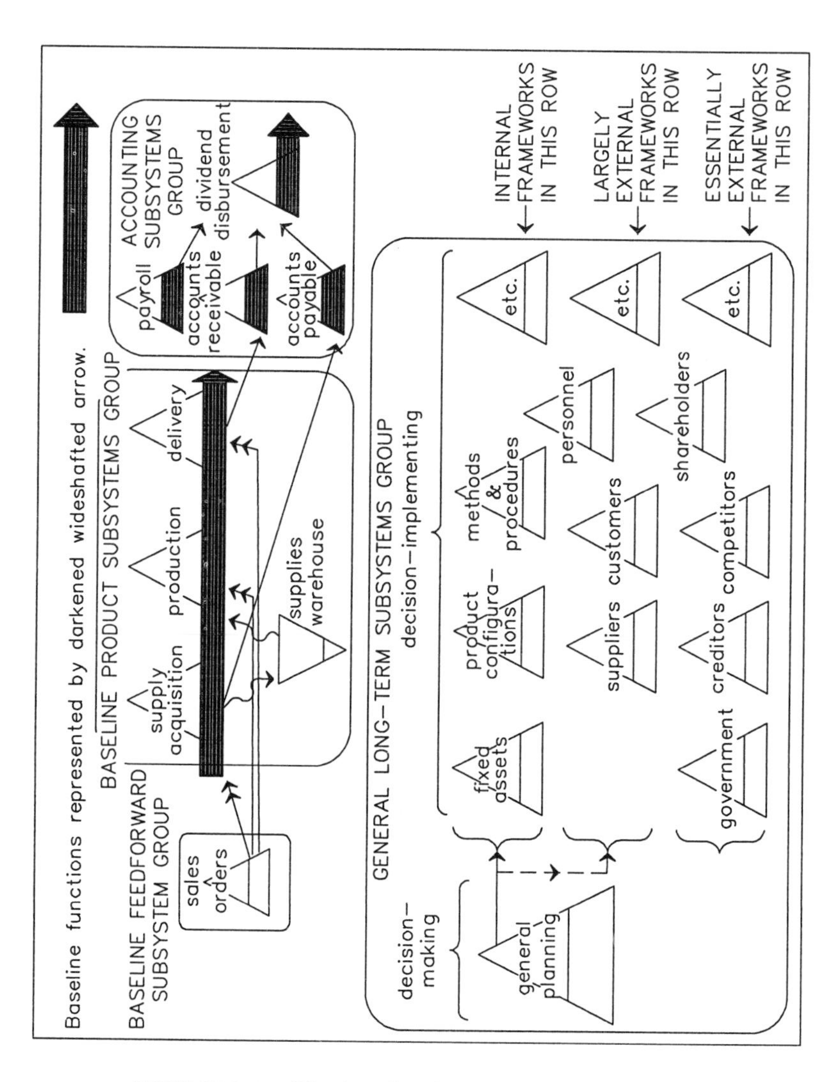

FIGURE 2-5: The functional model of the organization, with subsystems represented by triangles.

BASIC DESCRIPTION OF THE FUNCTIONAL MODEL

This sub-section describes the following:

—the major subsystems of the organization, i.e., the subsystems included in the diagrammed depictions of the model (see version given in Figure 2-5);

—"reservoir (buffer)" subsystems, i.e., a type of subsystem which is only implied in these diagrams;

—constraints present in the functional model.

Major Subsystems

Baseline Feedforward Subsystem

Sales orders, the baseline feedforward subsystem, performs current (short-term) feedforward functions. Simply, it obtains and enters customer orders. The baseline feedforward subsystem is informational in nature; i.e., it does not have a physical (in the sense of materials-processing) base of its own.

Baseline Product Subsystems

These subsystems directly achieve what the business sells on the marketplace. Because the example used in the present model of the organization is that of a manufacturing business, the baseline product subsystems are actually "physical" systems. (Here, **physical** is not being used in the sense of opposition to **logical**, but rather in the sense of "material.") However, there exist other types of organizations in the service industries whose baselines provide only services, i.e., may be informational rather than physical.

The following are characteristics of the so-called "product" subsystems of the manufacturing organization being modeled:

- **Supply Acquisition Subsystem** — obtains the transactional supplies to be transformed (or otherwise operated upon) by the production subsystem. For instance, it obtains raw materials and component parts to be used in manufacturing.
- **Supplies Warehousing Subsystem** — stores supplies prior to their being used in production.
- **Production Subsystem** — uses the transactional supplies to make the finished product(s) (either by transformation of the inputs, or by other means) ready for delivery to customers. For example, production would mean manufacturing in an organization that assembles finished goods from component parts.

- **Delivery Subsystem** — delivers the finished product(s) to customers.

Accounting Subsystems

The subsystems and their objectives are as follows:

—**Accounts Receivable Subsystem** — obtains money from customers in payment for products delivered.

—**Accounts Payable Subsystem** — disburses money to suppliers for goods and services provided.

—**Payroll Subsystem** — disburses money to employees for services rendered.

—**Dividend Disbursement Subsystem** — pays shareholders and others having an ownership interest.

Since the general accounting subsystem produces information used by high-level managers to plan business activity, this accounting subsystem is contained within the general planning subsystem, and thus is not exhibited in the figures by its own triangle.

Long-Term General Subsystems

The long-term general subsystems constitute the fourth group of subsystems. This group performs decision-making and decision-implementing operations which are long-term and general in scope.

General Framework Subsystems — The nature of the general framework subsystems, which carry out decision-implementing functions, depends upon the degree to which the framework is internal. The objectives of the internal framework subsystems are generally to acquire and install frameworks or to improve and/or replace them, such as by the acquisition of a new machine, the implementation of a new set of policies and procedures, the retraining of personnel, the start-up of a new product, etc. As for largely external frameworks, the subsystem objectives are more related to influencing the framework, e.g., winning new customers by a promotional campaign, improving relationships with a supplier, etc. Lastly, wholly external frameworks may not be directly affected by framework operations.

General Planning Subsystem — The general framework subsystems group also includes the general planning subsystem, which initiates and otherwise controls the general framework subsystems by rendering strategic-level decisions affecting general frameworks. The general planning subsystem is thus shown in the diagrams as being connected to the decision-implementing general framework subsystems.

"Federal" Role of the General Subsystems — While each triangle in the overall diagram contains its own local planning and framework operations, the general planning and general framework operations are placed in their own, separate subsystems to avoid conflicts among the various subsystems. Stated in another way, the roles of the long-term general subsystems versus the roles of other subsystems in the model is like the role of a federal government versus that of a province or state. A federal government does not necessarily have to be more powerful than a state or provincial one, since the only required differentiating criterion is that it perform control functions which have a more general scope of applicability than the ones the provincial or state governments perform. In fact, the categorization of subsystems as "federal" may simply result from the way the state and province subsystem boundaries are placed. For example, two very local areas of two different states may be affected by a function performed by the federal government. Thus, the categorization of something as "general" is not always a matter of **wideness** of scope, but can be a result of the scope lying over two different jurisdictions. Figure 2-6 portrays federalism, for whatever purpose, using a triangles chart. The idea of "federal" triangles applies in the present discussion to the placement of framework and planning operations which have a general scope of applicability in the functional

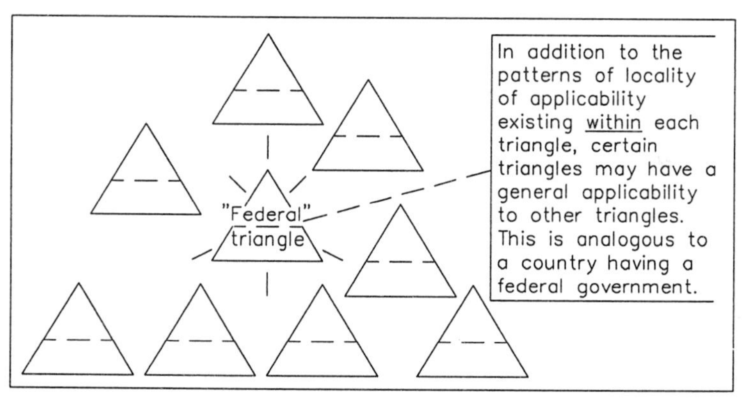

FIGURE 2-6: The idea of "federal" subsystems.
Each triangle represents a sub-system identified by functional cohesion within the same parent system as the other triangles. The part of each triangle below the broken line contains the baseline operations of that subsystem, while the part above it contains the control functions. Only one of the triangles is designated as "federal," while the remaining triangles represent non-federal jurisdictions.

model of the organization. Of course, as already mentioned, there also exist planning and framework operations **local** to each functional subsystem of the organization.

Buffer (Reservoir) Subsystems

Direct Interfaces versus Interfaces Via Reservoirs

The subsystems of the organization interface with one another, through both physical (material) and data (informational) flows. Although these flows sometimes occur randomly, on a transaction-by-transaction basis, they more frequently occur through reservoirs (stocks, or "levels"*), i.e., stores that accumulate things. Thus, data flows occur via files, in which data is accumulated, while physical flows involve the storage of physical supplies in depots, e.g., warehouses. For instance, data concerning customers is stored in the customer static data master file, data concerning customer orders is stored in the customer order transactions file,** supplies needed to fill customer orders are stored in a warehouse, and personnel needed to carry out the order filling are in a pool of employees available for assignment.

Subsystems for Operating Reservoirs

When a reservoir is involved, operations are needed to store items, to manage and control storage and stored items, and to remove items from the store when they are needed elsewhere. To illustrate, the supply acquisition subsystem sends acquired raw materials to a warehouse, where they are placed in bins, from which they are removed in response to requisitions for raw material from the ensuing production subsystem. While in the store, records are kept of bin location, transfers of bin location, spoilage or theft, etc. This data is used to control the store. Also, planning and framework operations are needed to set up the warehouse, to establish policies for running it, to provide methods for recovering erroneously removed contents, and so on. In other words, the existence of a store entails the existence of a subsystem needed to run the store, on both short-term and long-term bases.

---

* Forrester, J.W., Industrial Dynamics. Cambridge, Mass.: M.I.T. Press, 1961.

** See the related Ed Baylin book, Conceptual Prototyping of Business Systems—A Templating Approach to Describing System Functions, for a new scheme of classifying data files in computer business systems.

Diagramming of Reservoirs (Buffers)

When a reservoir is involved in the interfacing of different subsystems of the business, the reservoir belongs neither in one subsystem nor in the other (an incidence of the functional cohesion principle of avoiding conflict); that is, it acts as a buffer between the two. Similarly, the subsystem needed to operate the reservoir belongs neither in one subsystem nor in the other. The functional model of the business organization involves many such buffers and their associated operating subsystems, which are needed to interface the other subsystems. The diagramming of most of these reservoirs is left to more detailed charts showing more about the individual subsystems of the organization.

Constraints Present in the Functional Model

So far, the terms used in describing the functional model of the organization have mostly been those applicable to a manufacturing business. However, this model can in fact accommodate any type of industry. Table 2-2 gives the baseline subsystems of a manufacturer, a bank, a distributorship, and a hotel. In fact, any type of organization having a definable technical objective can be fitted with the functional model. Thus, the model could apply to a welfare agency, a discotheque, a management consulting firm, etc. Where the system base does not involve the transformation or transmission of materials, and the business produces services rather than goods, the baseline is of the informational, rather than of the physical type.

In addition to the constraint of terminology, the model, as developed so far, includes other constraints. The reader should be aware of this, since the functional model can perhaps be stated more flexibly. These constraints are as follows:

1- The objective of the organization is profit.

2- The business organization has only a single product line. However, in some cases, the same business may produce profit via more than one line. (Note: The perceiving of the **product**(s) as the system objective(s) would definitely be a **logical** constraint.)

3- The types of frameworks have been identified. Nevertheless, many more should perhaps be identified, the categories should be broken into sub-categories, certain categories should be eliminated, and so on.

4- Certain operations, namely, the general planning and framework ones, are executed less frequently, since their effects are long-term relative to the baseline operations. This dif-

	SUPPLY ACQUISITION	PRODUCTION	DELIVERY
MANUFACTURER	acquire raw materials and components	assemble and mold finished products	deliver finished products to customers
HOTEL	acquire cleaning and personal care supplies	clean and upkeep guest rooms with various supplies	provide rooms for guests
DISTRIBUTOR	acquire resale goods	store resale goods and remove them to fill orders	deliver resale goods to customers (i.e., retailers)
BANK	acquire monetary deposits	arrange for loans to customers	disburse loans to customers

TABLE 2-2: Baseline product subsystems of different industries.

ference in time frames means that a scheduling constraint has been incorporated into the model.

5- The supplies warehouse has been identified, as a buffer subsystem, meaning that supplies are stocked in advance of customer orders. In effect, batch processing is implied by the existence of this transactional store, similar to the difference between batch and unary (single-transaction) processing modes in computer systems.* This decision to have an accumulation point in the system flow means another scheduling constraint.

The above constraints have been embodied into the model of the organization, although one might find that these constraints are peculiar to given systems, i.e., are physical constraints rather than logical ones.

---

* See the related Ed Baylin book, Conceptual Prototyping of Business Systems—A Templating Approach to Describing System Functions, concerning batch versus unary processing in computer systems.

VARIATIONS OF THE FUNCTIONAL MODEL

The functional model of the organization, which was presented in a triangles chart, is now illustrated in groups of diagrams of two different types. In the first type, triangles are used to symbolize subsystems identified along functional lines. The base section of each triangle (in the shape of a rhombus) represents the baseline of the subsystem in question, while the upper remaining part of the triangle (which is a smaller triangle) represents the control operations of the subsystem.

The second kind of diagram is the structure-flow chart. Figure 2-7 is the structure-flow chart equivalent of the triangles chart in Figure 2-5. In the structure-flow chart, the base (baseline) is the only part of each subsystem which is actually diagrammed. That is, the control operations in each subsystem are, for simplicity, not illustrated in the particular structure-flow charts drawn here. Thus, while in the triangles diagram the baseline of the organization is the sum of the *baselines* of the organization's baseline subsystems, in the structure-flow chart the organization's baseline is simply the sum of the baseline subsystems. Regardless of which form is used, essentially the same subsystem set is identified in all diagrams, with the variations described next.

Variation Incorporating Finer Breakdowns of Subsystems

One of the triangles diagrams (Figure 2-8a), and an equivalent diagram using the structure-flow chart (Figure 2-8b), break down two of the business subsystems in more detail, to distinguish their initiative feed-forward operations from their other functions. This is accomplished by sub-dividing each subsystem into two sub-subsystems. The point of doing this is to show that a number of ways may exist in which the functional cohesion method can be applied to divide the business organization into subsystems, depending upon the level at which the system elements are perceived by a given analyst; e.g., one subsystem to one analyst may be seen as two by a second analyst, whereas these two subsystems would each be seen as sub-subsystems by the first analyst. In the examples given, of the sales and supply acquisition subsystems being each broken into two, it would appear that, in fact, these should be seen as sub-subsystems since their identification is based on a finer perception of system elements, i.e., perception of system elements at a more detailed level.

Decomposition of the Sales Orders Subsystem — For a finer breakdown, the sales orders subsystem is decomposed into the following sub-subsystems:

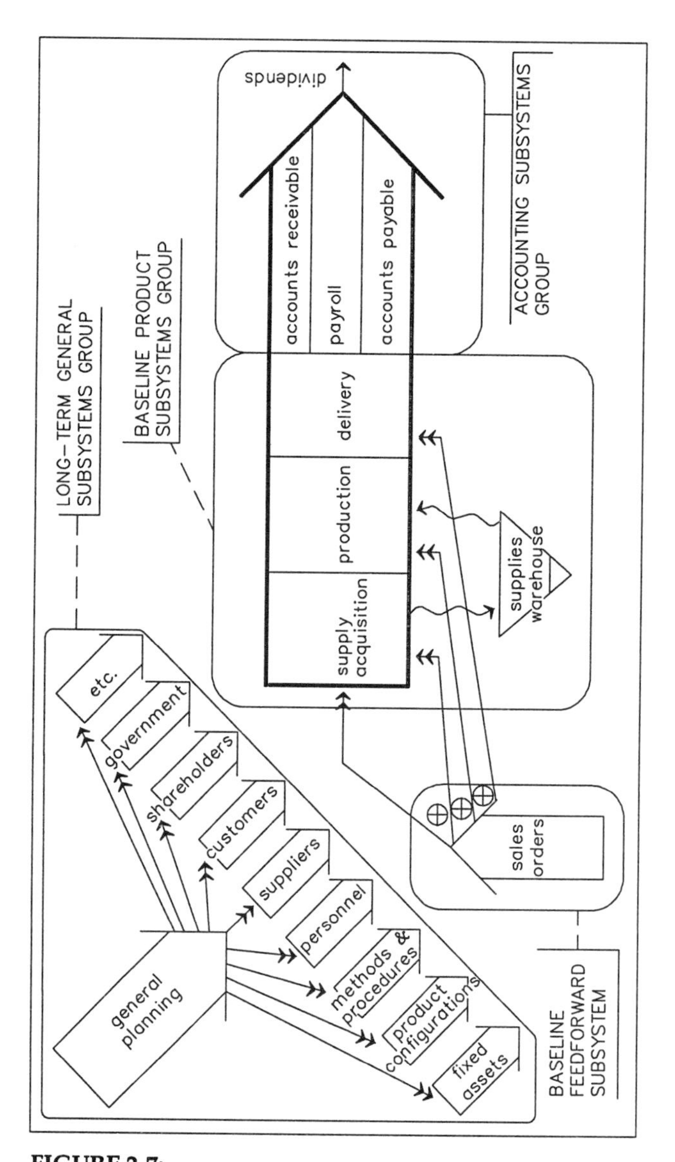

FIGURE 2-7:
Structure-flow chart representation of the functional model. *In this model of the business organization, the accounting operations are in the baseline, and the baseline feedforward consists of a single subsystem. This is the structure-flow chart equivalent of the triangles diagram in Figure 2-5.*

- **sales** — The baseline of this sub-subsystem is a procuration decision-implementing function of the sales orders subsystem. The sales sub-subsystem stimulates the arrival of specific customer orders, by influencing customers. It operates within a framework of customer relations established by the customers' general framework subsystem.

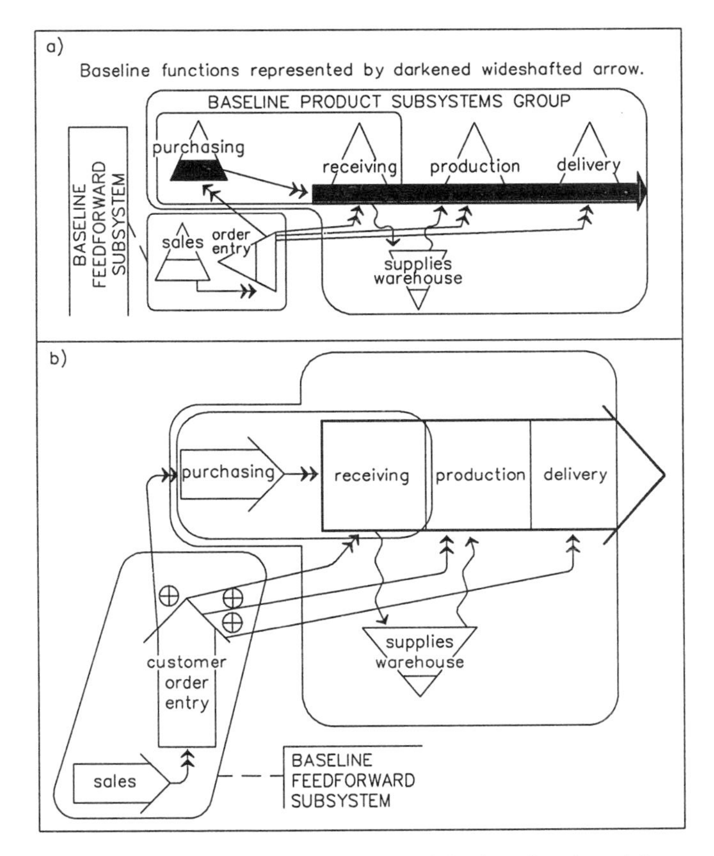

FIGURE 2-8: More detailed version of the functional model. *In this version of the functional model of the organization, the baseline feedforward subsystem, and the supply acquisition subsystem, each consist of two sub-subsystems. This version is illustrated in two different ways:*

a) representation using triangles to indicate subsystems;

b) representation using the structure-flow chart.

- **customer order entry** — The baseline of this sub-subsystem is activated either by sales activity, which acts as an initiative feedforward to order entry, or by customer demand in the absence of influence by current sales operations. The order entry sub-subsystem enters new customer orders. The next step depends largely upon whether inventory, finished or non-finished, is in stock. If, for instance, no inventory of either finished goods or raw materials is in stock, a new order will lead to initiation of the supply acquisition subsystem, via the latter's purchasing subsystem (sub-subsystem of the organization). In the other extreme case, if finished products are available in stock, new order entry will initiate the delivery subsystem, thereby bypassing supply acquisition and production.

Decomposition of the Supply Acquisition Subsystem — As for the finer breakdown of the supply acquisition subsystem (depicted in Figure 2-8), the purchasing sub-subsystem performs an initiative (feedforward) coordination function in relation to the supply acquisition subsystem's baseline, namely, the receiving sub-subsystem.

Variation Due to a Different Organizational Objective

Description and Consequences of the New Objective — Another variation of the functional model of the organization using a triangles chart (Figure 2-9a), and its equivalent with the structure-flow chart (Figure 2-9b), shows the accounting subsystems in a different perspective. This variation demonstrates a different view of what the business organization is all about (i.e., of its essence-identity), thereby changing the functional classification of the accounting operations, and hence the position of these operations with respect to other operations in the charts. This reflects the general principle that different perceptions of the logical constraints associated with a system may—sometimes surprisingly—alter the functional class of certain system operations.

Thus, in the variations of the functional model of the business organization represented by Figures 2-5 and 2-7, profit is seen as the only system objective. In contrast, the system objective in Figure 2-9 is the product sold by the business. Although the model illustrated in Figure 2-9 is essentially the same as the one in which profit is seen as the objective, the accounting operations are viewed as **framework upkeep** ones, rather than baseline ones.

Explanation of the Alteration in Class of the Accounting Functions — This change from baseline to framework upkeep can be explained as follows. When analysts see profit as a system objective, operations which collect and disburse money become part of the business baseline, for obvious reasons. On the other hand, if analysts do not see profit as the system objective, then to classify the accounting operations it is necessary to ask what would happen if bills from suppliers were not paid and if money were not collected from customers. Undoubtedly, it would not take long before the business' relationships with suppliers would deteriorate to the extent that supplies could not be obtained. That is, the framework of system relationships with sup-

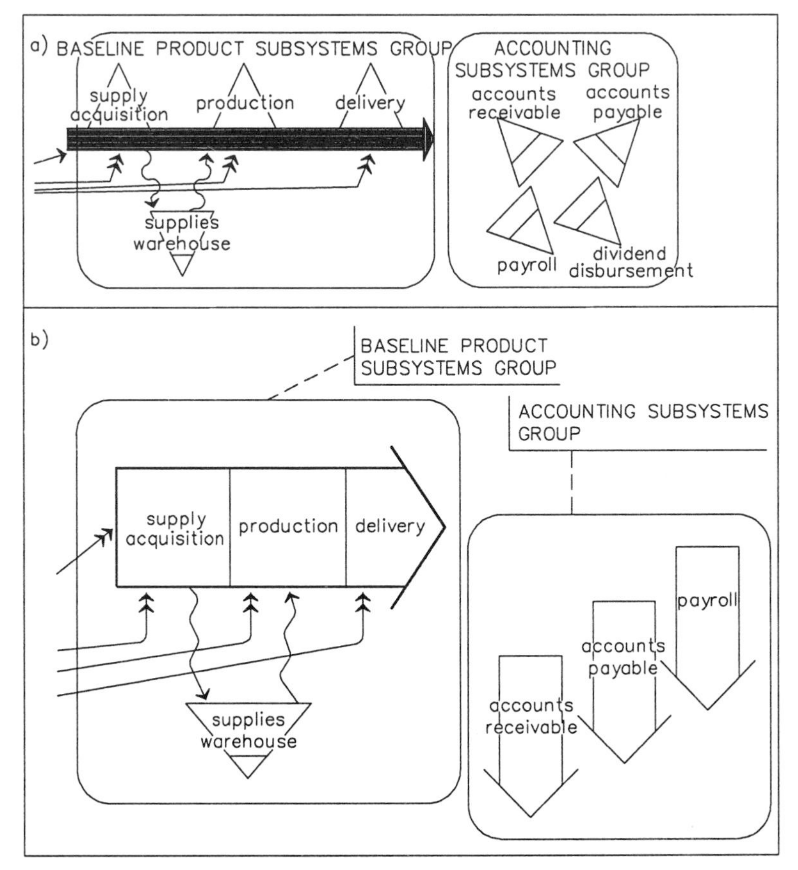

FIGURE 2-9: Version of the organizational functional model, with new roles for the accounting subsystems. *Here, the accounting operations are seen as framework upkeep ones, instead of baseline ones.*

a) triangles diagram representation;

b) structure-flow chart representation.

pliers and customers must be preserved by continually collecting compensation for products delivered, and making payments for supplies acquired. This establishes an energy interchange equilibrium* with the system environmental frameworks. Thus, in Figures 2-9 a and b, the accounting subsystems carry out framework upkeep functions.

A HYBRID MODEL IDENTIFYING INFORMATION-CONTROL SYSTEMS

Information systems are, from a macro perspective, subsystems within parent systems having "physical" (in the sense of material) bases, about which information is generated. Such information systems may be decomposed into subsystems corresponding to different levels of control, such that information subsystems higher in the control hierarchy control other information subsystems at the next level lower in the control hierarchy. These, *in turn*, at one or more levels lower in the hierarchy of control, ***directly*** *control systems having physical bases.*

This section develops a second model of the business organization, one which singles out the information-control subsystems, as well as the computerized parts of these (since not **all** information is produced by computers), thus reflecting the view of the computer/information systems analyst. The first step in identifying such computer sub-subsystems is to identify their parents, namely, the information-control subsystems. Only then should one take the next step of identifying those parts of these subsystems which are in fact computerized.

Two versions of the hybrid model are presented in this section:

—a simple version, which is developed using a mixture of the logical and procedural cohesion methods;

—a more complex version, which combines the functional cohesion method with the logical and procedural ones.

---

* See Chapter Five for more details on environmental equilibrium interaction.

The complex version of the hybrid model is cross-referenced to the functional model, thereby demonstrating two parallel paths which cohesion techniques may follow to divide the organization. This enables the hybrid model—the information systems analyst's view—to be compared to the functional model—e.g., the view of the computer information system user. This cross-referencing will be illustrated in Figure 2-12 below.

Cohesion Techniques Involved in Developing the Hybrid Model

Cohesion techniques discussed in the computer literature on structured analysis and design are usually applied to identify subsystems within computer information systems (groups of interrelated programs having a common purpose), or to identify subsystems within computer programs (since each program is a system at its own level). However, these same techniques can be used to analyze any system into subsystems, as has been done in developing the models of the organization presented in this chapter by applying these techniques to higher levels of subsystems than is usual.

The three cohesion methods involved are the following:

- *the functional cohesion method,* which involves keeping together related system elements and keeping apart unrelated ones, while at the same time avoiding both duplication and conflict;
- *the logical cohesion method,* which identifies subsystems by grouping system elements according to a conceptual classification scheme, e.g., by making subsystems for each of the planning, organizing, directing, and ensuring classes of operations;
- *the procedural cohesion method,* which identifies subsystems through grouping system elements belonging to a common procedural context. This idea may, for example, refer to the procedures connected with the particular instrument used to carry out operations.

A SIMPLE VERSION OF THE HYBRID MODEL

The computer/information systems analyst's view of the business is simplistically shown in Figure 2-10. Here, the large circle represents the business organization, while the other geometric figures portray subsystems of the business. In developing this simple model, only logical and procedural cohesion methods are used, while functional cohesion is not applied.

First-Level Decomposition — Using Logical Cohesion to Derive the Information-Control Subsystem

On first division into subsystems, the business is decomposed by the logical cohesion method into "information-control" and "non-information-control" (approximately called "operations") classes of operations, each surrounded by a dashed line in Figure 2-10. This division into two subsystems is explained with certain terms whose proper understanding requires a knowledge of the scheme for classifying functions (see Chapter Three), as follows:

- *Information-control operations* of a system are all its informational functions, **except** for baseline ones. More specifically, the information-control operations at all levels of that system consist of :
  - —decision-making (planning/framework-ensuring and coordination/validation) classes of functions;
  - —informational decision-implementing functions.
- *Non-information-control operations* are constituted by:
  - —the system's baseline operations—whether informational or physical (material) in nature;
  - —physical (material) decision-implementing control operations.

Thus, forming the information-control subsystem involves grouping parent system functions according to their **class**. This logical cohesion method produces very different results from those of the functional cohesion technique, since it does not guarantee that highly related system elements will be kept together in the same subsystem and that highly unrelated elements will be excluded.

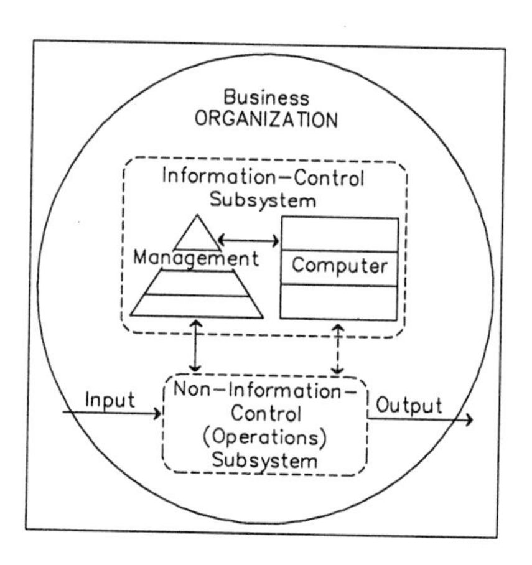

FIGURE 2-10: Simplified version of the hybrid model. *A computer analyst's view of the organization, in which only the logical and procedural cohesion methods are harnessed.*

Second-Level Decomposition — Using Procedural Cohesion to Derive the Computer Sub-subsystem

Next, the information-control subsystem is divided using procedural cohesion (according to method of operation), into:

—a "management" sub-subsystem—represented by the triangle in Figure 2-10—where the term "management" basically implies that the information-control operations are performed by humans, rather than by computers;

—a computer sub-subsystem.

Now it becomes possible to analyze the computer information-control sub-subsystem into further levels of subsystems, namely, programs and sub-programs (program modules).

Worthy of note is that the business computer sub-subsystem as a whole can itself be decomposed by logical cohesion into classes of computer systems in business. The related publication on conceptual prototyping of business systems discusses this classification.

A MORE COMPLEX VERSION OF THE HYBRID MODEL

A more complex pattern of hierarchical decomposition is shown in Figure 2-11. In this, each subsystem in the functional model of the organization is decomposed according to the above simple model.

Figure 2-12 tackles the problem of the complex hybrid model from a somewhat different perspective. Here, the left-hand side gives the decomposition of the business organization using only the functional cohesion method. The contrasting, alternative pattern of subsystem identification in the right-hand half uses logical, functional, and procedural cohesion methods at different levels of decomposition, a view familiar to the information systems specialist (albeit not in these terms). Logical and procedural cohesion only may be applied to the organization as a whole, as they are in the simplified hybrid model just described. However, they are usually applied in a way which mixes them with the functional cohesion method, thus relating the information-control operations being discussed to particular functional subsystems of the organization derived in the functional model of the organization described earlier.

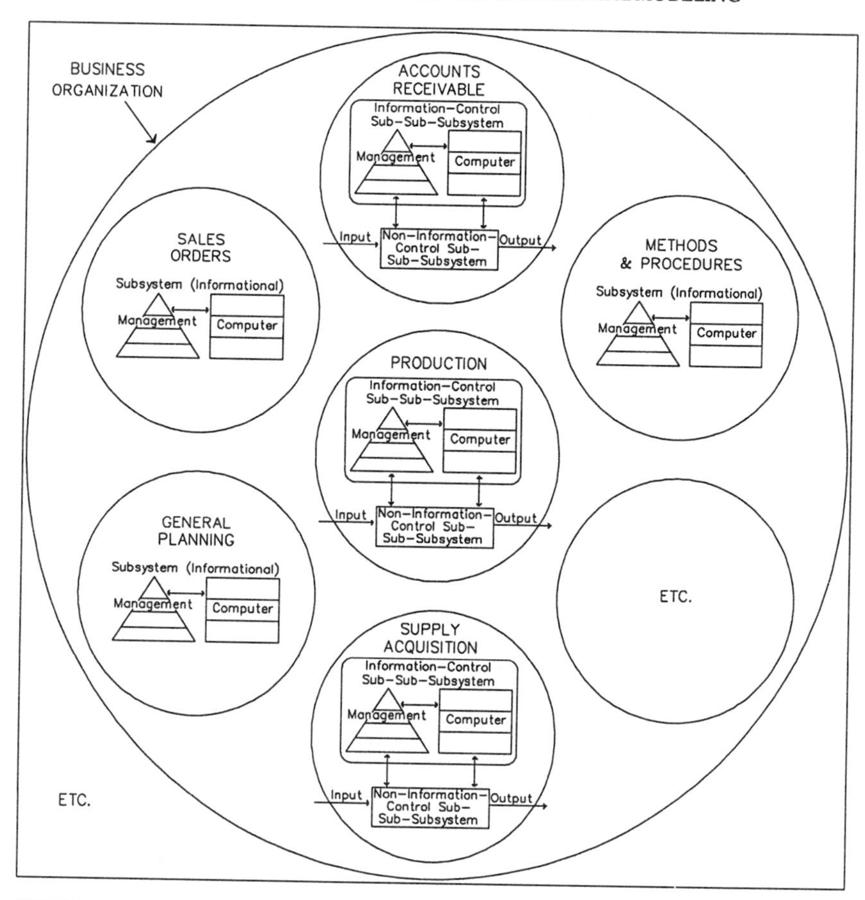

FIGURE 2-11: Hybrid model, complex version, composed from simpler versions. *Subsystems of the functional model of the business organization, are represented here by circles within a circle representing the business organization as a whole. Where the subsystem is not already of an informational base, an internal sub-subsystem is shown for its informational operations, while another one is shown for its remaining (non-informational) operations. Within each informational subsystem, or sub-subsystem, as the case may be, another level of internal subsystem is identified for the computerized informational operations.*

FIGURE 2-12: (facing page)
Hybrid model, complex version—one subsystem's perspective.
A view similar to that in Figure 2-11, but using a different format, and selecting only one of the information systems for further breakdown. Here, all the subsystems represented by triangles are identified by functional cohesion, while those symbolized by rectangles are identified by either logical or procedural cohesion, as the case may be.

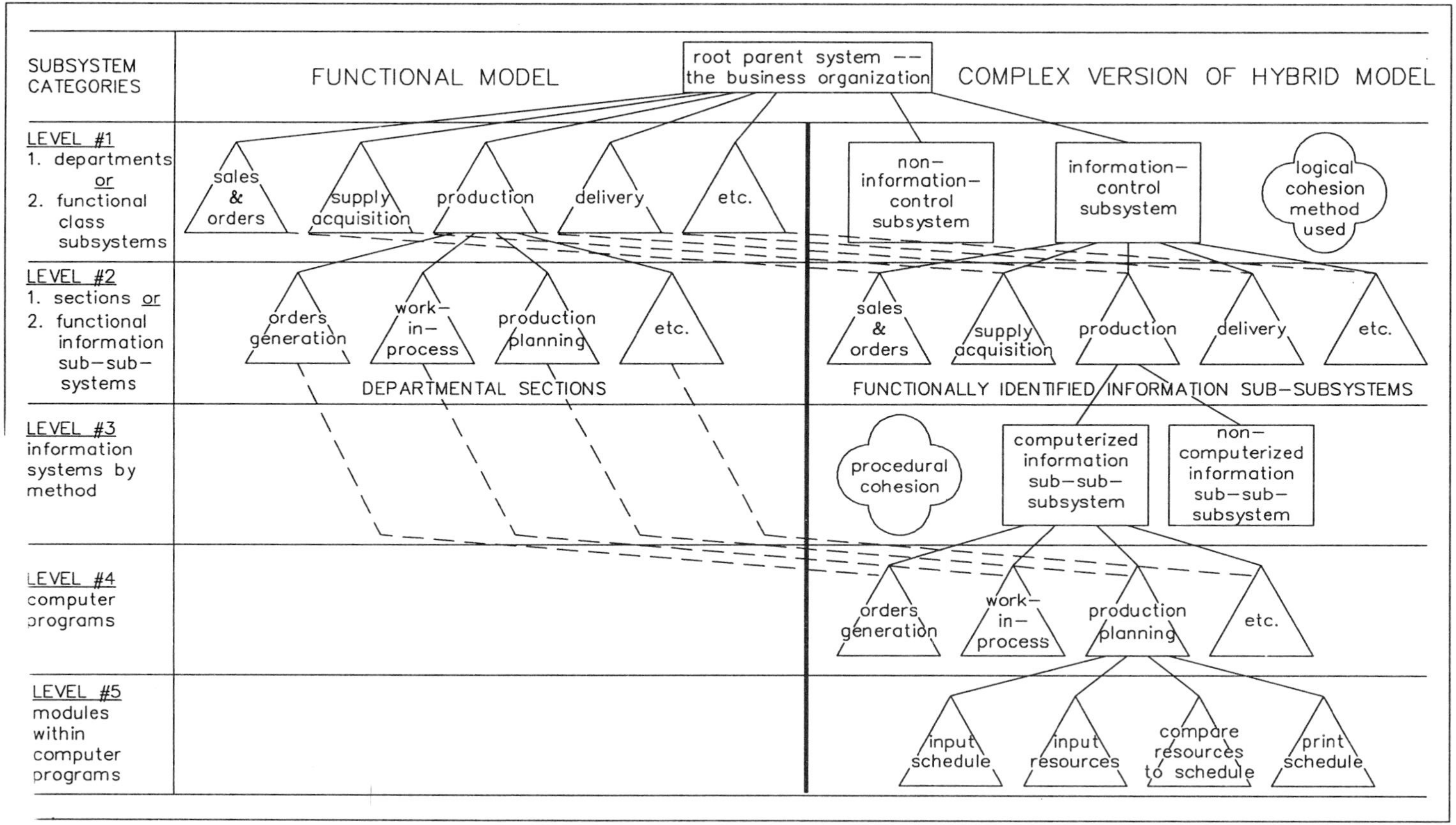

SUBSYSTEM CATEGORIES

FUNCTIONAL MODEL

root parent system -- the business organization

COMPLEX VERSION OF HYBRID MODEL

LEVEL #1
1. departments
or
2. functional class subsystems

sales & orders
supply acquisition
production
delivery
etc.

non-information-control subsystem
information-control subsystem
logical cohesion method used

LEVEL #2
1. sections or
2. functional information sub-sub-systems

orders generation
work-in-process
production planning
etc.

DEPARTMENTAL SECTIONS

sales & orders
supply acquisition
production
delivery
etc.

FUNCTIONALLY IDENTIFIED INFORMATION SUB-SUBSYSTEMS

LEVEL #3
information systems by method

procedural cohesion
computerized information sub-sub-subsystem
non-computerized information sub-sub-subsystem

LEVEL #4
computer programs

orders generation
work-in-process
production planning
etc.

LEVEL #5
modules within computer programs

input schedule
input resources
compare resources to schedule
print schedule

First-Level Decomposition — Using Logical Cohesion to Derive the Information-Control Subsystem

The first-level decomposition on the right-hand side of Figure 2-12 shows the breakdown of the organization as a whole using the logical cohesion method, exactly as in the simple version of the hybrid model (see Figure 2-10). To create the organization-wide information-control subsystem, all informational operations of the organization are included, except for any baseline ones (as is the case with an organization whose baseline is informational in nature).

The following points indicate which kinds of operations belonging to the different subsystems in the functional model of the organization (the left-hand side of Figure 2-12) are also part of the organization-wide information-control subsystem (the right-hand side of Figure 2-12):

- from the non-baseline subsystems identified in the functional model, *all informational operations.* More specifically,
  - —from the non-baseline **informational** subsystems (i.e., those having informational baselines, such as the sales orders, general planning, and methods and procedures subsystems) *both the baseline and the information-control operations;*
  - —from the non-baseline **physical** subsystems (i.e., those having physical baselines, such as the personnel and fixed assets subsystems) *only the information-control operations.*
- from the baseline subsystems (**physical or informational**) identified in the functional model, *only information-control informational operations.* (**Note**: Since the baseline operations of these subsystems constitute the baseline of the organization itself, they cannot be considered information-control operations of the organization.)

Second-Level Decomposition — Using Functional Cohesion to Derive the Information-Control Sub-Subsystem Corresponding to Each Major Subsystem Identified in the Functional Model

The second-level decomposition in the right-hand side of Figure 2-12 applies the functional cohesion method to the organization-wide information-control subsystem. This creates one information sub-subsystem corresponding to each subsystem of the functional model of the organization. The dashed lines linking the right and left sides of Figure 2-12 relate the functional model's perspective of the organization to that of the hybrid model. This cross-referencing reveals that the various information sub-subsystems of the organization (in the second level of the right-hand side of Figure 2-12) correspond to the functional departments (in the first level of the left-hand side of Figure 2-12), such that there is one information sub-subsystem for each department. This is true even though the single, organization-wide information-control subsystem has been identified at level one (right-hand side of Figure 2-12) on the basis of functional class, rather than by the functional cohesion approach to subsystem identification. The reason for this is that, within the information-control subsystem, various highly related system elements, which are clustered according to functional department, are present . Analogous to this is the human nervous system, which has nerve nets resembling the shape of the body, where the nerves form the information subsystem of the parent system, namely, the human body.

Third, Fourth, and Fifth-Level Decompositions

Third-Level Decomposition: Using Procedural Cohesion to Derive the Computer Sub-Sub-Subsystems — At the third level (right-hand side), the manufacturing information sub-subsystem, is subjected to procedural cohesion, which results in its being divided into computerized and non-computerized sub-sub-subsystems. (In the alternative subsystem breakdown into departments and sections on the left side of the figure on the third level, no further subdivision is shown.)

Fourth-Level Decomposition: Using Functional Cohesion to Derive the Computer Programs — The fourth level of subsystems (right-hand side in the diagram) shows the computer manufacturing information sub-sub-subsystem broken down into computer programs. If the computer programs are each identified as subsystems along functional lines, then it is possible, although not likely, that the configuration of computer programs will reflect the breakdown of the manufacturing department into sections (left-hand side), also along functional lines. This is demonstrated in the figure by the dashed lines which connect programs (right-hand side) to the departmental sections (at level two on the left side of the figure). Again, a situation exists in which a configuration of subsystems—this time the level four subsystems on the right side of the figure—**reflects** the functional breakdown of a preceding level of subsystems obtained by a different path of subsystem analysis on the left-hand side. Nevertheless, making this relationship at this fourth level may be difficult, since neither the information subsystem of the organization as a whole nor the computerized manufacturing information sub-sub-subsystem was identified along functional lines.

Fifth-Level Decomposition: Using Functional Cohesion to Derive Sub-Programs — Finally, the fifth level of subsystems in the figure is again identified using the functional approach to subsystem identification. This means that the subprograms represent groups of closely interrelated functions within the level-four subsystems, the computer programs. At this point, any relationship to the configuration of production department sections on the left side of the figure becomes irrelevant.

The View from a Particular Functional Subsystem of the Business

Figure 2-13 reformulates the perspective of the complex hybrid model to that of a single business subsystem, as identified within the organization by the functional cohesion method. In this, two levels of environmental entities are possible, namely, other entities of the organization, and entities outside the organization. The business subsystem itself is further subdivided by logical cohesion, into a base and a control sub-subsystem. The control (i.e., information) operations are then divided into computer and non-computer sub-sub-subsystems. The functional cohesion method is applied to divide the computer sub-sub-subsystem into further levels of subsystems, i.e., into programs and subprograms.

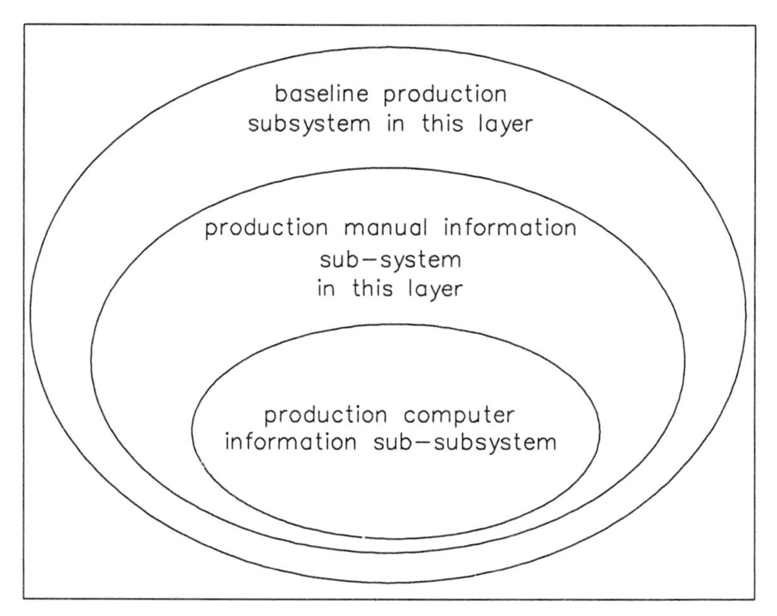

FIGURE 2-13: Hybrid model, complex version—yet another format. *This presentation, like that of the previous figure, takes the perspective of a single subsystem in the functional model of the organizat'ᵣn.*

CHAPTER SUPPLEMENT "A"— THE BLUMENTHAL MODEL OF THE ORGANIZATION

BACKGROUND

The Industrial Dynamics Model

Flows and Levels

The Blumenthal model ties together the Anthony scheme of functional levels (see below) with the Forrester Industrial Dynamics Model of system flow. The **Industrial Dynamics** Model* of flow in a systems starts with the assumption that flows involve what are here called "reservoirs." (Forrester calls these "levels," and Ahituv and Neumann call them "stocks.") "A stock is a buffer of any type of entities, e.g., raw materials inventory, fixed assets inventory, cash, manpower, etc. A flow is a sequence of activities which transfers entities from one stock to another, e.g. {to the} production line {from the} recruiting function {and from the} purchase of equipment function. {These} flows interlace."** Figure 2-14a indicates how flows from various processes can be combined, while Figure 2-14b demonstrates how the diagramming for the equivalent of Figure 2-14a would be done with the structure-flow charting method. In the Industrial Dynamics Model, the idea is to "observe various activities, combine them into processes, and then classify the processes."*** In fact, the classes of objective-defined functions described in connection with the structure-flow charting method provide a starting point for classifying the processes, since long-term framework flows may be distinguished from baseline ones.

The Industrial Dynamics Model also demonstrates the regulation of flows through "decision points," which adjust the flow rates. Each decision point receives information about resource levels, and outputs a decision affecting flow rates. The decision points and information

---

* Forrester (See references at end of this book).

** Ahituv, N. and S. Neumann, Principles of Information Systems for Management. Dubuque, Iowa, W.C. Brown©, 1982, p. 125. Reprinted by permission.

*** ibid., p. 124.

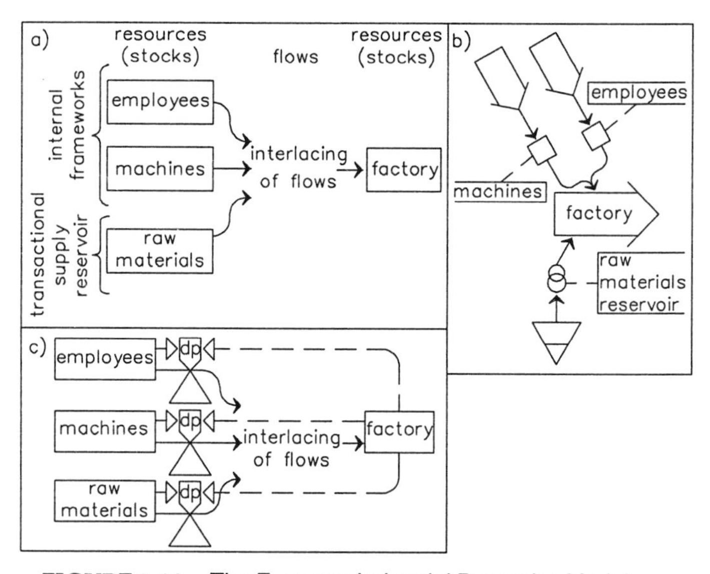

FIGURE 2-14: The Forrester Industrial Dynamics Model.
a) interlacing flows in the Industrial Dynamics Model;
b) equivalent of "a" using structure-flow chart;
c) flow regulation in the Industrial Dynamics Model.

flows applicable to Figure 2-14a are shown in Figure 2-14c. Of course, the representation of these information flows in these figures is somewhat simplistic, since all decisions are interdependent. That is, there would be other levels of managerial operations to coordinate the three independent decision points in these diagrams; e.g., the sales and customer order functions in a business might provide the needed source of coordination in deciding how much shelf stock of finished goods is required.

The links between the various subsystems are provided by the flow of information. The structure of the information systems is a reflection of the structure of the systems to which the information systems provide information. In fact, the information systems have their own stocks, maintained in data files.

Classes of Items Involved in Flow

The Industrial Dynamics Model does classify the various flows, such that the flow of machinery and personnel is in a different class than that of raw materials in the above example. In fact, this type of classification should be captured in a meaningful black box charting technique. For instance, the SADT diagrams* distinguish instruments (personnel, machinery) from inputs to be operated upon. In other words, instrumental framework inputs are distinguished from transactional ones in that the former come in from the bottom of the black box, while the latter enter from the left. This diagramming convention is used regardless of whether or not the framework is longterm. Thus, since a special chisel may be acquired for a specific system run, it is not a longterm framework. However, it still enters from the bottom, as it is an instrument.

The Anthony Scheme of Functional Levels

The Anthony scheme, represented by a triangle (see Figure 2-16a below), divides operations into the following four levels:

- **Strategic Planning**: "deciding on objectives of the organization, on changes in these objectives, on the resources needed to attain these objectives, and on the policies that are to govern the acquisition, use, and disposition of these resources."**
- **Management Control**: assuring "that resources are obtained and used effectively and efficiently in the accomplishment of the organization's objectives."***
- **Operational Control**: "assuring that specific tasks are carried out effectively and efficiently."****
- **Operations**: carrying out the low level tasks to achieve the objectives of the system. For example, in a business organization, inputs are physically transformed by the operations level, using personnel, capital, and machinery, into goods and services.

More details of the Anthony scheme can be found in Chapter Three.

---

* See an explanation of SADT in the related publication, by Ed Baylin, Procedural Diagramming for System Development—From a More Scientific Viewpoint.

** Anthony, Robert N., Planning and Control Systems: A Framework for Analysis. Boston: Division of Research, Harvard Business School©, 1965, p. 16.

*** ibid., p. 17.

**** ibid., p. 18.

A NEW MODEL BASED ON INTEGRATION OF SIMPLER MODELS

Integration of the Anthony and Forrester Models

Integration of Structure and Flow Dimensions

The Blumenthal model[*] is essentially an integration of the two different dimensions of analysis represented by the two simpler models; namely, the Anthony model of system functional levels and the Industrial Dynamics[**] one of system flow dynamics. Central to the Blumenthal model is the "functional unit" (see Figures 2-15 and 2-16b). In present terms, this model means a functional unit is a **subsystem** with a physical base identified along functional lines, in which only the physical base and control operations close to the subsystem baseline are included. "The most basic activities of an organization are handled by **functional units**. A functional unit receives information about stocks, makes some elementary decisions, and regulates physical flow rates."[***] According to Blumenthal, the functional unit consists of the physical internal framework and baseline functions and low-level control operations very closely connected to them. Following is a description of the Blumenthal functional unit, as seen in Figure 2-15a:

> At the bottom of figure . . . we can see the normal physical flow moving from one stock to the next. Information about a stock, combined with information from other functional units, is transferred to a **decision center** which, accordingly, instructs an **activity center** of actions to be carried out. The actions regulate the rate of physical flow. Note that the information is available also to the activity center, which may undertake some actions under its discretion whenever such actions are highly structured (almost fully programmed). The decision center is also guided by management

---

* Blumenthal, Sherman C., Management Information Systems: A Framework for Planning and Development. ©1969, Prentice-Hall, Inc., Englewood Cliffs, N.J.

** Forrester.

*** Ahituv, N. and S. Neumann, Principles of Information Systems for Management. Dubuque, Iowa, ©W.C. Brown, 1982, p. 128. Reprinted by permission.

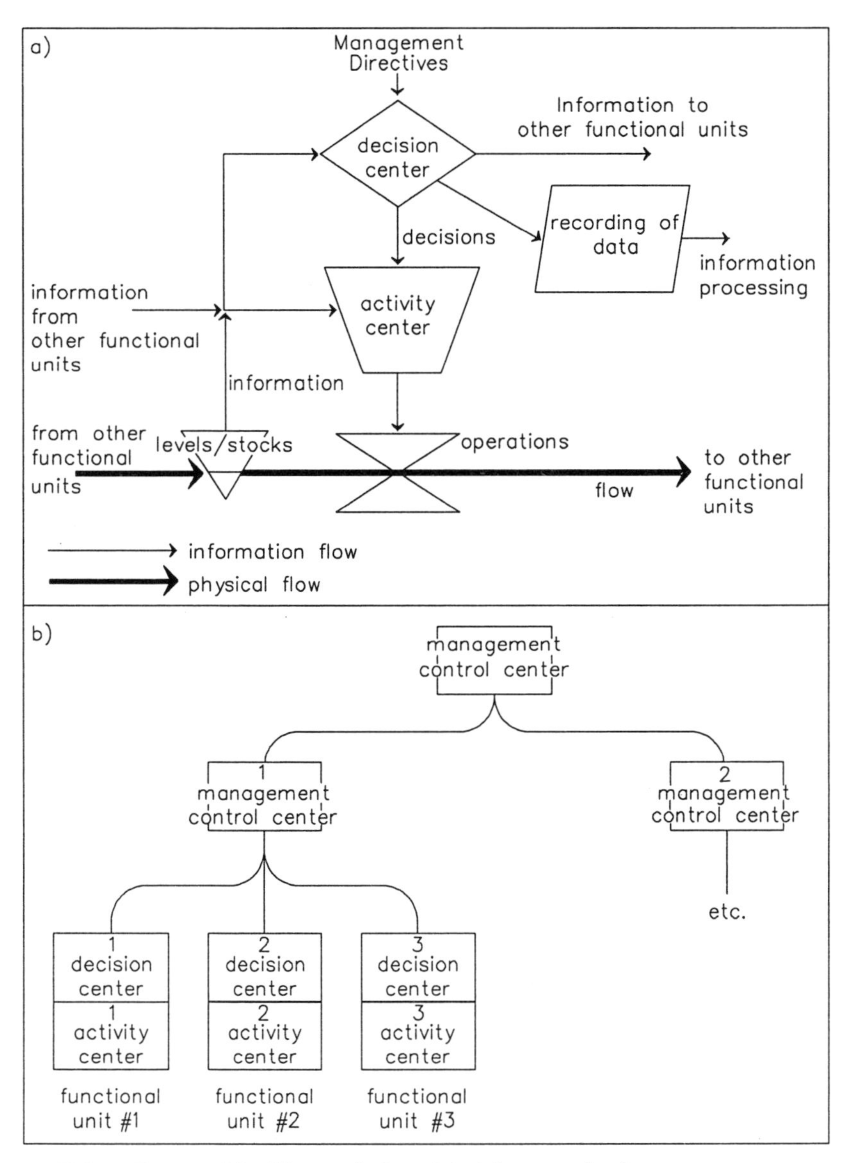

FIGURE 2-15: The Blumenthal model of the organization.

a) Blumenthal's functional unit. This diagram is comparable to that originally used to portray the ideas in Forrester's Industrial Dynamics book, as well as to variations of the original found in Blumenthal's Management Information Systems: A Framework for Planning and Development, and in Principles of Information Systems for Management, authored by Ahituv and Neumann);

b) Blumenthal's management control centers.

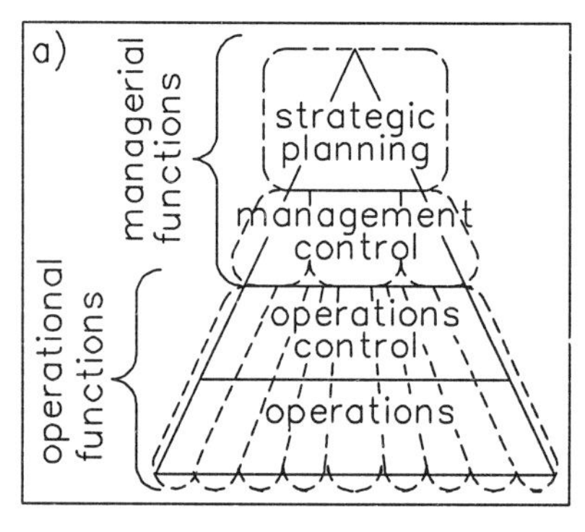

FIGURE 2-16: Blumenthal's idea of the "operational function."
a) Blumenthal's functions shown impressionistically in the Anthony triangle; {similar to a figure copyrighted by Hemisphere}
b) ***(facing page):***

directives determining goals, policies, and procedures. It may also convey relevant information to other FU's directly or via an information processing system (data capture). A functional unit is thus comprised of a decision center, an activity center, and an actions unit.*

Managerial Levels

Strategic planning and higher level control operations, although not contained within the functional unit of this model, is explained in the following quote:

> Functional units are the basic blocks composing the operations level of the Anthony hierarchical model.... They are directed by managerial functions composing the operational control level. These managerial functions are termed **management control** center (MCC) . . .. The distinction between a decision center and a functional unit and a MCC is that the former always supervises an activity center, while the latter always supervises other decision centers, either of FU's or of other MCCs. Figure {2-15b} illustrates the position of MCCs in the model . . .. Note that the term MCC refers to any managerial function regardless of its location in the organizational hierarchy. Certainly we may refine the classification of MCCs, associating each of them with a particular level in Anthony's model.**

In present terms, each management control center is separated into a functionally identified subsystem, since inclusion of it with any partic-

* Ahituv, N. and S. Neumann, Principles of Information Systems for Management. Dubuque, Iowa, W.C. Brown©, 1982, p. 128. Reprinted by permission.

** Ahituv and Neumann, pp. 128—129.

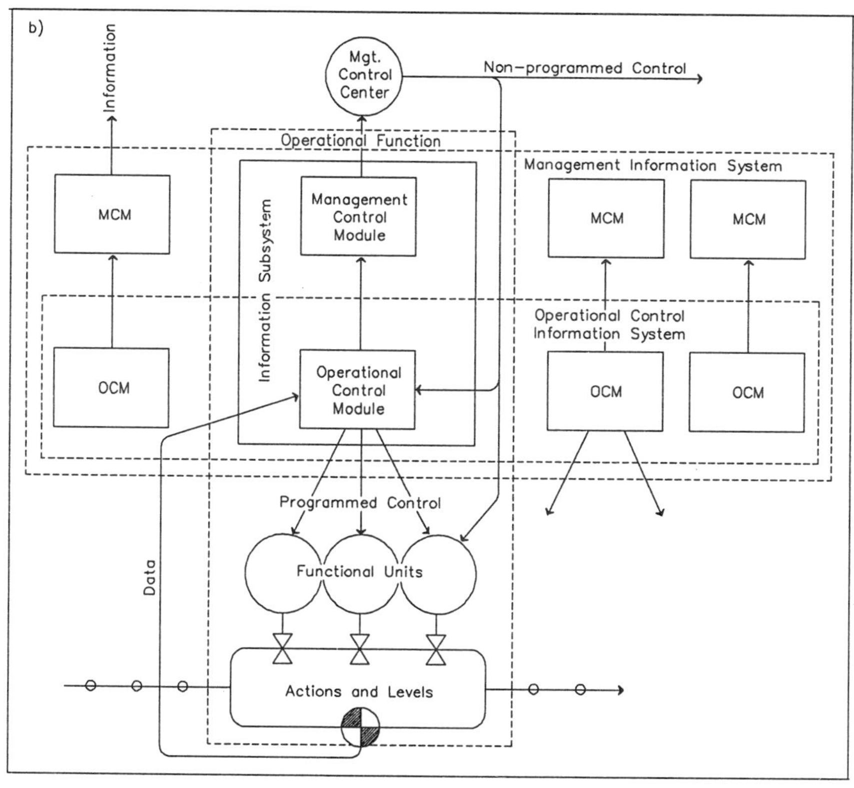

FIGURE 2-16b: Blumenthal's operational function. {Blumenthal, Sherman C., Management Information Systems: A Framework for Planning and Development. ©1969, Adapted by permission of Prentice-Hall, Inc., Englewood Cliffs, N.J.}

ular functional unit, or with any lower level management control center supervised by it, would create conflicts with other subsystems also supervised by it.

Operational Level Functions

Blumenthal apparently uses the word "function"—or at least what he calls an "operational function"—to mean what is here called a subsystem identified along functional lines. Thus, an operational function can be defined as:

> a class of any one or more types of actions that are interrelated by the flow of information between the levels {i.e.,

stocks or reservoirs} with which they are concerned and the decision centers which are part of the same functional units as those in which the actions take place, and between the levels and the management control centers accountable for the functional units in which the actions take place, and which regulate the inflow or outflow to or from sequences of levels as a group . . .. Operational functions can be conceived as having a "tree" structure made up of management control centers branching downward from a single focus of control and terminating in functional units.*

By a "class of any one or more types of actions," Blumenthal is simply referring to a group of operations considered to exist over more than a single run of the operations in question; that is, an "operations function continues to exist even when actual operations are not taking place."** Secondly, by the word "operations" as a modifier of the noun "function," he is referring only to functions below Anthony's management control level. In general, it would seem that Blumenthal's idea of a function is captured in Figure 2-16a, which is similar to the functional cohesion diagram, Figure 3-8b, in Chapter Three below. Figure 2-16b*** is a diagram of an operational function from the original Blumenthal book.

Avoidance of Subsystem Conflict Idea

Blumenthal appears to recognize the need for avoidance of subsystem conflict in identifying functions, by saying that each "functional unit may be part of several different structures, depending upon the number of operational functions in which it participates."**** That is, different subsystems identified along functional lines may call upon the same resources for their performance. This accounts for the idea of avoiding subsystem conflict for a common resource. Also, with management control centers, different functional units may be directed by the same decision-making operations. The latter provides the second form in which subsystem conflict is avoided (see functional cohesion explanation in the first section of this chapter).

---

* Blumenthal, Sherman C., Management Information Systems; A Framework for Planning and Development, copyright 1969, pp. 32—33. Adapted by permission of Prentice-Hall, Inc., Englewood Cliffs, N.J.

** ibid., p. 33.

*** ibid., p. 35.

**** ibid., p. 33.

CHAPTER SUPPLEMENT "B"— ORGANIZING BY FUNCTIONAL AREAS: THE TIME HAS COME*

by Brenda J. Shestowsky B.Sc.(N), M.Sc.**

The organizational structure of institutions such as hospitals has been that of a traditional hierarchial arrangement based upon division by class or speciality area (Figure {2-17}). This cohesion, whereby elements are often grouped merely upon their performing a certain logical class of functions, has resulted in an unnatural division among highly coupled (i.e., dependent) elements. This mode of hospital organization, as for example seen by the interconnecting lines in Figure {2-17}, results in fragmented speciality areas each with their own managerial and operational levels. Such fragmentation results in many problems, including inefficiency of operation, territorial disputes, and blurring of the organization's baseline objectives. A better organization is possible based upon functional cohesion. Here, elements are grouped according to their being a necessary part of a single function and essential to that function's performance while avoiding duplication and conflict among subsystems.

To some, the idea of functional cohesion, as just described, is highly impractical, since our technological society seems to require increasing specialization. However, specialization is often just one alternative, and, in this author's opinion, has already been carried too far, to the point where it may cause the system to explode. One of the difficulties with over-specialization is that it obliterates the "moral sense" of health professions, due to an overemphasis on technology. In addition, it "demoralizes" professionals, as it does not allow them to make the linkages among the various isolated tasks they perform and the overall mission of health care.

---

* Shestowsky, Brenda, Dimensions in Health Service, Vol. 65, No. 8, Nov. 1988, pages 19, 20, and 22. This article is published with permission of the Canadian Hospital Association, which has first publishing rights. It is the final version submitted to the journal, without any further pre-publication editing changes.

** Brenda Shestowsky is currently Director of Nursing at the Hotel Dieu Hospital, Cornwall, Ontario, Canada.

The Functional Cohesion Concept

As is commonly recognized in the literature on the development of computer systems, the functional cohesion method for identifying subsystems is the most desirable of methods, since it produces the most easily understandable and maintainable system modular structure (Yourdon and Constantine, 1979). This can be generalized to all types of man-made systems, since comprehensibility and maintainability are, in general, highly desirable in such systems. To explain this idea further, it is an accepted fact that systems accomplish their mission by meeting a number of intermediate objectives which lead to the fulfilment of the final, or system, objective(s). Roughly speaking, by the functional cohesion method, one subsystem is identified to meet each of these intermediate objectives, while a final subsystem is identified for each of the final (or system) objectives. Subsystems so identified are maximally independent of one another. Thus, it is easy to maintain subsystems, as a change to one can be made without the need for corresponding changes to another.

Before exploring this notion of organization based upon functional cohesion or functional areas, it is first necessary to explore the idea of function. Much confusion has surrounded the issue of how to identify functions so as to achieve functionally cohesive subsystems. While some authors have equated function with a definition of "specialty," others have defined it in terms of tasks leading to a technical objective without regard for duplication of effort or the resultant conflict (Baylin, 1986).

Baylin (1985) postulates that the most effective way to achieve functionally cohesive subsystems is by first defining functions via functional objectives, and then grouping the tasks closely related to these objectives, taking into consideration the need to avoid duplication and conflict. An objective-defined function is a subset of all the operations of a system whose cohesion as a group is based upon their all being **directly** necessary to the achievement of the same specific objective, or the same mutually contingent specific objective set, within a given functional class or level (Baylin, 1986). A useful functional classification scheme divides "adaptation" from "baseline" operations (Baylin, 1984). Adaptation operations are, in general systems theory, roughly equivalent to control operations, i.e., those operations which, according to Fayol (1949), plan, organize, direct, and ensure ("control") the baseline. Baseline operations are roughly equivalent to the operations level (Anthony, 1965), that is, those operations which directly carry out the purpose or mission of the system (organization).

It is important to note that the word "objective" in Baylin's definition of objective-defined function is usually taken to mean the technical

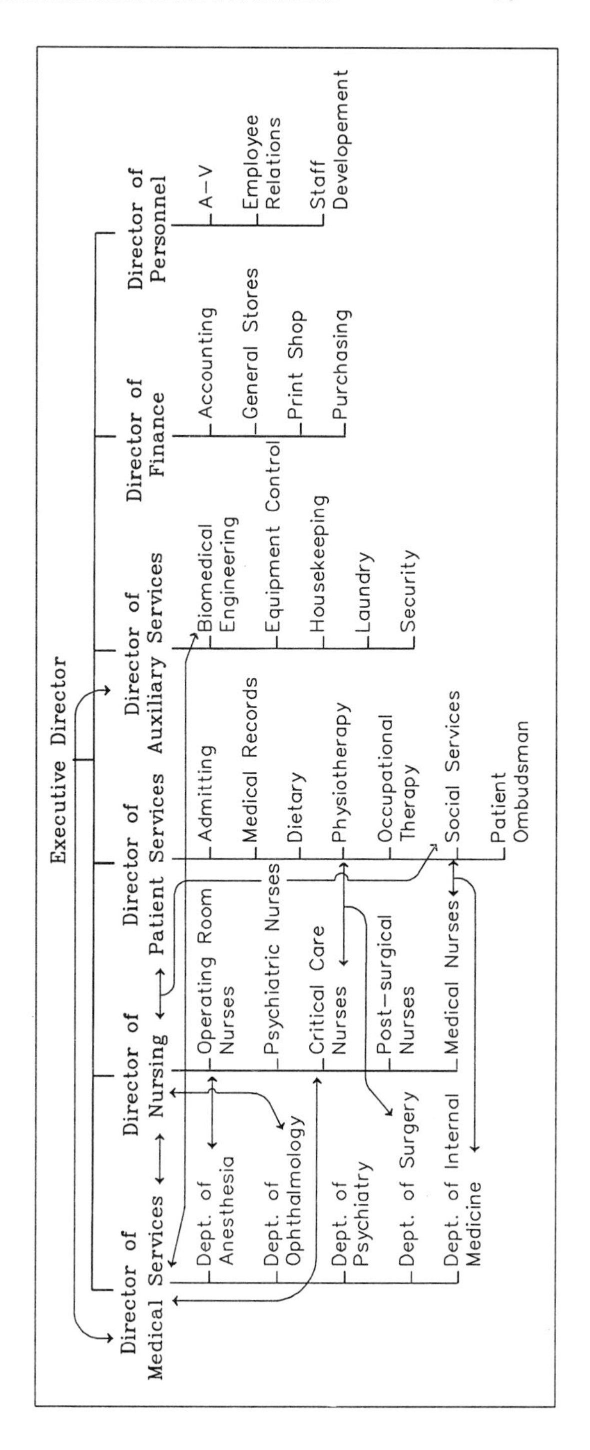

FIGURE 2-17: Examples of cross-communication and authority lines in traditional hospital structure.

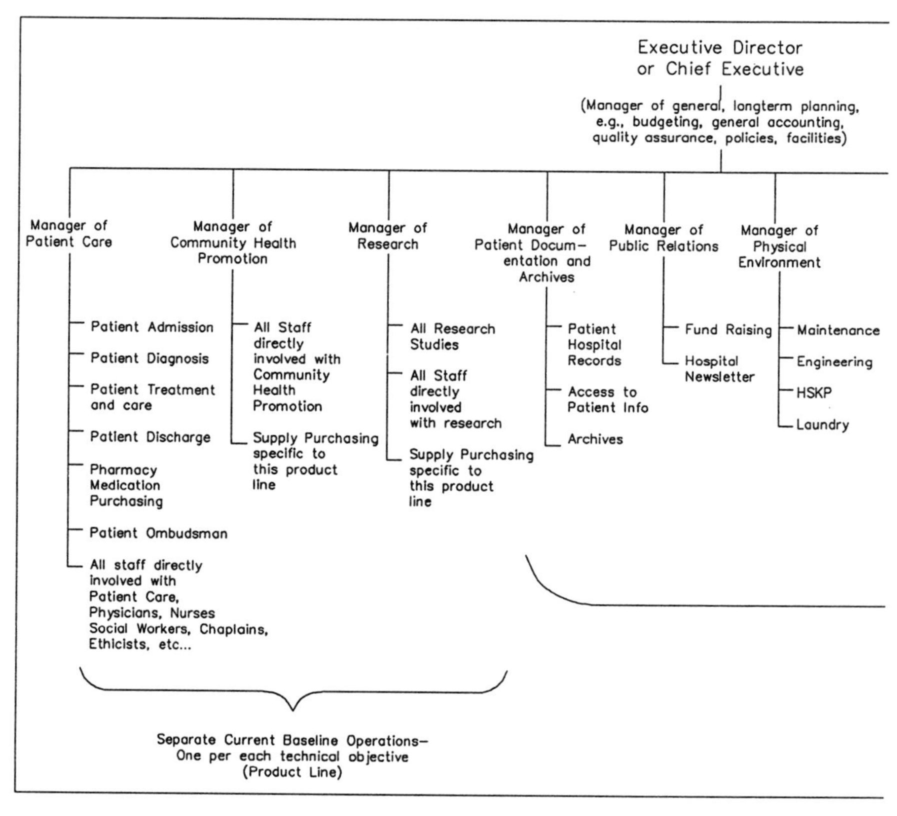

FIGURE 2-18: Functional organization of the hospital system. (Note: The right part of this figure is on the facing page.)

objective(s) and never includes the performance goals of the system or organization. For example, in a hospital, the technical objective can be specifically stated to be caring for the sick and promoting health, or more, globally, meeting the health care needs of patients. Therefore, professional objectives for advancement and power, staff objectives for fun and pleasure, or, institutional objectives for fame and profit are not technical objectives.

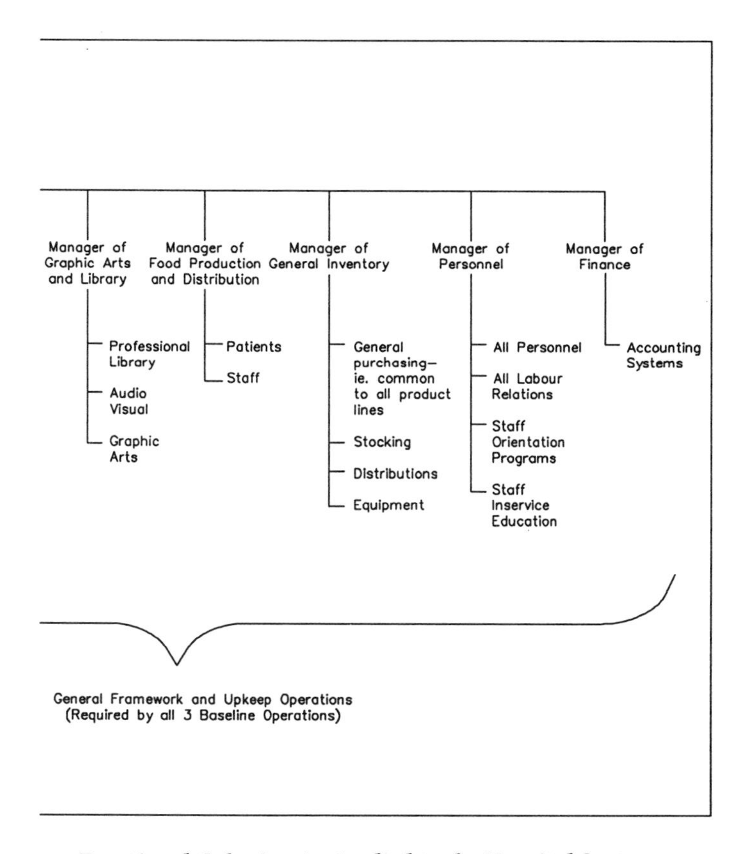

Functional Cohesion As Applied to the Hospital System

Given that the overriding mission of a hospital is the technical objective of meeting the health care needs of patients, we can begin looking at breaking down the hospital organization using functionally cohesive subsystems (i.e., systems identified along functional lines or along lines of objective-defined functions).

This process first involves the identification of the technical objectives which constitute the organization's mission. It is to be noted here that each technical objective is in fact a separate product line. For example, the following are separate product lines: meeting patient needs, promoting community health, and conducting research.

Although these are inter-related they nevertheless constitute a basis for identifying separate functionally cohesive subsystems.

Consider the following activities in an organization:

1. long-term planning;

2. creation and replacement of general long-lasting frameworks needed in daily operations, e.g., policies, standards, procedures, facilities, personnel;
3. enactment of current baseline operations (baseline operations directly fulfill the technical objective of the system/subsystem under analysis);
4. enactment of current framework upkeep operations (these are required by the baseline in fulfilling the technical objective and are, therefore, indirectly necessary to that fulfilment).

Looking at the hospital as a whole, management by functional cohesion would result in the following:

1. a manager of planning and long-term framework operations—i.e., the hospital president or executive director;
2. managers for each of the current operation (baseline plus framework) subsystems identified—i.e., a manager for each of the technical objectives.

Recalling, however, that the definition of functional cohesion specifically states that conflict and duplication among subsystems are to be avoided, current framework operations common to two or more subsystems should be contained within a subsystem of their own, thereby resulting in the organizational structure pictured in Figure {2-18}. Further subsystem breakdown adhering to functional cohesion principles is possible for each of the subsystems detailed in Figure {2-18}. It is essential to remember that at each subsystem level there exists current operations as well as planning and long-term framework operations locally applicable within that subsystem. The same principle of avoiding subsystem conflict is to be upheld. In other words, each subsystem identified by the functional cohesion method can be divided into subsystems of its own, again using the functional cohesion method.

Effect of Functional Cohesion on Different Levels of Personnel

It is perhaps easy to see how adaptation (control, i.e., planning, organizing, directing, and ensuring) operations within the hospital system can be organized in functionally cohesive ways. This would see the division of management based upon internally cohesive but independent subsystems corresponding to distinct baseline, framework and framework upkeep operations. Such a meshing of management functions would avoid the territorial conflicts and professional disputes now evident at management levels as there would exist one management hierarchy per baseline, framework, and framework upkeep operation. Management personnel would of course be selected on the basis of their skills rather than upon their membership in specific professional/worker groups. In this author's opinion, movement toward this end would eliminate many problems inherent in the organization of today's hospitals.

A cloudier, but nonetheless important, issue is the notion of functionally cohesive subsystems at the operations level. One might question why functional cohesive units at the operations level is something to be aspired to. As at management levels, problems with territory and inefficiency abound at operational levels. As well, another difficulty associated with organizations based upon speciality lines appears to be worker boredom. As workers begin to look for more interesting things to do, there is a tendency for them to leave the operational level and move "upward" into management and/or research. For example, nurses who have attained bachelors and masters degrees rarely stay at the bedside, and tend to move into management, research, and teaching positions. This produces dysfunctionality with the objectives of the hospital in society. Work organization based upon functional lines is one way of increasing worker role satisfaction through task variety and via an increasing sense of accomplishment which comes from working toward clear objectives within a mutually cohesive group. Here it becomes clear that the rigid boundaries established by professional and union classifications must dissolve. As at the management level, worker/professional assignment would be based upon the specific skills needed, and not upon membership in professional/worker groups.

If the above seems too utopian, and ignores a real need for specialization, then perhaps we should take a look at the already existing evidence, both negative and positive, which is pressing us toward a reorganization of structure within the health care industry. *On the one hand,* there exists a shortage of health care professionals, such as physiotherapists, pharmacists, and nurses. Part of the difficulty in attracting individuals to these groups comes from the disinterest and dissatisfaction created by the rigidly defined hospital environment, which divides work according to task specialty, rather than by task interrelationships. *On the other hand,* we already see areas in which functionally cohesive structures have been tried and been successful. Examples of this include primary nursing, where individual nurses are assigned all aspects of nursing care related to individual patients, and management structures in which there is a manager of all patient care services, including nursing, and, hopefully, in the future, medicine.

If these ideas continue to be implemented, we might begin to see less of a distinction between registered nurses and registered nursing assistants, nurses performing tasks traditionally relegated to physicians, and physicians performing tasks traditionally relegated to social workers. Such fluidity of role, although theoretically possible, is not presently feasible. A new orientation and a restructuring of professional schools, organizations and trade unions is necessary before such an intertwining of health workers all oriented toward the same technical objective is achieved. An organization truly based on functional lines is a goal for the 21st Century, towards which we must now begin to progress.

References Used in Article by Brenda Shestowsky

Anthony, R.A. Planning and Control Systems: A Framework for Analysis. Boston, Harvard University, 1965.

Baylin, E.N. "A Scheme for Handling Problems of Functional Classification/Time Orientation." Working Paper Series, Faculty of Commerce and Administration, Concordia University, 1984 (84-033).

Baylin, E.N. "Computer Systems Analyst's View of the Business Organization." Cybernetics and Systems: An International Journal, 16:305-323, 1985.

Baylin, E.N. "Identifying System Functions." International Journal of General Systems, 12:7-38, 1986.

Fayol, H. General and Industrial Management. Translated by C. Storrs, London, Sir Issac Pitman & Sons, 1949.

Yourdon, E. and Constantine, L. Structured Design: Fundamentals of a Discipline of Computer Program and Systems Design. Englewood Cliffs, N.J., Prentice-Hall, 1979.

CHAPTER THREE

UNDERLYING CONCEPTS OF FUNCTIONAL MODELING

This chapter deals with several complex constructs, which may not be fully understood until Chapters Four to Eleven have been completed.

- The first section introduces and critiques various current notions of identifying functions.
- The second section continues the critique and introduces new heuristics for identifying functions.
- The third section of this chapter describes and critiques existing schemes for classifying and levelling functions, and then moves on to the new scheme, introduced by this book.
- Finally, the fourth section discusses feedback/feedforward. It will be useful to refer forward to this fourth section at recommended points, especially when discussing the new scheme for functional classification.

The discussion of underlying concepts of functional modeling is extended in the next chapter, which deals with the principles of procedural (i.e., functional) diagramming.

The new heuristics for identifying functions and the new functional classification scheme are dealt with further in subsequent chapters, mainly in Chapters Eight and Nine, respectively. As well, Chapters Five, Six, and Seven are suggested for more discussion of functional levels and functional significances of inputs and outputs. These later chapters not only provide further aspects of the ideas involved, but also give many more examples and clarify a number of issues not thoroughly discussed in the present chapter. It is not unreasonable that the reader having particular further questions about materials in this chapter should glance ahead to these later chapters.

IDENTIFYING SYSTEM FUNCTIONS*

INTRODUCTION

Functions and Functionalism

What is a function? How are functions identifiable within a system? Although the word "function" has a fairly definite meaning in systems literature, many different ways exist to identify functions within a system. This applies equally to the mathematical and non-mathematical ideas of function. In fact, it can happen that functions are identified such that it becomes difficult to see the relationship between the functions in a system. Stated in another way, if the functions in the system are not first identified in certain ways, it becomes difficult to see the patterns by which highly related groups of operations in a system are related to the system mission. For instance, speaking of a "managerial function" in a business may not provide as much clarity as speaking of a "marketing" or a "production" function, because different managerial functions, applicable to different areas of the business, may have little relationship to one another, unlike the marketing or production

---

* The following is based on an article by Ed Baylin, "Identifying System Functions," International Journal of General Systems. New York: Gordon and Breach Science Publishers©, vol. 12, no. 1, 1986.

activities. In the case of the managerial function, it is internally bound more by so-called "logical" cohesion, in which operations are grouped merely by class, than by so-called "functional cohesion," where operations are grouped together by intensity of coupling, common objectives, and common processes.

Before discussing a new set of heuristics for identifying functions, this section treats the following matters:

- what is well agreed upon in the literature concerning the functional concept;
- the notion of functional**ism**, to which it would reasonably be expected that the function idea is closely related;
- current notions of function; first, the mathematical one, and then the non-mathematical ones (as in the study of organizations).

The Notion of "Functionalism"

While it may be agreed that function may be equated to a group of operations associated with a mission, a function does not need to be made up of highly related elements which as a group are optimally independent from the elements in other functions. For instance, the managerial function, the computer function, the field sales function, etc. all refer to functional contents which are either more restrictive or less restrictive than might be thought given the mission of the broader system in which these functions are components. Thus, while "sales" identifies a conceptual function," field sales" refers to the sales objective as achieved in the field only, which would appear to be unduly restrictive. Conversely, the "managerial" function may be associated with any number of different, often highly unrelated, elements, and therefore hardly forms a functionally cohesive function. Similarly, the "accounting" function is often identified in business organizations, although the different accounting subsystems hardly form a highly related cluster of functional elements. Perhaps, it might be said, such formulations of functions in systems are incorrect. Well, if so, the definition of function might be revised, since functions can be formulated in such "non-functional" manners.

However, it would seem that the idea of functional**ism** almost implies that functions have been identified in a functionally cohesive manner. One set of definitions of functionalism is as follows:

> (1) Doctrine or practice which emphasizes practical utility or functional relations. (2) Theory of culture which analyzes the interrelatedness and interdependence of patterns and institutions within a cultural complex or social system and

emphasizes the interaction of these forms in the maintenance of socio-cultural unity.*

Points of Common Agreement

First, without doubt, the function idea provides a conceptual viewpoint of system operations. According to W.T. Singleton: "Functional thinking has been developed to compensate for the limitations of thinking in terms of physical entities. This latter is essentially conservative or convergent and is also limited in range of comprehension."**

Second, a function identifies a component part of the total set of system operations. Functional decomposition and breaking a system down into subsystems are two common notions of systems analysis. Actually, the two ideas are often considered to be synonymous. However, in the author's opinion, these different types of analysis are highly interrelated, but not identical.

Third, it is generally agreed that the idea of function is fairly synonymous with a "black box" filled with operations leading to the achievement of an objective, be this objective the mission of a subsystem, or, what is the same thing, the target of a component group of system operations in a system called a "function." Certain other aspects of functions are, it seems, also generally agreed upon. Thus,

- The objective of a function does not include its own performance goals, such as the rate of execution of the function. Thus, the statement of objectives should not include a reference to the efficiency with which the mission is achieved.
- The mission does not refer to future plans for change in the mission of the function; nor does it refer to all the outputs of a function; i.e., it does not include outputs other than those which fulfill objectives.
- The mission is generally understood to have been stated in a fairly abstract (flexible, logical) way for the level of function under consideration. However, the whole discussion of "what" has to be accomplished versus "how" it is accomplished, i.e., of logical versus physical thinking, still has many ambiguities, so that it is perhaps better to think of achieving the most flexible **possible** statement of the mission of a subsystem or function, rather than of a purely logical versus a physical statement of mission. To illustrate, the following two definitions of function may be contrasted:

---

* Open Systems Group, System Behaviour, 3rd edition. London: Harper and Row©, 1981, Glossary.

** Open Systems Group, p. 126.

1) ... function is the mission, aim, purpose or primary concern of the system. **What** is to be accomplished describes this element, without concern for **how** it is to be accomplished. Thus, the function is the need which is to be satisfied . . .. Function is the only one of the . . . system elements that is intangible.*
2) . . . a general means of action by which the system fulfills its requirements. Functions are usually expressed in verb form (monitor, control) or participial form (monitoring, controlling). They are the first expressions of the **hows** of the system. They are expressed progressively more precisely. Ideally, functions are conceived apart from implementation by men and/or machines; in practice, they are usually expressed along with machine design implications.**

Putting aside the issue of logical versus physical, or of the whats versus the hows, the idea of system or function does refer to those outputs called "objectives," provided that these are not formulated in a way which is overly physical (insufficiently flexible).

DIFFERENT NOTIONS OF FUNCTION

The Mathematical Notion of Function

The ideas of function, including the mathematical one, are basic in all systems work. The mathematical idea is particularly important in connection with functional decomposition using so-called "fourth-generation" software for computer systems. Program code which is provably free of errors, provided that the program objective is correct, has been claimed to exist largely on the basis that the functional idea has been rigorously used in the mathematical sense. The algebraic notion of function recognizes that functions involve targets, or outputs, that is,

* Nadler, G., Work Design: A Systems Concept, 2nd edition. Homewood, Illinois: copyright G. Nadler, Irwin Publishers, pp. 207—208.

** Open Systems Group, chapter by De Greene, K. B., p. 96.

the left-hand side of the equation. To generalize this notion, it may be said that;

> A function has one or more objects as its input and one or more objects as its output..... An object may be a data item, a list, a table, a report, a file, a data base, or it might be a physical entity such as a circuit, a missile, an item undergoing manufacturing scheduling, a train, tracks, switching points, etc.*

Here, the outputs may be multiple, as in the case of some of the examples from the HOS (Higher Order Software)** method. Figure 3-1, borrowed from a description of HOS by Martin and McClure, shows multiple outputs, x_t, y_t, and z_t, in one of its examples of the concept of function.

Unfortunately for the relationship between the mathematical function and the functionalism of the system to whose operations it is applied, functional cohesion of operations is not guaranteed where multiple objectives are involved. It all depends upon the relationships between these objectives and the coupling of the operations for achieving these various ends. Thus, the use of mathematics simply guarantees that the system is partitioned using abstract concepts, rather than being based upon physical entities. It is still possible that the operations in a mathematical function are bound together in non-functionally cohesive ways, e.g., by procedure, time, communications channel, class, etc.

Critique of the HOS Method of Identifying Functions

Based on the above, it may be premature to say that the HOS method fulfills the type of criteria needed for the most advanced possible species of functional decomposition, as described by Martin and McClure.*** However, the criteria used in HOS perhaps establish rigor within the frame of reference of the HOS method itself. But, do they provide a clear relationship between the functional breakdown and the overall patterns of achieving functionalism in the system? In this vein,

* Martin J. and C. McClure, Structured Techniques for Computing. Englewood Cliffs, N.J.: Prentice-Hall©, 1985, p. 357.

** Hamilton, M. and S. Zeldin, Integrated Software Development System/Higher Order Software Conceptual Description. TR-3, Higher Order Software, Inc.©, 1976.

*** See functional decomposition charting methods in the related Ed Baylin book, Procedural Diagramming for System Development—From a More Scientific Viewpoint.

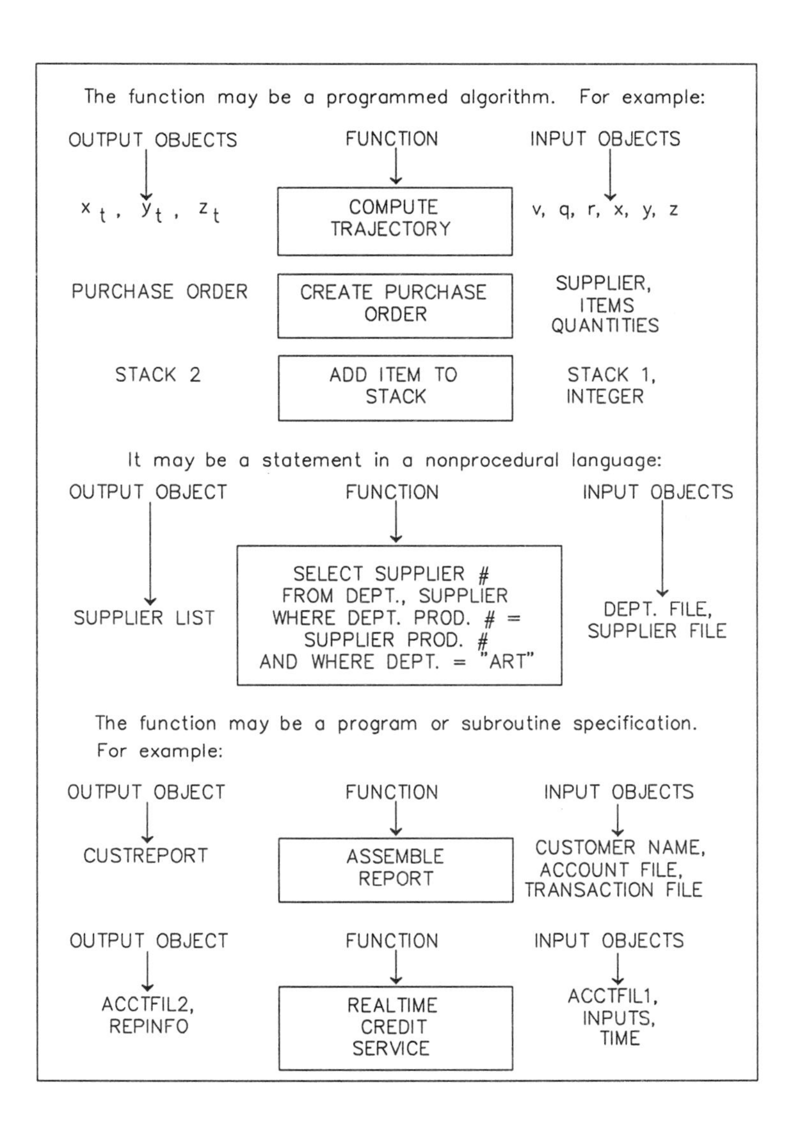

FIGURE 3-1: Examples of functions in the Higher Order Software method (From Martin and McClure, p. 358).

it is interesting to note the following comment made somewhat earlier by another well-known computer systems writer: "I'll bet you can't even define the word function except in a purely mathematical sense."* Is it possible that further species of functional decomposition are needed, such as species IV, V, etc., i.e., species which are still more advanced than the species III fulfilled by HOS?

As to whether the **HOS** method shows the patterns of functional cohesion, it may be concluded that **this** mathematical method perhaps **does** handle the problem of relating functions to objectives, although not all the operations in the system flowing towards these objectives need have been included (see below on these problems). However, objectives-orientation is only the first step in defining functions in ways which facilitate seeing the patterns of functional cohesion in a system. Some further thoughts on this are provided in the following paragraphs, although their comprehension will be greatly aided by completing the second section of this chapter.

First, in the case of a function/sub-function having multiple objectives, does good functional decomposition require certain constraints, such as a set number of types of possible relationships between functional objectives? The latter idea is essentially similar to the present idea of "mutual contingency." As seen, the importance of the relationships between different objectives of the same function is not evident in the definition of HOS functions. However, unless certain constraints are placed upon the relationships of these functional objects to one another, it is possible that, with a function having multiple objects, the outputs and the operations making up the function could be related by nothing more than, say, class or time, as in logical or temporal cohesion, respectively.

Second, it is not apparent in HOS that a function's objects might not all be themselves fulfillments of the functional objectives, since some of these simply fulfill roles such as feedback for control purposes, symbiotic interface with other functions, offshoots of no consequence, etc. In this book's method (see next section), these different types of functional outputs must be clearly identified.

Third, and equally important as tying functions to objectives, is the notion of limiting the operations in a function to those which flow **directly** towards the same objective(s). The constraint imposed herein that each function contain **all** and **only** those operations in the function's direct line of flow within the system functional class/level/time-orientation implies that, in most cases at least, different functions

---

* De Marco, T., Structured Analysis and System Specification. Englewood Cliffs, N.J.: Prentice-Hall©, 1978 and 1979, pp. 41—42.

cannot be combined (imploded, concatenated) into a single function. This notion perhaps further enhances the making of a solid starting relationship between the system functional breakdown and the system mission.

The overall reason for the limitations of HOS for functional decomposition may be that mathematical consistency hardly guarantees the best possible functional breakdown, i.e., the best bridge between the breakdown of the system into functions and the relating of this breakdown to the system's functional**ism**.

Non-Mathematical Ideas of Function

A similar problem, but from the reverse perspective, is often present in non-mathematical formulations of the functional idea, which distinguish functional thinking from objectives-oriented thinking. Thus, it has been said by Mintzberg, "we have the fundamental distinction between grouping activities by ends, by the characteristics of the ultimate markets served by the organization . . . or by the means, the functions (including work processes, skills, and knowledge) it uses to produce its products and services."[*] That Mintzberg may at times be referring to something like procedural or logical cohesion, as opposed to functional cohesion, is further confirmed by the following extract: "For example, one lathe operator may have to consult another, working on a different product line.... In effect, we have interdependencies related to specialization, which favour functional grouping."[**] Based on this, it would appear that Mintzberg thinks of a function as a grouping by class of activity, rather than by objective. He states: "Functional structure also encourages specialization, for example, by establishing career paths for specialists within their own area of expertise."[***]

The Mintzberg notion of function has apparently been influenced by the Frederick Taylor idea of "functional organization," of which the following gives a summary:

> **Functional organization** structures, in which groups are organized by function, were established to try to remove some of the line structure's dependence on a few key persons. Originated in the early 1900s by Frederick Taylor, this type of structure calls for a combination of line and staff activities grouped into distinct areas of specialization, each dealing

* Mintzberg, H., The Structuring of Organizations. Englewood Cliffs, N.J.: Prentice-Hall©, 1979, p. 114.

** Mintzberg, p. 122.

*** Mintzberg, p. 124.

> with specific functions. The staff group is concerned with organization maintenance, that is, service and support activities.
>
> There are eight areas of specialization in functional structures: time and credit, routing, instruction, gang (work direction tasks), disciplinarian, speed, inspection, and repair and maintenance. Each one was designed to support the workers and control them in their jobs . . . it results in more complex relationships between supervisor and staff because of multiple reporting and authority lines. The workers have eight functional supervisors over them plus the general foreman.*

The Taylor** idea of "functional organization" is perhaps based upon identifying functional elements within the framework of a general classification scheme for these. However, it is not clear from Taylor's ideas that these functional elements are intended to be reassembled into functionally cohesive subsystems. Moreover, the classification scheme used (scheduling, quality control, routing, etc.) is of questionable value, and cannot claim to cover all categories of system functions.

The apparent equation between specialty area and "function," based on the Taylor ideas, comes through in many books on organizational structuring,*** often in connection with contrasting the "project" approach to the "functional" one. Two major problems are present in this contrast:

1) The functional viewpoint, compared to the project one, is thought to emphasize levels of control and "line" (as opposed to "staff," or "support") organization, as well as a formal, more highly structured mode of operation. In contrast, this author's view of function transcends these types of distinctions, whose discussion should instead be relegated to the separate matter of system evolution over time.

---

* From Senn, James A., Information Systems in Management, second edition, copyright 1982 by Wadsworth, Inc., used by permission, p. 70.

** Taylor, F.W., The Principles of Scientific Management. New York, N.Y.: W.W. Norton, Co. Inc.©, 1911.

*** e.g., see Rakich, J.S., Longest, B.B. Jr., and K. Darr, Managing Health Services Organizations, 2nd edition. Philadelphia: W.B. Saunders©, 1985, pp. 154—157.

2) The functional viewpoint is thought to be equivalent to the perspective of specialty areas. In a hospital, for example, nursing, social work, dietary, physical therapy, etc. are often referred to as different "functional" areas.* In the present view (also expressed by Brenda Shestowsky in Supplement B to the previous chapter), these different areas are different **classes** of functions more so than distinct functionally cohesive subsystems.

Integration of Different Notions of Function

Supplement A to Chapter Two gave the Blumenthal idea of function, which, it seems, does a better job of relating the different aspects of functions and functionalism. Blumenthal handles both the problem of identifying functions based on functional mission, and the problem of identifying functionally cohesive subsystems across different functional missions. In other words, he integrates the ideas of Mintzberg with those of HOS. This is, in effect, what the present book attempts to do in its idea of functional cohesion (see the first section of Chapter Two).

TOWARDS A BETTER WAY OF IDENTIFYING FUNCTIONS

SEARCH FOR BETTER WAYS OF IDENTIFYING FUNCTIONS

The subject of rigor in identifying functions is discussed frequently in the computer literature. What follows is meant to provide further progress in obtaining that much needed rigor. As shown below, the first step is to introduce "directivity" into identifying functions. That is, functions should be identified such that they contain all and only those operations in the same direct line of flow towards objectives within the same functional class/level/time-orientation. With this starting point, it is often not possible to concatenate (implode, join together) two functions into a larger function within the same system, since this would cause the inclusion of indirect operations.

---

* See Rakich, Longest, and Darr, p. 157.

The next step in introducing more rigor is to provide more explicit understanding of the relationships between different functional outputs, as in the following two senses:

- show gating (flow routing) logic, i.e., AND (parallel branch) and OR (decision branch) relationships between functional outputs;
- identify various types of outputs, e.g., outputs corresponding to functional objectives, feedback outputs, symbiotic outputs (produced only for the maintenance of a non-directly related function), offshoot outputs, etc.

Having thus identified functions, further decomposition into functional elements takes place. The functional elements may then be reassembled to construct the subsystems of the parent system in which the functional elements are identified, that is, according to the principle of "reconstructability analysis." This mixes top-down and bottom-up approaches to system structuring. Simply stated, this approach goes from level 1 to level 2 to level 3, and then a new configuration of level 2 is built by a bottom-up reconstruction of the level 3 elements.

THE OBJECTIVE-DEFINED FUNCTION CONCEPT*

Essentials of a New Definition

An objective-defined function is defined in this text as a sub-set of the operations of a system, consisting of only those operations directly necessary to the achievement of the same specific objective, or the same "mutually contingent" set of specific objectives. It also contains all the operations directly necessary for this purpose, unless the operations fall into different functional classes or are at different functional levels. In these cases, different objective-defined functions must be identified corresponding to the difference in basic class or level. As well, differ-

---

* Parts of the following sub-section are based on the following articles by Ed Baylin:
1) "Functional Modeling of the Business Organization," Cybernetics and Systems: An International Journal. Washington, D.C.: Hemisphere Publishing Corp.©, vol. 15, nos. 3—4, 1984;
2) "Computer Systems Analyst's View of the Business Organization," Cybernetics and Systems: An International Journal. vol. 16, 1985, pp. 305—323;
3) "Identifying System Functions," International Journal of General Systems. New York: Gordon and Breach Science Publishers©, vol. 12, no. 1, 1986.

ent time-orientations may serve as the basis for distinguishing distinct objective-defined functions within a single direct line of flow.

Following are some further details to clarify the above definition:

- "Directly" means that the operations which plan, organize, direct, and ensure operations in that function are not included. These control types of operations have only an indirect—albeit necessary—bearing on the functional objectives, as they do one of:
  - —setting up frameworks for the current operations;
  - —directing (coordinating) and ensuring the carrying out of these current operations.
- "Mutually contingent" means that the functional objectives may be related by the various Boolean logic flow routing understandings, namely: AND (both required), EXCLUSIVE OR (mutual exclusion, i.e., either one but not both), and perhaps even INCLUSIVE OR (and/or, i.e., either one or both). The inclusion of the OR in the definition of mutual contingency expands what certain others* call "mutual contingency."
- It is assumed throughout that the word "objectives" is used in senses other than performance goals. Also, objectives are always stated as logically, conceptually, and flexibly as possible, without losing specific meaning.
- The differences possible in functional class, time-orientation, and level are introduced later in this chapter. As will be seen there, the primary criterion used in establishing the functional class is based on the very distinction of direct versus indirect relationship to objectives. This establishes two basic classes, namely, baseline and adaptation operations. The adaptation operations are further divided into sub-classes which may cause different objective-defined functions to be identified for adaptation operations otherwise having a common direct relationship to objectives.

In general, there is much more to the notion of function than initially meets the eye. Since the functional notion is not only very important, but takes many more pages of text to explain in detail, many of the important issues are simply skimmed below, with further details being left until Chapter Eight. The functional concept is perhaps the single most important idea in this book, since it underlies both the new charting method and the explanations of the cohesion methods for subsystem identification.

---

* e.g., Martin and McClure, p. 323.

Further Points about Objective-Defined Functions

Flow Functions

In addition to "objective-defined" functions, defined above, there are also what are here called "flow" functions (see Figure 3-2). These are conceptual components of an objective-defined function, and comprise one or more functional elements.

Functional Elements

Objective-defined functions are composed of serial/parallel operations referred to as "functional elements" (the elements which are grouped by the functional cohesion method as described in Chapter Two). Figure 3-2 shows some of the relationships of functional elements which might, for instance, occur within an objective-defined function. A functional element can be considered to be any part of an objective-defined function, a flow function, or even a whole objective-defined function. However, a functional element cannot be a grouping of more than a single objective-defined function.

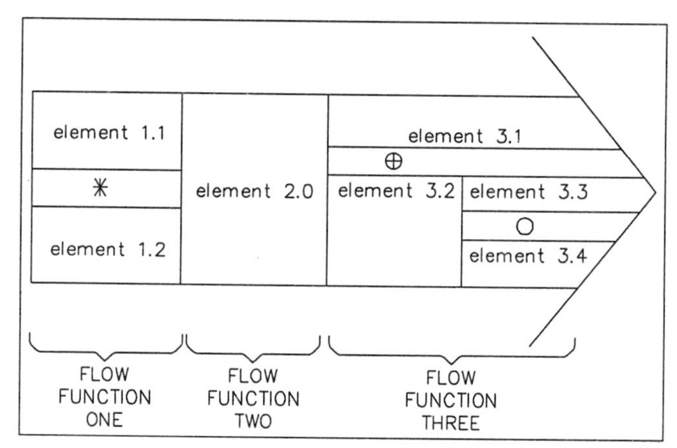

FIGURE 3-2: Objective-defined function and its functional elements (copyright by Hemisphere Publishing Corp.).
The asterisk (AND Boolean symbol) between functional elements 1.1 and 1.2 indicates a parallel relationship. The EXCLUSIVE OR symbol between functional element 3.1 and the other two functional elements in the third flow function indicates alternative paths. Finally, the INCLUSIVE OR (and/or) symbol separating functional elements 3.3 and 3.4 symbolizes that one or both may be performed in a given run of the objective-defined function (represented by the wideshafted arrow as a whole).

Final Versus Intermediate Objectives

Unless what is being dealt with is the last flow function in an objective-defined function, flow functions define "intermediate" objectives, as opposed to "final" ones. The last flow function of the objective-defined function(s) which directly achieve(s) the **system** objective(s) may also be said to have the final objective(s) of the system as a whole. As just implied, certain objective-defined functions do not have final objectives with respect to the system as a whole. That is, they themselves do not directly carry out the system objective(s). In other words, many cases exist in which the purpose of an objective-defined function is, like that of a flow function, actually nothing more than a step along a line of flow towards a final goal, which will eventually be achieved by a later objective-defined function. According to the presently developed functional classification scheme (see the third section of this chapter), the final system objective(s) is(are) achieved by "baseline" (or "system base") objective-defined functions, while those objective-defined functions having only intermediate system objectives are classified as "adaptation" ones. Adaptation functions have an indirect relationship to the system objective(s).

Primary Versus Non-Primary Objectives

Since an objective-defined function can have more than one specific goal, the question of establishing the priorities of the various objectives is raised. While the **scheduling** of objective achievement can, in certain cases of mutual contingency, be scientifically addressed, assessing the **importance** of the objective is really a subjective matter. In fact, in the rare instance where an objective of an objective-defined function may be produced by a flow function occurring earlier than the last flow function, it would be tempting to say that the objective achieved later is more important. However, even here, choosing the priority involves a value judgement.

Specific Versus General Objectives

Functional objectives must be "specific," that is, they must be identified in a "reasonably" narrow way. The subject of specific versus general objectives involves issues such as tolerance ranges within which it is assumed that an objective has been reached, parallel explosion of detail, objectives as being slices of one or more other objectives, and objectives as being stated such as to cover different mutually exclusive options.

Non-Objective Outputs

Many functional outputs do not correspond to either primary or non-primary objectives, since they are not objectives at all. For instance, they may be connected with the means by which objectives are attained, or they may simply be produced in connection with interfacing with other objective-defined functions or systems. Confusing all outputs of a function with functional objectives is common, since it is connected with an overly simplistic application of the black box idea. Figure 3-3a shows that, in such an application, the functional significances of the inputs and outputs are not symbolized in the diagram. However, inputs could be transactional supplies to be operated on, guidelines to control the operations, instruments which carry out the tasks, etc. Different types of output include those representing objectives, those used to obtain feedback for purposes of control, symbiotic outputs to interface with other functions/systems, and so on.

Figure 3-3b also shows a black box, i.e., the wideshafted arrow, but one in which the functional significances of both the inputs and the outputs are displayed in the diagrammatic symbols. The symbols in

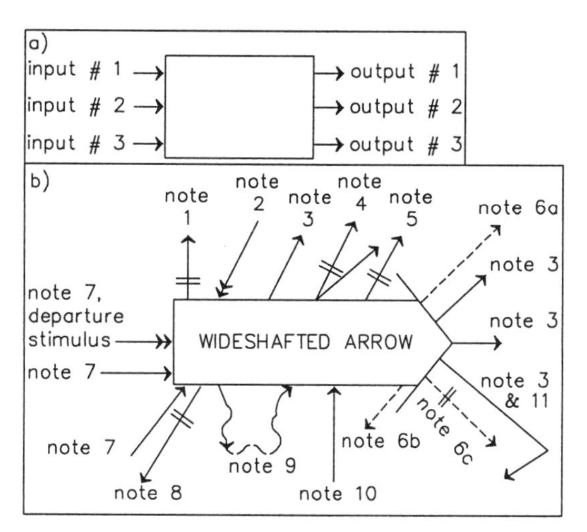

FIGURE 3-3: Functional significances of inputs/outputs in wideshafted arrow (adapted from a figure copyrighted by Hemisphere Publishing Corp.). The wideshafted arrow is seen here as a black box where the functional significances of the inputs and outputs are symbolized:

a) *a black box where the functional significances of the inputs and outputs are not indicated at all by the symbols;*

b) *(a black box in which the functional significances are, in contrast, clearly indicated by the symbols. The meanings of the reference numbers in this figure are given in Table 3-1 (opposite page).*

REF. #

1- An offshoot, i.e., something provided, or made available, outside the function, but which may not be considered to be one of the other output types; e.g., exhaust gas, waste product, an output considered to be of no relevance to the emitting function.
2 - A retarding or terminating stimulus.
3 - An output which represents achievement of a specific objective (four examples of number 3 occur in the figure, one prior to the end of all operations in the function).
4 - An output with a hybrid significance, combining the meanings of both notes 3 and 5; this is just an example of a possible hybrid.
5 - A symbiotic output sent to a cooperating function or entity, but of no direct relevance to this run of the emitting function.
6a- Future-oriented feedback data, i.e., start of a loop affecting **future** runs of the emitting function; data which is not a symbiotic output, but one used for purposes of control of the emitting function.
6b- Past/current-oriented feedback data, i.e., start of a loop leading to control of operations with errors to be corrected by redoing something in the **current** run of the emitting function.
6c- Future-oriented feedback data passed by one objective-defined function on behalf of another, and viewed as symbiotic output by the function which passes it on behalf of the other function.
7- An input used during functional execution to promote further execution of this run. This could represent many things, although only three examples are present in the figure; e.g., the following:

- input providing transactional supply, or a guideline, an instrument, depending upon what other symbols are drawn on the tail end of the arrow (see Figure 3-4 for these);
- an initiation stimulus (if enters at start of run, e.g., see arrow with double headed tip), or an encouragement or continuation stimulus (if arrow with double headed tip enters after start of run).

8- An extracted/competitive/involuntary output; e.g., something taken by another function, but not "voluntarily" provided by the emitting function.
9 - Something output in this run of the function which returns at a later step in the very same run of this function, other than feedback affecting the current run.
10- Non-specific (to-be-defined) type of input.
11- Hooked objective output, i.e., output sent to the system environment with the intent of initiating the sending of an input from the environment to another function of this same system, in a chain of linked functions.

General Note: Once entry or exit of any type of arrow is complete, the small arrows may go in any direction, depending upon the circumstances. Also, the small arrows may split and branch during their flow, as the situation requires.

TABLE 3-1: Explanation of reference numbers in Figure 3-3b.

Category	Symbol	Meaning
INPUTS & OUTPUTS	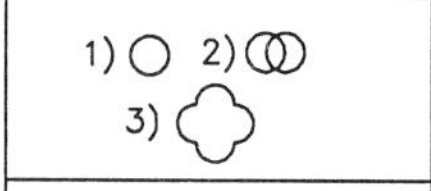 1) 2) 3)	1) transactional input, 2) a batch of transactional inputs, and 3) a synthesis of transactional inputs, respectively. Transactional input is what is operated upon by, or what causes, the run.
	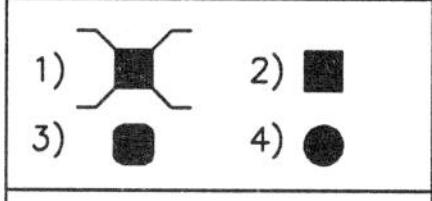 1) 2) 3) 4)	methodological frameworks – i.e., guidelines in carrying out operations – of 1) highly stable, 2) medium-stability, 3) temporary, and 4) one-shot types, as numbered
	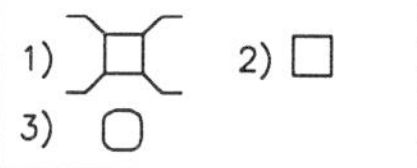 1) 2) 3)	instrumental frameworks – i.e., agents/instruments who/which carry out the operations – of the highly stable, 2) medium-stability, and 3) temporary types
FLOWS	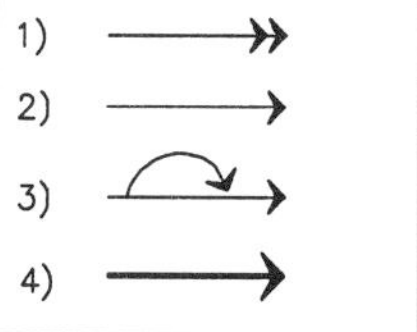 1) 2) 3) 4)	1) stimulus (signal or impulse) representing a transaction, which may initiate (activate), terminate, continue, slow-down, speed-up, etc. a serie of events, 2) a flow route segment, but without stimulus, 3) an input passed along for use by a later function or functional element, 4) material (physical) flow, as opposed to data (or information) flow
	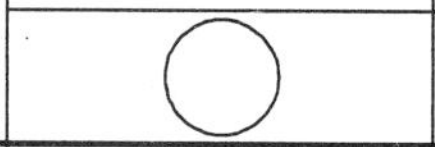	on-page connector
INPUTS & OUTPUTS & FLOWS	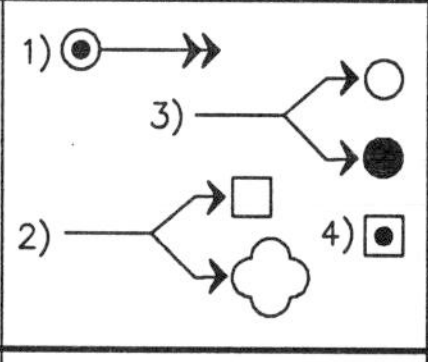 1) 2) 3) 4)	hybrids of the above symbols, i.e., 1) a combination of one-shot guidelines and transactional input along with the stimulus which begins the transaction, 2) a combination of a synthesis of transactional inputs and a medium-stability instrumental framework, 3) a combination of a one-shot instrumental framework and transactional input, 4) etc.
STORES	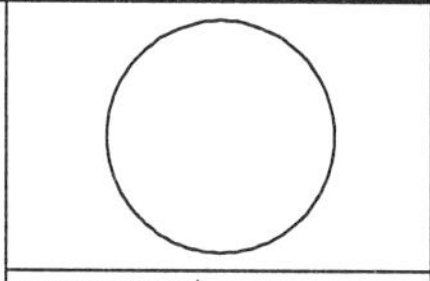	deposit point, as opposed to a reservoir, i.e., a store, but not necessarily a store used for accumulation
	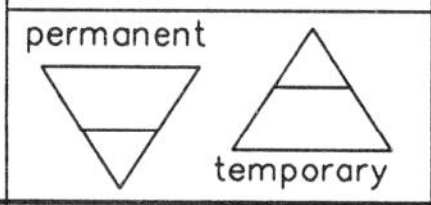 permanent temporary	a reservoir, or a collection of like entities, likely in a deposit point (store) used for accumulation
ENVIRONMENTAL INTERFACES	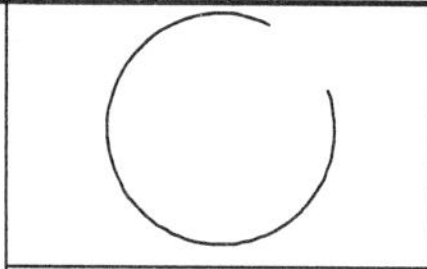	environmental framework, i.e., a source or destination with which the system interacts
	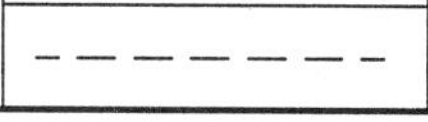 – – – – – – –	system boundary

FIGURE 3-4:
Other functional significance symbols of wideshafted arrow.
The symbols shown here are used on the tails and tips of the small arrows associated with the wideshafted arrow. Some of these ideas actually represent points on a spectrum.

Figure 3-3b (along with explanations in Table 3-1) are used in conjunction with the details in Figure 3-4 to obtain a fairly refined view of the functional significances of inputs and outputs in the new structure-flow charting method explained in Chapter Five. A major section of that chapter is devoted to discussing the inputs and outputs of the wideshafted arrow.

Further Points About Non-Objective Outputs

Meanwhile, until Chapter Five is read, "non-objective" functional outputs can be exemplified as follows. A worker hands material over to a second person on the assembly line. Since the final transmission of this material is the only objective of this worker's baseline function, the question is how to classify some of the other outputs transmitted by this worker to the next worker. For instance, the worker has social conversation with the second worker. If this conversation ultimately results in the workers' functioning as a more cohesive team in the achievement of the designated task, it can be called a "symbiotic output" of the first worker to the second. This output has no direct role in the production of the output in a given run of the worker's function. Nor is it feedback for purposes of control. However, it indirectly affects the worker's production.

Similarly, the general ledger postings in an accounts payable system are symbiotic outputs, since failure to make such postings will eventually result in distortion of the company's financial statements, thereby leading to detrimental effects on the accounts payable system. The general ledger postings have nothing **directly** to do with the achievement of the payment of money to suppliers for goods delivered, that is, the achievement of the accounts payable system objective.

Another example of non-objective output may be derived from the worker's conversation. If this has no ultimate effect, one way or the other, on the achievement of the worker's designated task, this social conversation is "off-shoot" output. However, if the social conversation in fact **worsens** the relationship between the two workers, this would be diagrammed as a "competitive" output, since it has the opposite effect of a symbiotic output. When one country has its secrets stolen by another country in a spy operation, the exporting country is said to have an "extracted," or "involuntary," output. The latter are diagrammed in the same way as competitive output.

Feedback and "out-back-in" outputs are two other output types. First, feedback output may be sent to a control function for purposes of receiving either correction, or the authorization to continue current operations, or, in the case of future-oriented feedback, to produce control actions affecting future runs of the feedback-originating function. For instance, future-oriented feedback output may be sent to a planning function for use in long-term planning which will eventually affect the feedback-originating function through new guidelines.

Second, out-back-in output exits the objective-defined function in the current run, for accumulation in a reservoir or for interface with an environmental entity. It comes back later in the same run, after storage in the reservoir and/or generation of a response from the environmental entity. This is distinctly different from the feedback idea, since this output is not used for purposes of control. It is also different from the idea of symbiotic output, in that the return to the originating function is directly related to current functional operation.

Unfortunately, feedback and out-back-in outputs are often both called feedback, because both involve a loop in which an output returns to its originating function. True, the same output may sometimes serve both purposes. Feedback, as seen here, is, however, associated only with control of operations of the **originating** function. More is said about this terminological problem in the fourth section of this chapter.

Another problem, but of a very different origin, is that a function being restricted to a direct line of flow results in the possibility of representing the same functional output as either feedback or symbiotic output, depending upon whether the output is seen as going from the function being controlled or the function doing the control, respectively. For instance, a worker has two supervisors, one in charge of initiating the worker's task, and the other in charge of validating the results of the task and causing it to be redone in case errors are discovered. Data about the actions of a worker used to control future tasks of that worker may be sent to the task initiation supervisor, either directly by the worker himself, as feedback output, or via the task validation supervisor, as symbiotic output of this latter supervisor. This is symbolized by item #6c in Figure 3-3b.

Example of an Objective-Defined Function

So far, the following dimensions for analysis of functional objectives have been discussed:

—system/final versus intermediate objectives;
—primary versus non-primary objectives;
—specific versus general objectives.
—objectives versus non-objectives;

The function of entering customer orders illustrates the latter two of these dimensions. The order entry objective-defined function may be decomposed into the following flow functions:

1 - enter the order data;
2 - decide whether to accept or reject the order, i.e., perform a credit check on the customer, and see if inventory is in stock;
3 - either print the workorder (for picking of the order by the warehouse) and the shipping documents (for shipping), or communicate refusal of the order to the customer.

In Figure 3-5a, the order entry function may be viewed as having two specific, mutually exclusive objectives, namely, the printed order or the order refusal. The Boolean symbol between the two arrows representing these objectives indicates their EXCLUSIVE OR relationship. The two objectives of order entry jointly fulfill a more general objective, namely, "information about the order entry outcome," or "initiative control of the business baseline."

A symbiotic output, the new order data, going into the open order file, has nothing directly to do with initiating further processing of the order if the order is accepted. However, it will assist further control functions needed in the fulfillment of the customer order, and could thereby eventually affect further orders, since new orders may depend on how well the current order is filled.

The symbiotic output's relationship to the order acceptance objective output is indicated by the Boolean AND symbol between the two neighbouring arrows representing these outputs. In Boolean logic, the AND symbol is applied before the OR, as in the following equation: ORDER OUTPUT equals (PRINTED ORDER AND FILE UPDATE) OR ORDER REFUSAL.

Indeed, still more specific objectives may be appropriate for the level of system under consideration. Thus, Figures 3-5a to Figure 3-5e explode the printed order objective into the workorder, which goes to the warehouse, and the shipping order, forwarded to shipping. The differences between what is represented in each diagram include the relative scheduling of the production of these two outputs, and the degree of similarity of the two outputs. These distinctions, indicated in the descriptions of the diagrams in question, are just as much an art as a science.

Figures 3-5f and g use the idea of showing EXCLUSIVE OR logic with he decision (diamond) symbol, as in the transaction-centered design n structure charts. These diamonds may be used in conjunction with Boolean logic symbols. Research with the structure-flow chart has shown that **both** Boolean **and** structure charting flow route branching symbols for decisions need to be used in hybrid fashion if many of the subtleties of the different permutations and combinations of flow routing logistics are to be effectively captured.

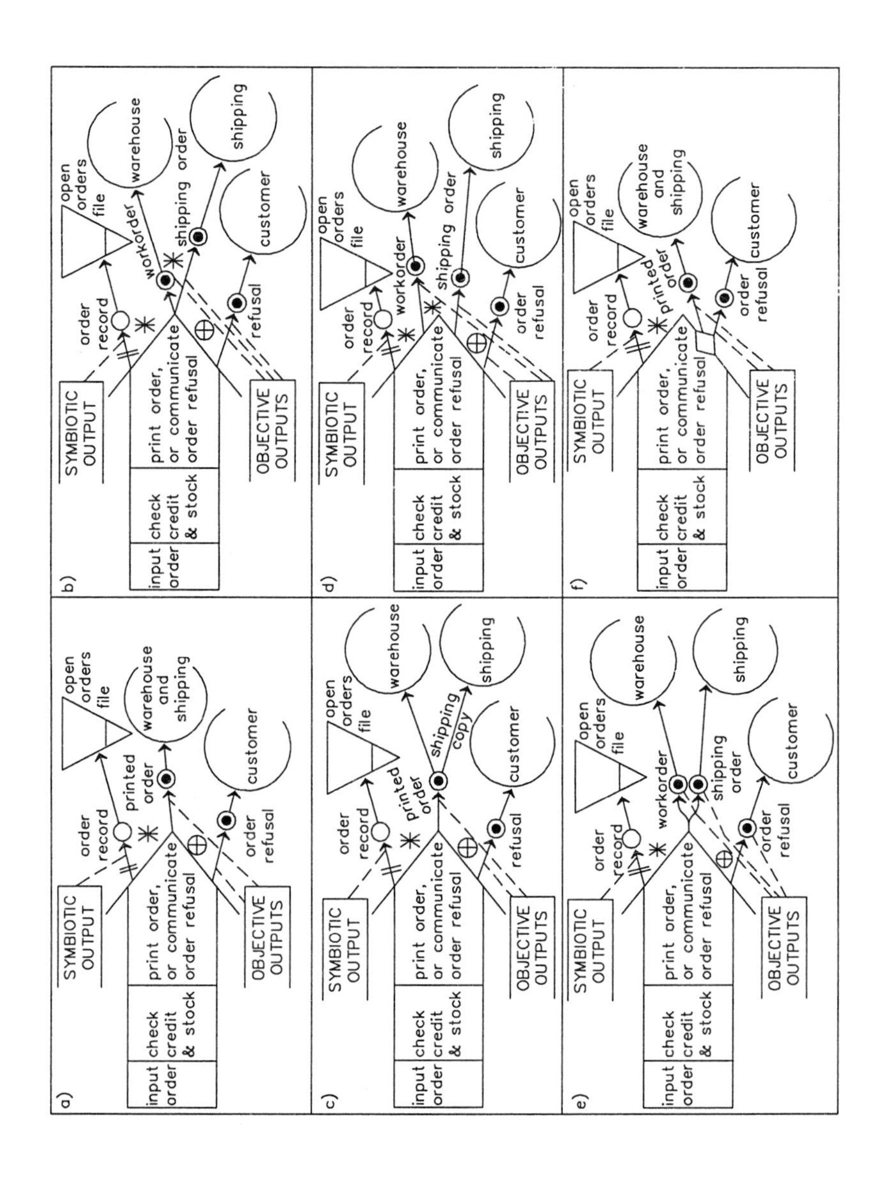
a)
input order
check credit & stock
print order, or communicate order refusal
SYMBIOTIC OUTPUT
OBJECTIVE OUTPUTS
order record
printed order
open orders file
warehouse and shipping
customer
order refusal
b)
input order
check credit & stock
print order, or communicate order refusal
SYMBIOTIC OUTPUT
OBJECTIVE OUTPUTS
order record
open orders file
workorder
warehouse
shipping order
shipping
customer
refusal
c)
input order
check credit & stock
print order, or communicate order refusal
SYMBIOTIC OUTPUT
OBJECTIVE OUTPUTS
order record
printed order
open orders file
warehouse
shipping copy
shipping
customer
refusal
d)
input order
check credit & stock
print order, or communicate order refusal
SYMBIOTIC OUTPUT
OBJECTIVE OUTPUTS
order record
open orders file
workorder
warehouse
shipping order
shipping
customer
order refusal
e)
input order
check credit & stock
print order, or communicate order refusal
SYMBIOTIC OUTPUT
OBJECTIVE OUTPUTS
order record
open orders file
workorder
warehouse
shipping order
shipping
customer
order refusal
f)
input order
check credit & stock
print order, or communicate order refusal
SYMBIOTIC OUTPUT
OBJECTIVE OUTPUTS
order record
printed order
open orders file
warehouse and shipping
customer
order refusal

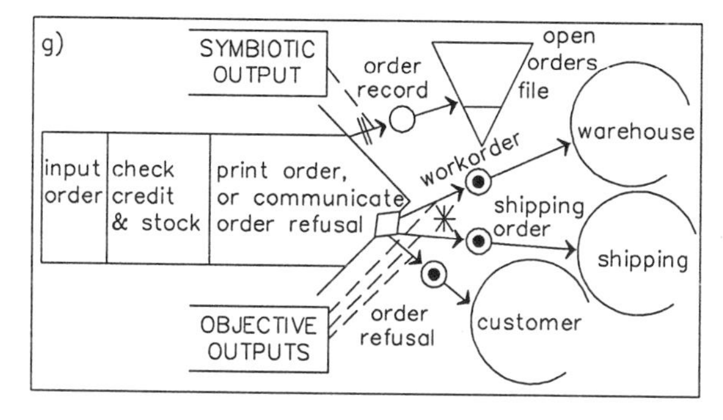

FIGURE 3-5:

Various versions of order entry objective-defined function.

Different versions of the outputs are portrayed in these figures:

*a) **(opposite)** version #1, one symbiotic output and two mutually exclusive objective outputs, the symbiotic one being produced only in connection with accepted orders;*

*b) **(opposite)** version #2, the same as "a", except that the printed order is shown as two different outputs, having essentially (but not exactly) the same data, but going to different destinations; the common origin of the arrows representing these two outputs symbolizes the essential similarity of their data;*

*c) **(opposite)** version #3, the same as "b", except that exactly the same data goes to both the warehouse (for picking) and to the shipping department; this is really just one specific functional objective transmitted in two different directions;*

*d) **(opposite)** version #4, the same as "b", except that the data on the workorder is seen as being printed at what may be a different time than the shipping order; also, the shipping order data is quite different from the workorder data, such that two different arrow origins are used for these outputs;*

*e) **(opposite)** version #5, similar to "d", except that the workorder for the warehouse and the shipping order for the shipping department are produced at essentially the same time; as in the preceding figure, these two outputs are substantially different from one another;*

*f) **(opposite)** version #6, the same idea as "a", except that the diamond symbol is used for mutually exclusive options instead of the Exclusive Or symbol; this diamond symbol comes from the transaction-centered design in structure charts;*

*g) **(above)** version #7, similar to "b" or "d" or "e", depending upon how it is interpreted; this provides a hybrid between the Boolean gating logic symbols and the symbols used in structure charts for flow routing.*

FUNCTIONAL ANALYSIS VERSUS SUBSYSTEM ANALYSIS*

Essential Relationship

What is often called a "function" in the literature is equivalent here to a "subsystem identified along functional lines." If the concept of the objective-defined function is used, functional analysis is one way of breaking down the system's operations, while subsystem analysis is an alternative way. However, a connection exists between these two manners of decomposing system operations. Thus, having first identified functions, i.e., objective-defined functions, and then identifying functional elements within these using a top-down approach (e.g., by using a hierarchical decomposition charting method, such as HOS, to decompose the objective-defined function symbolized by the wideshafted arrow), the functional elements may then be reassembled in a different way within and across different objective-defined functions to form subsystems via a bottom-up approach. The term which could be applied to this approach for determining the make-up of subsystems is "reconstructability analysis." That is, the system is decomposed into functional elements so these may be reassembled to form a subsystem within the system. As seen in Figure 3-6, the goals of reconstructability analysis are achieved by going from the level of the system as a whole to the level of objective-defined functions, and then to the level of functional elements, and then to reassembling the functional elements into a different second level of detail, namely, subsystems.

To repeat, common usage of the term "function" refers to what is called here a "subsystem identified along functional lines." For example, the accounts payable system is often referred to as a "business function," rather than as a "subsystem of the business." (In fact, reference is often made to the "accounting function" in speaking about organizations, although the various so-called accounting subsystems are so loosely coupled to one another that it could hardly be said that they form a subsystem identified along functional lines.) Further details about this usage of the word function are contained in Supplement A to Chapter Two, on the Blumenthal notion of function and of modeling of the organization.

* This sub-section is based on materials appearing in the article by Ed Baylin, "Computer Systems Analyst's View of the Business Organization," Cybernetics and Systems: An International Journal. Washington, D.C.: Hemisphere Publishing Corp.©, vol. 16, 1985, pp. 305—323.

In contrast, with the objective-defined function concept of identifying functions, it becomes important to avoid confusing pure functional analysis with subsystem analysis along functional lines. For instance, in the case of a stereo system for the reproduction of sound, each component is designed so that, in current terminology, one would say that it performs a particular "function" designed by engineers, i.e., amplification, tuning, sound emission, loudness control, etc. Nevertheless, each component contains more than one **objective-defined** function, at least insofar as it must have certain internal control functions related uniquely to its own objectives. Thus, although this component subsystem may perform a function of the stereo system as a whole, in the sense of a subsystem identified by the the functional cohesion method, it contains various objective-defined functions as well as the base function relative to this particular subsystem. These non-baseline functions must at least include those which are "inseparable" (Figure 3-7a) from the baseline function, or, in other words, **inevitably** internal (or "self-administered"). In sum, the idea of directness in the concept of objective-defined function greatly changes the use of terms.

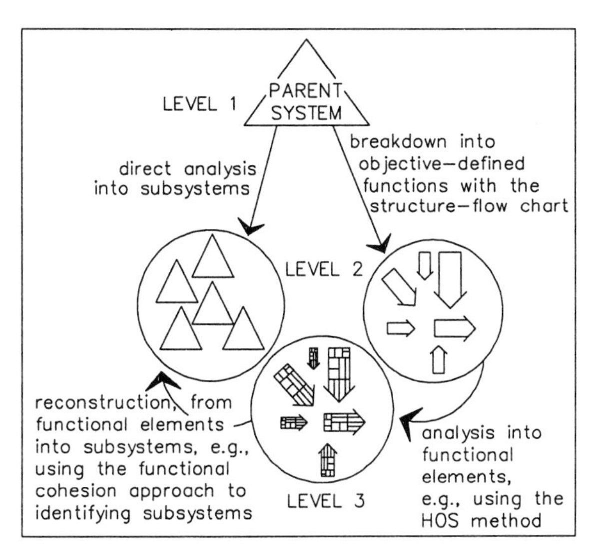

FIGURE 3-6: The idea of reconstructability analysis (copyright held by Hermisphere Publishing Corp.). *This shows how to go from functional to subsystem analysis via reconstructability analysis.*

Functional Cohesion Versus Logical Cohesion

Subsystem analysis along functional lines, i.e., functional cohesion, is often confused with analysis into subsystems according to functional class, a method called "logical cohesion" in the computer literature. However, while subsystem analysis using functional cohesion results in the grouping of highly coupled functions, using logical cohesion often results in both the dissociation of highly coupled functions, and the grouping of loosely coupled functional elements into the same subsystems. (Because of this, logical cohesion should perhaps be referred to as "illogical cohesion.")

The essential difference between logical and functional cohesion is illustrated in Figure 3-8b (whose comprehension will be helped by also looking at Figure 3-8a). The subsystems in each triangle are encircled by dashed lines. Each triangle in this diagram is also divided into two by a solid line which creates both a smaller triangle, in the upper part, and a rhombus at the bottom. The smaller, internal triangle represents the decision-making—i.e., informational—class of system operations,

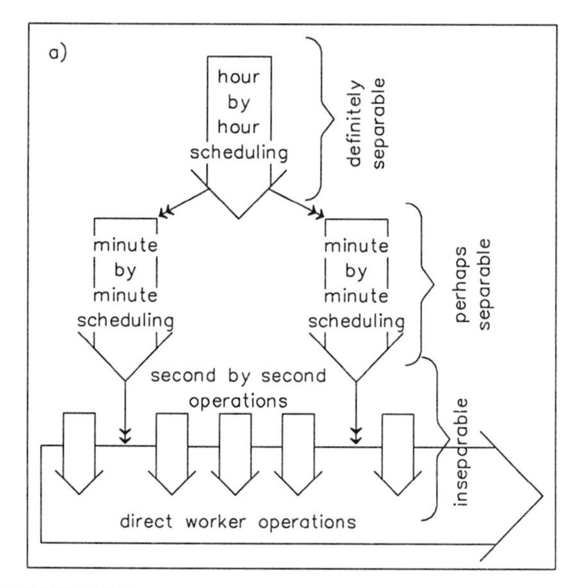

FIGURE 3-7:
The idea of inseparable internal adaptation functions (copyright held by Hemisphere Publishing Corp.).

a) ***(above)*** *idea of "inseparable" (inevitably internal) adaptation operations, with these impressionistically represented as embedded within the shaft of the base function to which they apply;*

b) ***(opposite)***

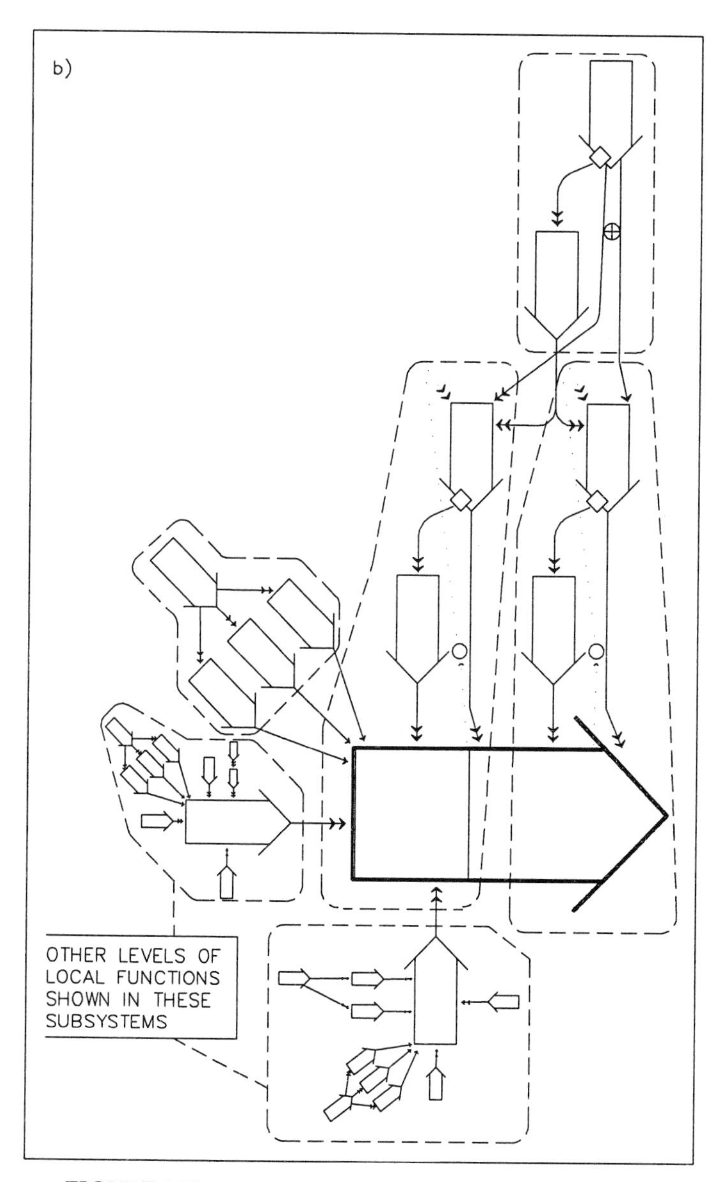

FIGURE 3-7b:

b) levels of objective-defined functions in the internal sense using the structure-flow chart. The more logical level structure-flow charts often do not show the internal adaptation functions within the umbrella of functional elements.

while the rhombus represents the remainder of the system operations, i.e., the non-decision-making operations. When logical cohesion is used, as seen in one of the triangles in the figure, the subsystems correspond to the division of the triangle provided by the solid line. In contrast, when functional cohesion is used as the method of subsystem identification, as seen in the other triangle, individual decision-making operations are grouped with the non-decision-making ones to which they specifically apply. Thereby, the subsystems identified by functional cohesion are basically along a vertical, rather than a horizontal dimension, with the exception that the horizontal dimension is used at the top of the triangle showing functional cohesion. This symbolizes grouping operations at the strategic planning level to avoid subsystem conflicts, which would otherwise arise since operations at this level are highly coupled to all the vertically identified subsystems in the triangle.

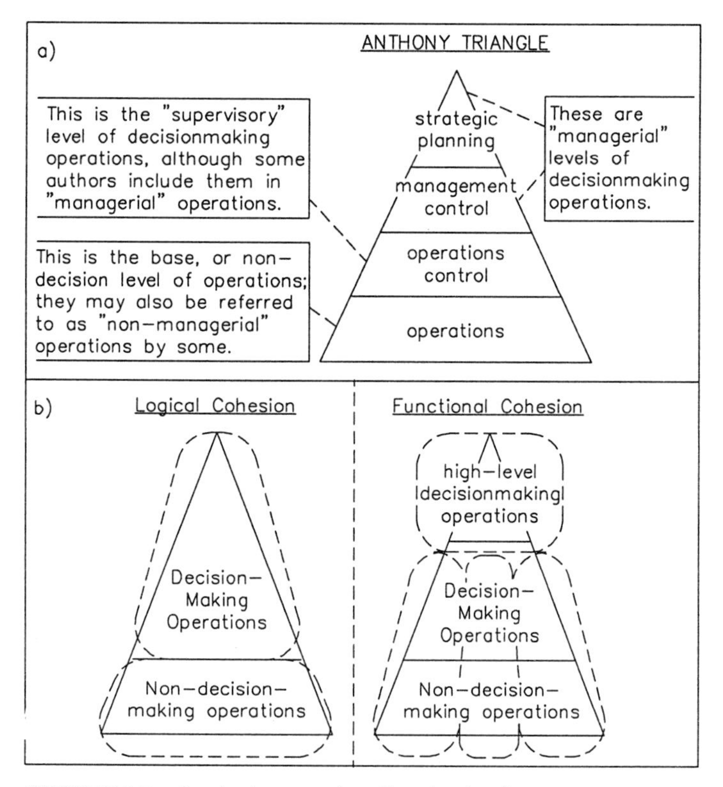

FIGURE 3-8: Logical versus functional cohesion—
an impression captured using the Anthony triangle (copyright Hemisphere Publishing Corp.).
a) the Anthony triangle; b) logical versus functional cohesion.

FUNCTIONAL CLASSES, LEVELS, AND TIME-ORIENTATIONS

In order to provide a basis of comparison with something already well-known, the new functional classification scheme, presented in this section, is discussed after describing the Fayolian scheme. Preceding the Fayolian scheme is that of Anthony, which may be interrelated with the Fayolian one. Explaining the new scheme of functional classes introduced here also requires a new interpretation of some of the terminology of feedback and feedforward, which are discussed in the fourth section of this chapter.

Before beginning, it may be best to attempt to deal with the inconsistent way in which the all-important word "control" has been used by different writers. Following are some of these different uses:

- The Fayolian scheme effectively relates control to negative feedback affecting the redoing of past operations in the current run.
- Anthony distinguishes between management control and operational control associated with supervisory work, both of which fit below the strategic planning level of managerial work. These uses of the word "control" clearly do not have anything to do with whether the orientation is future or past (current), although it may be understood that most managerial level control is concerned with long-term, future-oriented measures, while supervisory control involves both types of time-orientations, with about equal frequency.
- The word" control" in cybernetics has a still broader meaning, including both current and non-current adaptation operations. With respect to current decision-making (i.e., coordinating/validating in this book's scheme) operations, control operations provide stimuli to initiate, cancel, speed up, slow down, continue, stop, restart, rerun, etc. operations in the current run, in either a future or a past/current-oriented direction. Such control operations may begin by validating past operations, or by modeling the next operations in the current run, following which corrective or preventive measures, respectively, may be taken. As for non-current control operations, called planning/framework-ensuring and framework operations in this book's functional classification scheme, these are also control operations, but in a different time frame from the current ones. (The difference between the term "time frame" and "time-orientation" will be described below.)

To deal with the above problem of semantics, the following terms are adopted here:

- "Ensurance" refers to a time-orientation in which a deviation-reducing feedback loop is involved and corrections are past (current)-oriented, that is, completed by redoing something already done erroneously in the current run, before continuing with operations in the current run.
- "Feedforward" refers to future time-orientation, which may or may not involve a loop. Feedforward may be of either the initiative or interruptive varieties.
- The word "adaptation" is used instead of "control" for this author's functional classification scheme, since control already has too many meanings. Regardless, the word" control" appears frequently in this text in order to describe the various workings of adaptation operations, especially those which perform decision making, and also to describe different levels of adaptation operations (as in levels of control). When used in this text, unless otherwise specified," control" takes on its widest possible meaning, as it does in cybernetics.

A REVIEW OF EXISTING SCHEMES

Anthony's Scheme

Anthony's Levels of Functions

The Anthony scheme of control levels, represented by a triangle (see Figure 3-8a earlier), divides operations into the following four levels:

- **Strategic Planning**: "deciding on objectives of the organization, on changes in these objectives, on the resources needed to attain these objectives, and on the policies that are to govern the acquisition, use, and disposition of these resources."*
- **Management Control**: assuring "that resources are obtained and used effectively and efficiently in the accomplishment of the organization's objectives."**
- **Operational Control**: "assuring that specific tasks are carried out effectively and efficiently."***

---

* Anthony, R.A., Planning and Control Systems: A Framework for Analysis. Boston: Division of Research, Harvard Business School, Harvard University©, 1965, p. 16.

** Anthony, p. 17.

*** Anthony, p. 18.

- **Operations**: carrying out the low level tasks to achieve the objectives of the system. For example, in a business organization, inputs are physically transformed by the operations level, using personnel, capital, and machinery, into goods and services.

STRATEGIC PLANNING	MANAGEMENT CONTROL	OPERATIONAL CONTROL
• Choosing company objectives	• Formulating budgets	
• Planning the organization	• Planning staff levels	• Controlling hiring
• Setting personnel policies	• Planning working capital	• Controlling credit extension
• Setting marketing policies	• Formulating advertising programs	• Controlling placement of advertisements
• Setting research policies	• Selecting research projects	
• Choosing new product lines	• Choosing product improvements	• Scheduling production
• Acquiring new divisions	• Deciding on plant rearrangement	
• Deciding on non-capital expenditures	• Deciding on routine capital expenditures	• Controlling inventory
	• Formulating decision rules for operational control	• Measuring, appraising, and improving worker's efficiency
	• Measuring, appraising, and improving management	

TABLE 3-2:
Examples of activities at different levels in Anthony scheme.
Adapted version, from Anthony, Robert A., Planning and Control Systems: A Framework for Analysis. (Boston: Division of Research, Harvard Business School, Harvard University, 1965), p.19.

Changes In Going From the Bottom to the Top of the Anthony Triangle

Table 3-2 provides examples of operations from each of the top three classes of operations in the Anthony triangle. These illustrations are based on business organizations, as the Anthony scheme finds its widest use in organizational analysis. A study of the changes in moving from the bottom to the top of the Anthony triangle suggests that the principles upon which the Anthony scheme is based include the following:

- Operations become less directly involved in achieving the system objectives. While this idea of directiveness may not have been rigorously applied by Anthony, it is applied in the same rough sense in which accountants distinguish between direct and indirect costs.
- The decision-making time frame becomes increasingly long-term, while its time-orientation becomes more future-oriented.
- The scope of decisions increases, in the sense that decisions have a wider applicability within the system. The scope factor seems to hold fairly consistently, since this is really the most basic principle of the triangle. Decision making becoming increasingly long-term is often overridden by the fact that managerial levels handle short-term matters on an exception basis. Also, in exceptional circumstances, managers may become fairly directly involved with levels as low as operations.
- As for information used in decision making, it becomes increasingly external in its sources, more summarized in form, more periodically and/or randomly provided, as well as less frequent and less current (more historical) because of delays.*

The Fayolian Scheme

This scheme,** initiated early in the century as a method of classifying functions (i.e., operations) normally performed by managers, developed the following original classes: planning (foreseeing), organizing, commanding, coordinating and controlling. These were evolved by English-speaking organizational theorists into a somewhat different

---

* Findings of Gorry, G.A. and M.S. Scott Morton, "A Framework for Management Information Systems," Sloan Management Review, vol. 13, no. 1, 1971, pp. 55—70.

** Fayol, H., General and Industrial Management, trans. C. Storrs. London: Sir Isaac Pitman & Sons©, 1949.

set of semantics, thereby giving rise to what is referred to here as the "Fayolian" scheme. The classes in this scheme are as follows:

—**Planning**: deciding what is to be done in the long-term;
—**Organizing**: developing the appropriate structure to accomplish the plan;
—**Staffing** (derived from Fayol's "command" category): obtaining the appropriate personnel;
—**Directing** (derived from Fayol's coordinating): commanding the personnel in the direction of plan accomplishment;
—**Controlling**: ensuring that the plan objectives are met.

This Fayolian scheme is used for what some loosely call "managerial" functions, which include supervisory functions and decision-making functions performed by those decision centers at the bottom level of the Blumenthal scheme (see Supplement A to Chapter Two) and which may include some of the operations level of the Anthony triangle. In contrast, the Anthony scheme, as well as the one to be described, cover **all** the operations in a system.

Cross-Cutting the Anthony and Fayolian Schemes

The Fayolian scheme of functional **classes** may be cross-cut by the Anthony scheme of functional **levels**.* Table 3-3 shows this cross-cutting in the form of a matrix. A point of view often put forward** is that certain classes of Fayolian functions are more frequent at certain levels of the Anthony triangle than at others. Specifically, the higher the level, the more planning and organizing and the less controlling and directing occur, with the greatest changes occurring in the extents of planning and controlling. Cross-cutting the Fayolian scheme with the Anthony scheme provides another dimension in analyzing functions. As well, as will be explained below, certain other types of levels, not accounted for by the Anthony triangle, are needed.

---

* e.g., See Ahituv, N. and S. Neumann, Principles of Information Systems for Management. Dubuque, Iowa: W.C. Brown, 1982, p.123.

** e.g., Thierauf, R.J., Effective Management Information Systems. Columbus, Ohio: Charles E. Merrill, 1984, p. 8.

	CONTROL LEVELS		
TYPE OF FUNCTION	Strategic Planning	Management Control	Operational Control
Planning			
Organizing			
Staffing			
Directing			
Controlling			

TABLE 3-3: Cross-cutting the Fayolian and Anthony schemes.

A REPLACEMENT SCHEME FOR FUNCTIONAL CLASSIFICATION

Derivation Principles

Figure 3-9 and Table 3-4 will be used to illustrate the set of derivation principles for this author's scheme, which divides system functions

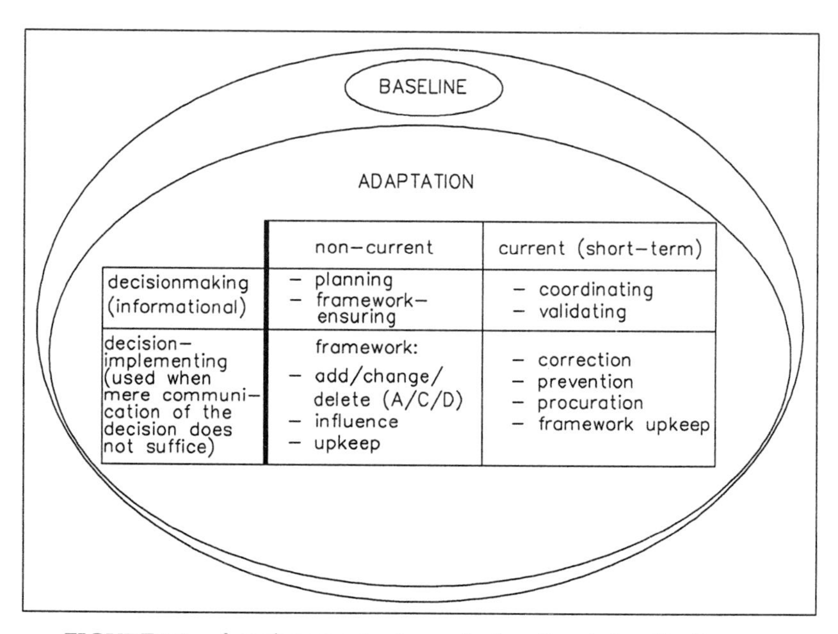

FIGURE 3-9: A replacement scheme for functional classification. *(an adapted version of a figure originally copyrighted by Hemisphere Publishing Corp.)*

(simply referred to as "operations" at many points) into various levels of subsets, from highest to lowest, as follows:

Level #1—Separation of Baseline from Adaptation

This distinguishes operations (i.e., functions) on the basis of whether or not they directly achieve the system objectives. It separates the overall set of system operations into two basic subsets, called the baseline (or system base) and the adaptation operations. This division is represented by the two inner ovals in Figure 3-9. Direct versus indirect is **not** being used here in the sense of line versus staff, as in traditional organizational theory. Rather, it refers to the distinction between operations which do the work versus those which plan, direct, and otherwise manage, and set up the frameworks for the work. The adaptation operations correspond to what some theorists call "control" operations, when the word control is used in its broadest sense.

Level #2—Primary Distinctions in Adaptation Operations

Figure 3-9 contains a matrix, whose columns and rows provide the following distinctions:

- **between current (short-term) and non-current (long-term)**: This separates the adaptation operations according to how current they are in their effect, as represented by the columns in the matrix in Figure 3-9. In other words, this separates adaptation operations according to time frame, which is a flow scheduling detail.
- **between decision-making (or "informational") and decision-implementing**: This is used to cross-analyze each of the non-current and current categories in the matrix in Figure 3-9, according to whether they make or carry out decisions. Decision-implementing operations only become necessary when mere communication of the decision fails to produce the necessary implementation.

Level #3—Secondary Distinctions in Adaptation Operations

A number of optional further categories may be provided for adaptation objective-defined functions. These cross-cut the classes already presented in Figure 3-9. At the option of the analyst, these distinctions may or may not be used when making different objective-defined functions by partitioning functional elements belonging in the same direct line of flow. These secondary distinctions, illustrated in Table 3-4, are the following:

- **between adaptation operations having internal and external focus:** This further separates the decision-implementing operations according to whether or not decisions are implemented via the system environment. Those classes which operate via the environment are "framework influence" and procuration current decision-implementing. Decision-**making** operations are not sub-divided into internal versus external, although their effect may be either.

- **between ensurance and feedforward current adaptation operations:** This distinguishes feedforward (future-oriented) and ensurance (back to make corrections in the present run before the products of the functional run are delivered to further operations) time-orientations. Time-orientations refer to the flow routing relationships between functions, as opposed to the flow scheduling relationships given by time frames. Further discussion of the different time-orientations is given in connection with the analysis of the concepts and terms of feedback and feedforward, contained in the fourth section of this chapter.

Baseline Operations

The operations directly related to the attainment of the system mission are referred to as "baseline" (or "system base") ones, while those operations indirectly related to this achievement are the system's "adaptation" ones. Although the specific baseline objectives may sometimes be considered modifiable frameworks through which the system mission is attained, the baseline operations may still be considered to be those which directly achieve the system mission.

TIME-ORIENTATION	TIME FRAME	D.M./ D.I.	← FOCUS →	
			INTERNAL	EXTERNAL
ensurance	current	d.m.	validating	
		d.i.	correction	procuration
	non-current	d.m.	framework-ensuring	
feedforward	current	d.m.	coordinating	
		d.i.	prevention	procuration
	non-current	d.m.	planning	
		d.i.	framework a/c/d	framework influence

TABLE 3-4 Time-orientation/focus vs. class of adaptation function. *In the above, "d.m." refers to decision-making operations, while "d.i." indicates decision-implementing operations.*

The baseline operations directly produce and deliver those outputs corresponding to the system mission. Terminologically, the word "baseline" combines both the notion of the "base" level of operations in the Anthony triangle (i.e., the "operations" level of this triangle) and that of the "line" (as opposed to staff) part of the system organization (e.g., see the discussion of levels-of-control charting in Chapter Seven).

Examples of Baseline Operations

In the following examples of baselines in different industries, the mission of each business organization is assumed to be its technical objectives. For instance, the mission of a firm in the warehousing business is to distribute goods, rather than to make its owners rich and/or to provide a good work atmosphere for its employees. In such a perspective, the baseline operations would be described in the following ways, for each mentioned type of industry:

Manufacturing

—receipt of raw materials;
—transformation of raw materials into finished products;
—delivery of finished products to customers, according to the parameters of the customer orders.

Warehousing

—receipt of goods needed to fulfill customer orders;
—picking, assembling, and packaging of stocked goods in response to customer orders;
—delivery of goods to customers, as per the customer order.

Hospitality—Hotels

—receipt of goods and services;
—preparation and maintenance of rooms for guests;
—provision of prepared/maintained rooms to guests.

Hospitality—Restaurants

—receipt of food and perishable supplies;
—preparation of food, beverage, and tables used to satisfy customer needs;
—serving of food and beverage to customers.

Transportation

—preparation of transport vehicles and personnel for specific voyages;
—provision of transport services.

These examples itemize activities in their sequence of occurrence in response to particular customer demands. They include only current activities which directly fulfill the customer requirements. They do not include indirect activities such as training of personnel, purchase and upkeep of long-term assets (buildings, machinery, etc.), establishing standards and procedures of a general nature, initiating activity via sales and receipt of customer orders and ordering of supplies, and other tasks involving decision making, such as task coordination and validation. All the latter types of operation fall into the adaptation category of operations.

These illustrations of baseline are related to what is usually understood as "direct" operations. However, what is direct and what is indirect is all relative to the essence-identity of the system. In fact, the mission of a system may be the objective of providing adaptive support and direction to other operations. For example, an organization such as a government may have as its chief aim to plan, coordinate, and otherwise control. That is, the baseline activities of this system are in fact adaptation operations of a larger system, since the government really belongs to the larger, societal system, in which it is a subsystem. As well, the government has its own adaptation operations, which are at least one or more levels "removed from" the adaptation operations which the government performs relative to the society in which it plays its role. Thus, the government has its own administrative operations, which do not directly act upon the population, since the population is directly administered only by the government's baseline operations.

That adaptation operations of a parent system may be perceived as baseline operations from the point of view of a subsystem within this system is part of the complex subject of the relativity of functional conceptualization. The conceptualization of adaptation functions may change in other ways as well, as may that of flow functions when looking at the operations of a parent system from the context of one of its subsystems.

Adaptation Operations

Adaptation operations enable, support, direct, etc. the baseline ones, either directly or indirectly. That is, the ones which in fact really enable, support, direct, etc., **other** adaptation operations, **indirectly** enable, support, direct, etcetera the baseline operations. Also, adaptation operations can act on the baseline or other adaptation ones indirectly in the sense that the interaction occurs via the system environment. Figure 3-10 provides an impression of the general relationships between the adaptation and the baseline, and among adaptation operations themselves. Since the adaptation operations are not the ones which perform the system base (baseline) functions, their objectives are by definition intermediate in terms of the final system objectives, achieved by the baseline. Moreover, while baseline functions may have agent-based objectives, such as the power or profit of the system operators, the adaptation objectives are viewed as being technical in nature, as are all intermediate objectives.

Table 3-5 provides some generic details of the flow involved in the various classes of objective-defined functions. The long-term (non-current) operations make decisions on, and implement, long-term frameworks. A "framework," a word with a special meaning developed in this book, may be defined in a general sense as a guideline for operations, or an instrument used to carry out operations, or an environmental entity. In other words, the framework notion refers to the resources of the system, except for those resources which are the transactional inputs (operands and stimulants) of the execution of current system operations. Thus, a customer, an organizational policy, and a piece of machinery are different types of frameworks in a business, but not so a stockpile of raw materials. Since the latter is operated upon by current business operations, it is not a framework, even though large stocks of raw materials are, in effect, of long-term existence relative to the duration of the current business operations.

The framework operations, except for framework upkeep ones, implement (add/change/delete) long-lasting internal frameworks, and influence largely external frameworks of the system. The basic framework operations of the system may be combined with the baseline operations for certain types of analysis. No special term has been

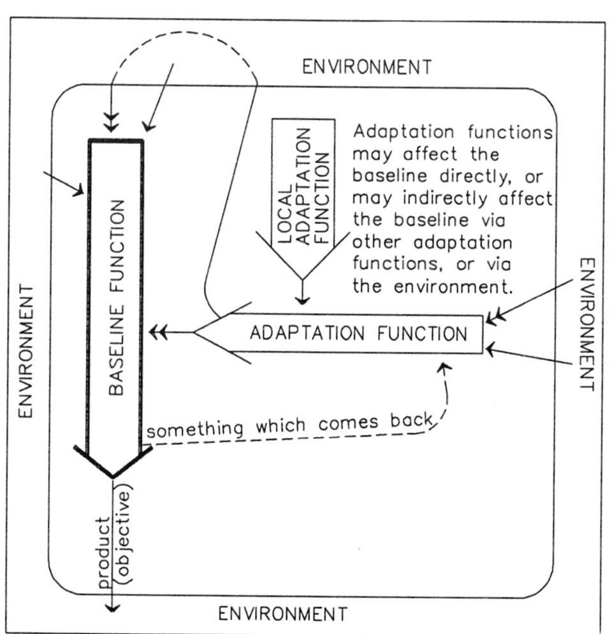

FIGURE 3-10: Baseline-adaptation relationship—an impression.

FUNCTIONAL CLASS	PHASE OF FLOW		
	INPUT	PROCESSING	OUTPUT
BASELINE	acquire transactional supplies	produce finished products	deliver finished products
COORDINATING /VALIDATING	detect problem or opportunity	perform further study, consider alternatives, and make choice	communicate decision for implementation
CURRENT IMPLEMENTING	receive decision of decision-making function	prepare implementation of actions needed to correct, prevent, etc.	implement actions
PLANNING/ FRAMEWORK-ENSURING	detect problem or opportunity	perform further study, consider alternatives, and make choice	communicate decision for implementation
FRAMEWORK	receive decision of planning function	prepare implementa- of add/change/delete or influencing actions	implement actions

TABLE 3-5: Generic details of flow in different functional classes.

devised to indicate this type of grouping, although the phrase "operating core"* does come to mind.

COMPARISON TO FAYOLIAN SCHEME

Fundamental Similarities

Figure 3-11 cross-references the Fayolian scheme to the replacement one developed here. In reviewing the cross-reference lines, one can identify certain the following approximate similarities.

- Except for framework-ensuring, the non-current decision-making class is fairly equivalent to the Fayolian planning, including both strategic and tactical levels of planning. Similarly, the non-current decision-implementing operations roughly include organizing. However, organizing also involves decision making. This would

---

* Mintzberg.

be equivalent to the operations performed by lower level planning functions in the new scheme.

- The current decision-making operations, coordinating and validating, are roughly equivalent to directing and controlling. Different **levels** of scope of these operations are not distinguished in either scheme. (Scope differentiation is done via the Anthony scheme.)
- The baseline operations definitely belong outside of the Fayolian scheme, which is only for operations. (However, they belong in the operations level of Anthony's triangle.)

Essential Differences

A number of differences between the present scheme and the Fayolian one are apparent. Although not great differences, they may add a few clarifications. These are as follows:

- The Fayolian scheme does not formally distinguish between decision making and decision-implementing, as does the new scheme, although it would appear that organizing and staffing belong in non-current decision-implementing, while planning would be restricted to decision making. However, in the current operations, i.e., directing and controlling, it is clear that no such distinction has been made between decision making and decision-implementing. In all fairness to Fayol, he perhaps did not consider decision-implementing to be a managerial task, and so it may not have been included in the Fayolian scheme for this reason.

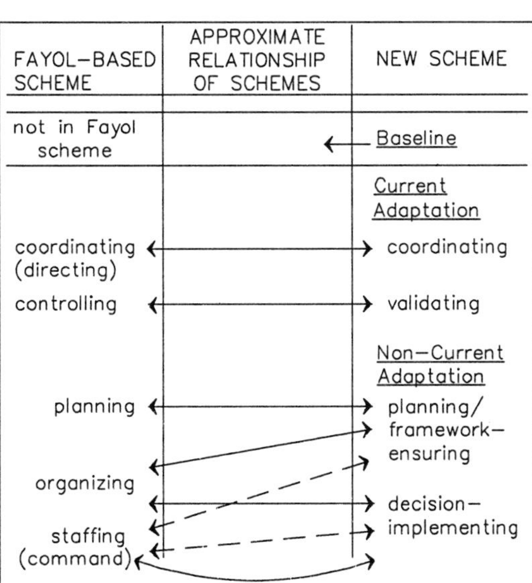

FIGURE 3-11
Cross-reference of replacement scheme to Fayolian one.
Note that a solid relationship line means that the categories tend to be quite similar, while the dashed lines represent a weaker degree of overlap of the categories on the different sides of the table. In all cases, the equivalences are based merely on guesswork and intuition.

- Although internal adaptation operations of the various Fayolian classes can exist at very local levels, below what would normally be called the tactical level, the Fayolian scheme applies to the organization as a whole. Otherwise, if the secretary of the department does planning when he or she decides to go out for a coffee break, while the president of the company does controlling when she or he ensures that the coffee break is not missed by the secretary, it might be said that the secretary does more planning than the president, while the president does more controlling. In contrast, the new scheme explicitly allows planning and other adaptation operations in either the internal or the external senses, as different types of levelling dimensions may be used to cross-cut the new scheme. This explicit distinction may have been implicit in the Fayolian scheme.
- The time-orientation distinction between coordinating (feedforward) and validating (ensurance) current decision-making adaptation operations is on the surface the same as that as the Fayolian distinction between directing and controlling. However, to which of the two Fayolian classes (directing and controlling) does the interruptive feedforward time-orientation (see the fourth section of this chapter) belong? Does interruption ensure that the system meet its objectives, as in controlling, e.g., by preventing problems; or does interruption push the system in a new direction, as in directing?

TIME-ORIENTATIONS AND FEEDBACK

Both the Fayolian scheme and to a degree the present one include the concept of different time-orientations, which are flow routing details. Thus, controlling, in the Fayolian scheme, and framework-ensuring, validating, and correction decision-implementing, in the scheme introduced here, share a time-orientation which can be characterized by the word "ensurance." On the other hand, planning and organizing, in the Fayolian scheme, and planning, coordinating, framework decision-implementing, and prevention decision-implementing, in the present scheme, have a "feedforward" time-orientation, as they affect operations yet to be attempted.

The Terms Feedback and Feedforward

Since the meanings of feedback and feedforward have been developed in different planes, dealing with these words can be frustrating. Hopefully, the following will somewhat clarify their meanings, at least as they apply to this book.

In charting system operational flow, the following separate problems exist:

- —time-orientation: past (current run) or future;
- —loop presence: loop present or not;
- —amplification effect on the system:
  - › positive (deviation-amplifying), or
  - › negative (deviation-reducing).

Feedback, in many senses of the word, and feedforward, in all cases, are control mechanisms, of which feedforward affects the system in a future-oriented direction, while feedback may or may not affect the system in a future-oriented direction. Directing (coordinating) the system, as in the Fayolian scheme, is just one way of affecting the future of the system. Similarly, the Fayolian scheme's other functions of planning, organizing, and commanding (staffing) are future-oriented. All of these may, perhaps, be grouped into the notion of feedforward.

Feedback is not opposite to feedforward, despite the apparent opposition of "back" to "forward." The word particle "back" in feedback simply implies that a loop exists, while a loop may or may not be involved in feedforward. If a loop is involved with feedforward, then it may be said that this is a "feedforward feedback loop," as confusing as that may sound.

Insofar as a feedback loop is connected with system control, it is a loop in which data leaving the system function originating it results in something returning to this same function in the form of a controlling effect. For example, the data leaving the originating function passes via decision-making functions which then affect the data-originating function. If the decision-making function is a first-level control function (e.g., the operational control level in the Anthony triangle), this loop is called a "first-order" feedback loop. When the decision making is passed on to other levels (say, the management levels in the Anthony triangle), this becomes a multiple-order feedback loop, especially where long-term decisions come back to affect operations in the originating function.

Figure 3-12a portrays positive feedback. Here, the effect of the increase in wages is to augment inflation, which in turn boosts wages, and so on, in a deviation-amplifying direction. In contrast, Figure 3-12b exhibits a negative feedback loop, in which an increase in government spending increases the rate of economic activity, which then leads to a decrease in government spending, since the government has no further need to stimulate the economy to this extent. In this deviation-diminishing type of feedback loop, the positives cancel the negatives, whereas a positive feedback loop has an unequal overall balance of positive and negative effects.

In the above examples, it would appear that feedback is connected with control, although the control is not clearly visible in the charts. One should be careful in seeing feedback everywhere a loop is present, as has been discussed in connection with non-objective functional outputs earlier in this chapter. When a loop is not clearly for control purposes, the output starting the loop is referred to as something other than "feedback." Rather, it is symbiotic, offshoot, competitive, or objective output, according to the scheme established in the symbols for objective-defined functions in the second section of this chapter.

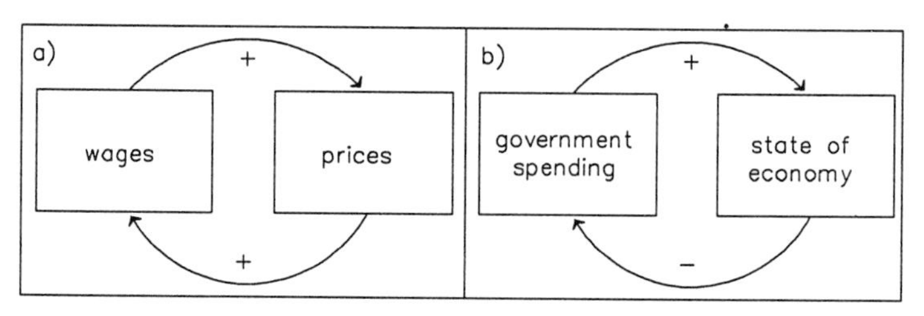

FIGURE 3-12: Positive versus negative feedback loops. *a) a positive feedback loop; b) a negative feedback loop.*

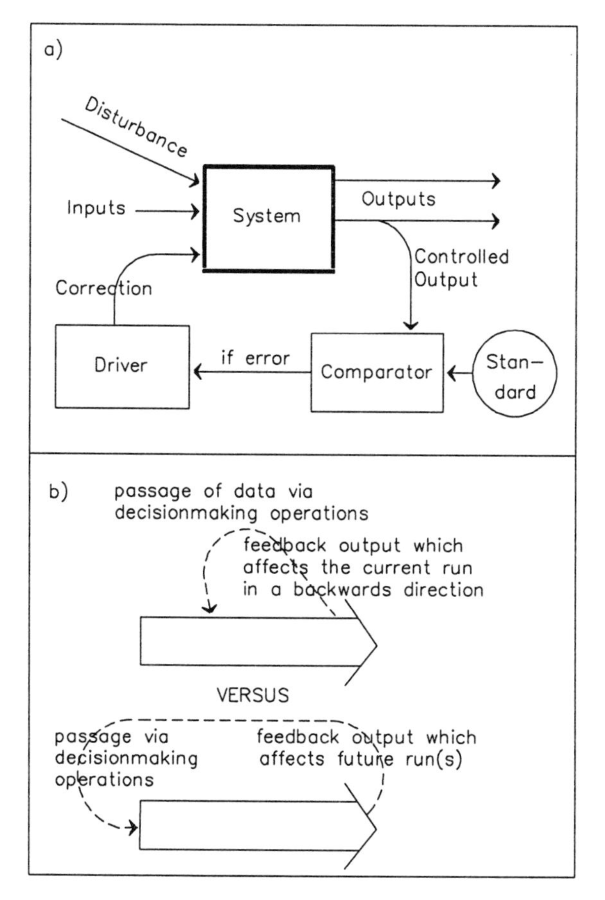

FIGURE 3-13: Feedback associated with system control.

a) negative feedback visibly associated with control;

b) current versus future-oriented feedback.

Note the following:

—that the angle of exit from the wideshafted arrow of the feedback arrow for current feedback is different from that of the feedback arrow for future-oriented decision making;

— that the loop must return to the originating function for it to be feedback;

— that the loop may pass via a decision-making function on its way from the originating function back to the originating function controlled by this feedback.

In both Figures 3-12a and b, no easily visible association is present between feedback and control. In contrast, Figure 3-13a exhibits a negative (deviation-deamplifying) feedback loop where system outputs are compared to a standard, so that errors can be detected. If errors are discovered, the driver is invoked to make corrections. This loop is clearly associated with control.

Examples of Feedforward

Figure 3-14a portrays the idea of feedforward. Because of external disturbances, errors here are detected in advance of system operations, or, at least, before the current system run has completed. Detection of the anticipated problem or opportunity relies on some sort of model of how the system functions. Action may then be taken to cancel or otherwise interrupt the system run. Actually, feedforward may result in a whole range of future-oriented effects on the system, including initiation, cancellation, stopping (pause), continuation, slowing down, speeding up, prevention, etc. Initiation and cancellation may occur before the next system run begins, while cancellation, stopping (pause), continuation, slowing down, etc. may take place in the sense of system interruption after the run begins.

Following are examples of feedforward:

1- A credit check may be performed on a customer before deciding to accept the order; or the order may be cancelled, changed, delayed, or expedited, based on modifications of the customer stance with respect to an in-progress order within the business organization.

2- The purchase of excess supplies of stock in a business is initiated when sudden increases in market demand are foreseen.

3- The execution of the subsequent operations is sped up. To illustrate, the delivery of a customer order is expedited based on a request from the customer.

4 - The execution of ensuing operations is cancelled; e.g., the customer cancels the order. The cancellation may also involve communicating with other operations, to undo the side-effects of previous steps. For instance, actual demand, updated in a sales forecast file as soon as the customer order was entered to the computer, must then be revised in the file as a result of the cancellation of the customer order.

5- The execution of later operations in a sequence is slowed down.

6- Changes are inserted before the next step, without changing the results of any of the previous steps. To illustrate, the customer changes certain line items on the order. The previous update of the customer order file is not undone, since it executed correctly, but it is revised before continuing. This is "prevention," as opposed to "correction," since correction is a word used here only in connection with ensurance (past-oriented) control.

Examples of Feedback, and Where it Merges (or does not Merge) with Feedforward

Like feedforward, feedback may affect the system in a future-oriented direction. On the other hand, unlike feedforward, feedback can affect past operations in the current run of the originating function. To illustrate, a quality check of manufactured goods before delivery to customers reveals a defect. If this defect is corrected **before** shipping occurs, the **past** operations in the current system run are corrected in some fashion prior to the point where the current system run may be complete. This notion is illustrated in Figure 3-13b (above).

The idea of feedforward, as presently understood, has nothing to do with the existence of a loop, but it may be associated with either a negative or a positive feed loop which affects the future of the system. In this case, feedforward is part of the loop which returns to the system. Therefore, one may be tempted to refer to a "feedforward loop," as strange as this terminology may appear. Of course, this terminology is only necessary when feedforward is somehow connected with a looping phenomenon.

A future-oriented feedback loop, i.e., a "feedforward feedback loop," occurs in the following two examples:

1- A paddler takes a canoe trip across a lake towards a specific location on the other side of the lake. The paddler looks up every so often

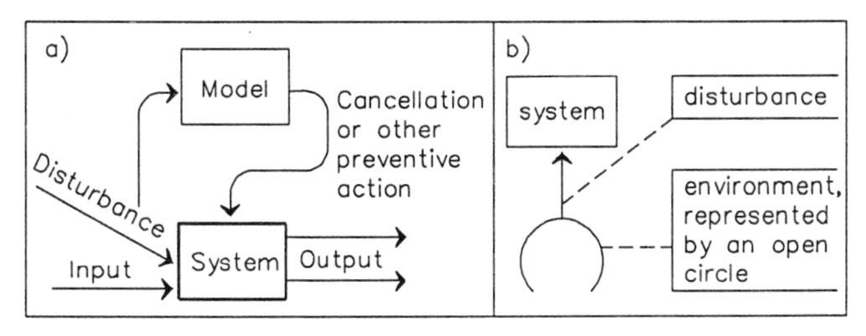

FIGURE 3-14: Feedforward control.
a) use of a model to make decisions;
b) the uni-directional aspect of much feedforward, whereby the source cannot be traced in the form of a feedback loop.

to see if the canoe is on course. Correction is made in a forward direction, i.e., without going back and restarting from a point at which a previous validation of trajectory was made. That is, the correction is made in the **next** step of the journey.

2- The canoeist must cross the lake five times, in an iterative process. The objective is considered to have been achieved each time the lake is crossed. Looping back to restart is not to be understood in the sense of ensurance (past-oriented) feedback.

Thus, feedforward may be connected with an information loop starting from past system experiences, as long as it affects operations not yet attempted. In this sense, feedforward forms part of a future-oriented feedback loop. However, unlike future-oriented feedback, feedforward may also result from changes or disturbances in the system environment, which may not be connected with a looping phenomenon, i.e., may not be connected with feedback. For example, as a result of environmental disturbance or change (Figure 3-14b), operations are interrupted (cancelled, modified, sped up, slowed, etc.), as in Figure 3-14a, feedforward does not emanate from within the system, and is not connected with feedback. Such feedforward may also take place where there is a decision-making process leading into the start of a new system run.

CHAPTER FOUR

PROCEDURAL DIAGRAMMING FUNDAMENTALS*

The present chapter starts by discussing the following classifications involving procedural diagramming methods:

—procedural versus non-procedural charts;
—flow versus levels-of-control charts;
—formatted charts;
—multi-level (hierarchically shaped) charts;
—single-phase versus multi-phase charts.

Next, there follows a discussion of various aspects of procedural charting symbols.

Finally, this chapter introduces various criteria for evaluating procedural diagramming methods, based on certain materials which have

* Various parts of this chapter are based on an article by Ed Baylin, "A Comparative Review of Popular Information System Charting Methods," Auerbach Information Management Series, Auerbach Publishers©, Oct., 1985.

been extracted from the fuller treatment of this subject in the related publication on procedural diagramming.

Procedural Versus Non-Procedural Charts

The two most fundamental categories of charting methods used in the systems field are procedural charts, which show operations performed, and non-procedural charts, which show relationships of non-action (non-procedural) entities. Non-procedural charts can be exemplified by the entity-relationship diagrams used in logical data modeling for relational databases of computer systems. A second example of a non-procedural chart is one which relates attributes, such as a circle labelled "red" being connected by an arrow pointing to another circle labelled "aggression." A third example is an office layout drawing. Procedural charts usually do contain a number of non-procedural entities, to represent inputs and outputs, flows, stores, and environmental interfaces. However, these are linked via procedural entities.

Although **only procedural charts are dealt with in this book**, certain forms of procedural charts may also find usage as non-procedural charts. Specifically, these are the charts which in some fashion or other take on a hierarchical shape, or, in other words, can show more than one level. These hierarchically shaped charts are discussed later in this section.

BASIC CATEGORIES OF PROCEDURAL CHARTS

FLOW CHARTS VERSUS LEVELS-OF-CONTROL CHARTS

The following deals with a particularly useful way of categorizing procedural charting methods, namely, one which emphasizes the system's flow features as opposed to its levels of control ones. Based on this difference, two basic categories of these charting methods may be established, to which almost all such charting methods belong.

Brief Description

The first basic method, the flow chart, has a sequence of steps, usually going in one general direction. In Figure 4-1a the steps are represented by squares. The second basic type of chart is the levels-of-control chart. Here, levels of control operations related to a given process are shown in the form of a hierarchy. Actually, there may theoretically be an infinite number of control levels. Figure 4-1b is a simple, two-level levels-of-control chart. Here, the sequence of steps is obtainable by following the small arrows accompanying the lines which connect the

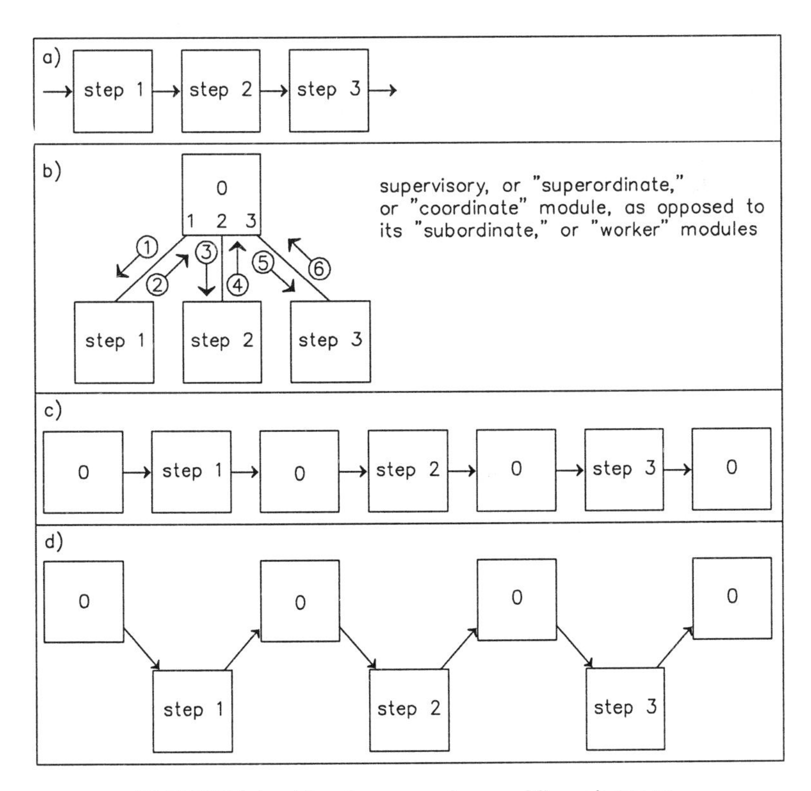

FIGURE 4-1: Simple comparisons of flow charts to levels-of-control charts. [Auerbach©]

a) a simple flow chart;

b) a simple two-level levels-of-control chart;

c) a flow chart which illustrates the same flow as in the levels-of-control chart in part "b";

d) different rows for control vs. controlled operations in a flow chart, using the same modules as in part "c."

different modules, represented by squares. In this levels-of-control chart, the essential flow is from left to right, except that the flow is initiated by the superordinate (the top square), and each module at the subordinate level (the bottom squares) reports back to the superordinate after completing its operations. Figure 4-1c is a flow chart which demonstrates the same sequence of operations as the levels-of-control chart in Figure 4-1b. In Figure 4-1c, the module labelled "0" is equivalent to the top module in the levels-of-control chart.

Fundamental System Flow Concepts

Flow Routing Versus Flow Scheduling — The difference between flow routing and flow scheduling is important for the understanding of flow. As anyone who has ever scheduled vehicles in the transportation industry will know, flow scheduling cannot be achieved until the flow routes are established. Flow routing in a procedural diagram involves laying down paths going toward the same end objective(s), without regard to scheduling details, based on the constraints established by the relationships between the inputs/outputs of different operations. It is useful here to think in terms of networks (as in PERT or CPM charts for project control) which fan out and then fan back in at various points in the flow towards an objective. For example, the following kinds of routing can exist:

- **sequence,** where the output of one operation is needed as an input to another;
- **concurrency,** or **parallelism,** where the output of one operation is not an input needed by the other operation, although their inputs originate from the same point or their outputs have the same destination; i.e., they branch from, or into, the same node in the network of flow routing paths;
- **alternative** (mutually exclusive) paths stemming from, or going to, the same node in the network;
- **optional** path(s);
- **return,** via part of a feedback loop, to a previous point in a sequence to retry operations in the event of errors;
- routes used for **cancellation** of further activity in the event of unrecoverable error;
- routes used for **interrupting** activity from outside the network, to initiate, or cancel, or restart activities (but not to reschedule activities, e.g., by speeding them up or slowing them down).

Scheduling, on the other hand, allows arrangements such as the following to take place:

- **prioritization/synchronization** of operations occurring in parallel routes, so that one route is executed in a specified (or unspecified, as the case may be) time relationship to another path;
- **speed-up** or **slow-down** of operations in a sequence;
- scheduling an **accumulation** or **depletion** of things to be done by iteration(s) of operations, instead of by a single run of operations;
- establishment of **relative time frames**, e.g., long-term versus current.

Flow routing and *scheduling* are easily confused, largely because the same symbols may often be used for both. Thus, flow lines to show a sequence can also be used to show a priority relationship in the scheduling of parallel branches (which thereby appear to be in a sequence). Similarly, iteration to retry operations in case of errors may be diagrammed in the same way as iterations for purposes of accumulation/depletion.

Iterative Branches — Iteration, or repetition of operations, involves simple decision branching, to return to the start point of a block of operations, following which, basic flow routes are used to repeat steps. It should be noted at this point that three essential reasons exist for iteration:

1) to accumulate or deplete;
2) to redo operations to correct errors;
3) to represent the existence of different time frames, whereby some operations occur less frequently than others, as they establish frameworks for execution of the more frequent operations.

Only iteration to redo involves flow **routing**. The other two reasons have to do with flow **scheduling**.

Direct and Indirect Flow Toward The System Objective(s) — Both direct and indirect flow toward the system objective(s) occur in any system. **Direct** flow towards an objective is also referred to as "baseline" flow in this text. It consists of all the system elements which flow towards that objective, except for those elements which play a controlling role in this flow. In other words, operations to plan, organize, direct (coordinate), and ensure the operations that are in the direct line of flow, constitute the **indirect** flow towards the objective.

Some indirect flow, which results from errors or interrupts, is referred to here as the "current contingency" type. Within this flow category exist both "basic" flow routes as well as special routes which are unique to current contingency indirect flow.

Defining Characteristics of Flow and Levels-of-Control Charts

As seen in Figures 4-1, some procedural charting forms can be made to capture both flow **and** levels of control, with different proficiencies. However, in order for a procedural charting method to be categorized as a flow chart, that method must **inherently** illustrate flow. In other words, it should not be possible to compose the chart without showing flow. Similarly, a procedural charting method must **inherently** illustrate levels-of-control, in order to be categorized as a levels-of-control chart. It should be noted that one current form of chart, the structure chart, is inherently both a levels-of-control and a flow chart, although it is much more proficient in demonstrating levels of control than it is in showing flow. This method is described in Chapter Seven, which analyzes levels-of-control in the structure-flow chart versus the structure chart. The structure-flow chart, introduced in the next chapter, is also, as its name suggests, both a flow and a levels-of-control (structure) chart. However, the structure-flow chart **excels** both in showing flow and in demonstrating levels of control.

Some Limitations of Flow and Levels-of-Control Charts

While flow can be made visible in the levels-of-control charts when a simple left to right flow pattern is involved (e.g., see "transform-centered design" discussion in Chapter Seven on levels-of-control charts), almost all of these charts are weak in flow situations involving parallel branches, optional steps, corrective loops, etc. On the other hand, most flow charts are very convenient for capturing flow, while the capturing of levels of control, etc. is rather poor. This is clearly noticeable by comparing Figures 4-1b and 4-1d. While levels of control are represented in a specially formatted flow chart in Figure 4-1d, these control levels are still far less clear than in the levels-of-control chart in Figure 4-1b. This is true despite the fact that a different **row** is used in Figure 4-1d for the control modules as compared to the "worker" modules.

The different types of limitations of flow and levels-of-control charts is a complex subject, whose aspects require a long discussion. The preceding merely mentions a number of the major issues involved. Some further treatment of these matters appears at various points in this book.

FORMATTED CHARTS

A formatted chart, be it a flow chart or a levels-of-control chart, is a diagram in which pre-defined page layouts are used, so as to highlight or standardize certain features, although this formatting may be otherwise restrictive. These layouts are either inherent in the charting method, i.e., constitute the one basic format of the chart, or are special, i.e., beyond those page layouts inherent in the charting method. Many types of such special chart page alignments may be devised. An example is the flow chart given in Figure 4-1d.

Both levels-of-control charts and flow charts can be formatted. In fact, **any** charting method covered in the related publication on procedural diagramming either can be, or is inherently, formatted. Some types of procedural diagramming methods are actually **designed** to have rigorously arranged page layouts, which are required by the method. These types of charts are considered to have an **inherent** (i.e., basic, or overall) format. The single-phase flow charts (see below) are about the only procedural methods which are definitely not inherently formatted, although they can be formatted in ways selected by the chart user. In this latter case, these flexibly arranged charts are said to be **specially** formatted. Actually, even those charting methods which have a basic inherent format can undergo formatting in ways selected by the chart user. In this situation, the additional formatting does not override the inherent format, but must occur within the bounds established by the latter. Thus, this additional formatting is referred to as **special sub-formatting**.

Some of the formats inherent in given charting methods may be emulated, with varying degrees of success, by other charting methods capable of being formatted in more than one basic way. As well, certain formats are applicable only to particular charting forms, in which circumstances the charting forms to which they apply may or may not also be capable of being specially formatted in other ways. For instance, computer program flowcharts have a unique special format often referred to when speaking of "structured flowcharting" in computer work. However, computer program flowcharts can also be specially formatted in alternative ways. In contrast, the hierarchical functional decomposition formats are not only unique to the hierarchical functional decomposition methods, but are also the required, inherent, formats for these, in whose framework special sub-formats may be selected.

Certain formatted flow charts have hierarchical shapes, and are thus particularly easy to confuse with levels-of-control charts. This type of flow chart demonstrates hierarchical functional decomposition along with flow. Various forms of inherently formatted flow charts to show hierarchical functional decomposition have been devised, and are particularly popular in information systems work. These include Warnier-Orr, Action, HOS (Higher Order Software), and HIPO methods.

Although HIPO as a whole is a hierarchical functional decomposition method, it should be noted that the so-called "Table of Contents" level charts of the HIPO (Hierarchy plus Input-Processing-Output) method is neither a flow nor a levels-of-control chart, since it merely demonstrates the tree structure of a multi-level functional decomposition, while the more detailed levels in HIPO, the IPO charts, are just single-level flow charts, i.e., are not hierarchical. These various tools are described in the related publication on procedural diagramming, along with one other form of inherently formatted flow chart, which happens to be used for purposes other than hierarchical functional decomposition, namely, the SADT method.

MULTI-LEVEL (HIERARCHICALLY SHAPED) CHARTS

Hierarchical Decomposition Vs. Levels-of-Control Charts

Dichotomy Involving Procedural Charts — Functional decomposition refers to the breaking down of system operations into modules. **Hierarchical** functional decomposition breaks the system operations down into different levels of detail; e.g., modules are decomposed into sub-modules, and sub-modules into sub-sub-modules. While hierarchy charts are often used to show hierarchical functional decomposition, they may instead be used to show levels of control, thereby not breaking down the system operations into **multiple levels** of sub-sets. That is, when used to show levels of control, hierarchy charts show **single**-level only functional decomposition, with the higher level modules containing those operations used to control the lower level ones, perhaps along with some "baseline" operations not delegated to the lower levels.

The Dichotomy Involving Non-Procedural Charts — To complicate the matter even further, the hierarchy chart can also be used as a **non**-procedural chart to show either multi-level decomposition or levels of control. For instance, a hierarchy chart having the president at the top, the vice-presidents at the next level, and so on, down to the workers, is a non-procedural levels-of-control chart, while a chart having the company at the top, the company major divisions at the next level, and so on, down to the small work units, is a non-procedural hierarchical functional decomposition chart.

Differences Clarified With An Example — The difference between levels-of-control uses and multi-level functional decomposition uses of a hierarchically shaped chart may be captured from a procedural view-

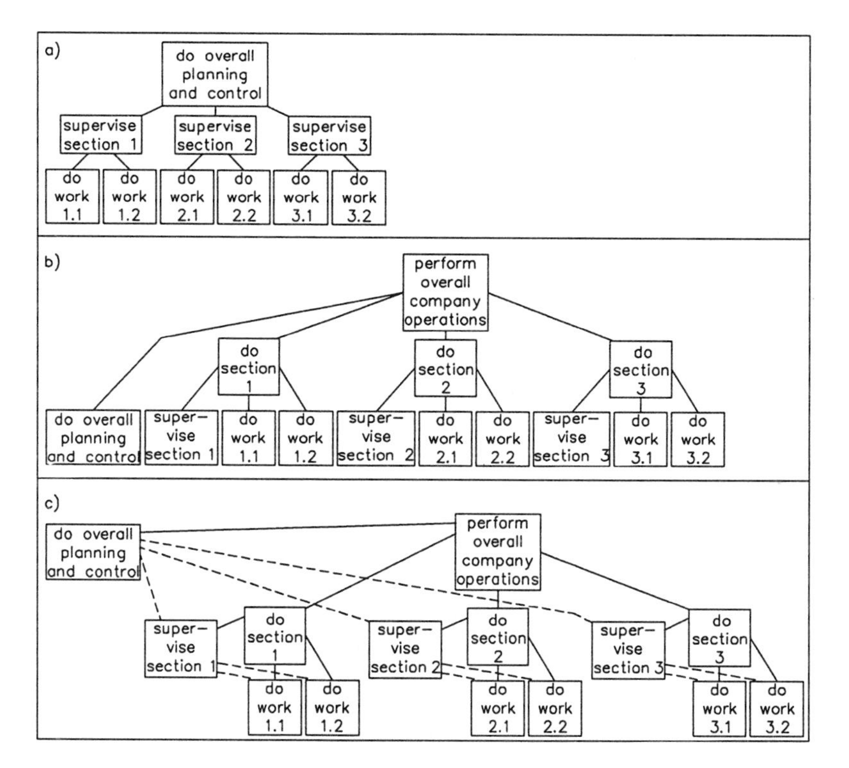

FIGURE 4-2:

Levels-of-control chart versus hierarchical decomposition chart.

a) levels-of-control chart showing levels of control, in which the total company operations equals the sum of the operations in all the boxes in the chart; [Auerbach©]

b) a hierarchical functional decomposition chart, in which the sum of the company operations is equal to any one of the following: [Auerbach©]

- *the operations represented by the box at the top of the hierarchy;*
- *the sum of the operations in the boxes which directly connect to the box at the top of the hierarchy;*
- *the sum of the operations in the bottom row of the chart.*

c) Figure "b" reformatted to facilitate making the relationship with Figure "a." The dashed lines add levels of control to the hierarchical functional decomposition.

point in the comparison of Figures 4-2a and 4-2b. The first of these figures, a levels-of-control chart, represents different levels of control operations, while the second figure, used for hierarchical functional decomposition, decomposes sets into sub-sets **at various levels**— a very different idea.

The confusion between levels-of-control and hierarchical functional decomposition is rampant in most literature on charting. It is, however, quite understandable, since the design of the control levels in a system often parallels that of its hierarchical decomposition. Thus, comparing figures 4-2a and 4-2b, the general control operations correspond to the company as a whole, the next level of managerial operations correspond to the various divisions of the company, the supervisory operations are along lines of the next level of sub-division of the company, and so on, down to the bottom of the hierarchy. Interestingly, it is possible to include the levels of control connections in the same chart with hierarchical functional decomposition. This is accomplished with the dashed lines in Figure 4-2c (which somewhat reformats Figure 4-2b to facilitate making the relationship between Figure 4-2a and 4-2b). Although the hierarchy used for multi-level decomposition can be

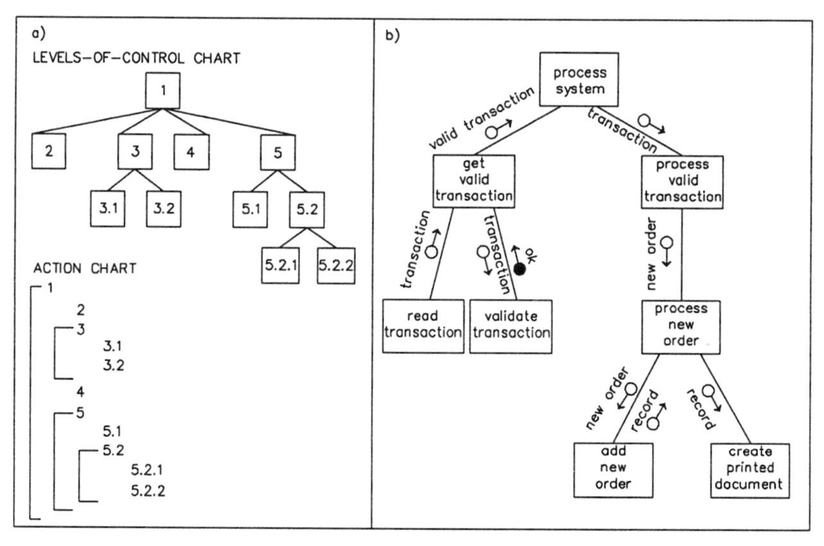

FIGURE 4-3: Examples of questionable use of hierarchy charts.

a) equation of action chart, for hierarchical functional decomposition, to levels-of-control chart;

b) a levels-of-control chart having questionable descriptors. For instance, "get valid transaction" would read "control getting valid transaction" if this were clearly a chart used to show levels of control.

made largely congruent with the levels-of-control hierarchy, the two can never be identical. The essential reasons for this can, for instance, be seen in comparing parts "a" and "b" of Figure 4-2.

Examples of Confusing Use of Hierarchy Charts — Specific examples of confusion between hierarchical functional decomposition charts and levels-of-control charts are given in Figures 4-3. Thus, in Figure 4-3a,

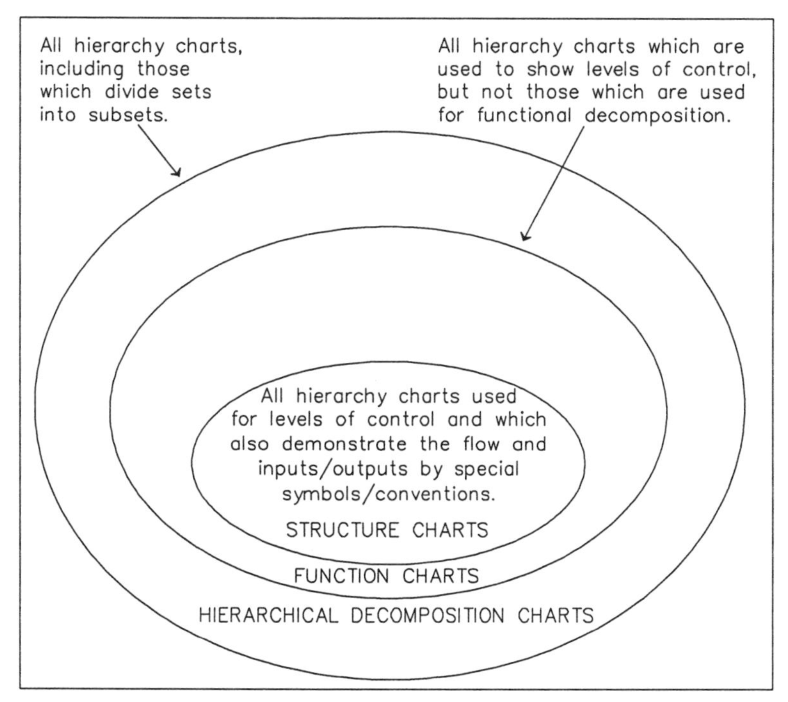

FIGURE 4-4: Different extents of meaning of "structure chart."
The entire area confined within each oval represents a meaning sometimes given to the term structure chart. Each oval corresponds to a different meaning, which is given by the non-capitalized text. The structure chart described in Chapter Six of this book is associated with the innermost oval. The other meanings sometimes given to the term "structure chart" are symbolized by the entire areas of the larger ovals. As well, in this diagram, each oval contained within another oval represents a sub-set of the next larger oval. The all-capitalized text in the diagram identifies what is contained in a layer, i.e., an oval minus any inner oval, like the strata of a tree trunk. Thus, the middle layer corresponds to the function chart, also to be described in Chapter Six, while the outermost layer represents the use of hierarchical charting forms for multi-level functional decomposition.

an action chart (which is one of the forms of flow chart inherently formatted to show hierarchical functional decomposition) is equated to the levels-of-control chart at the top of the same Figure 4-3a. Strictly speaking, this equation is incorrect.

Also, Figure 4-3b shows how ambiguity arises from failing to make the distinction between levels of control, with its single-level only functional decomposition, and multi-level functional decomposition. If this is indeed a levels-of-control chart, as its label indicates, then the "get valid transaction" module in this chart should really be labelled "**control getting** a valid transaction." The latter represents a very different set of operations from those operations **directly** involved in actually **getting** a valid item. Further study of Figure 4-3b reveals that the module in question should have been labelled as a controller, since the flow of data (little arrows accompanying the hierarchical relationship lines) in the chart is proof that, despite the erroneous label, the idea being captured here is, in fact, levels of control, rather than division of sets into sub-sets at different levels. Otherwise, why would data flow between different **levels**?

Conclusion — It is quite evident that the use of hierarchically shaped charts is a major source of confusion, as it is easy to slip from one use into the other, thereby creating inconsistencies. Major attention must be given to clarifying the intended use of such charts. Table 4-1 summarizes the focal points of the above discussion.

Different Meanings of the Term "Structure Chart"

Computer-oriented literature has usually applied the term "structure chart" in a restricted fashion, to include only one type of levels-of-control chart. This particular form also demonstrates inputs and outputs and flow routing logic (decision branches, loops, etc.) with a measure of success. This is in contrast to the other common levels-of-control

CHART CATEGORY	LEVELS-OF-CONTROL	MULTI-LEVEL (HIERARCHICAL) DECOMPOSITION
Procedural	e.g., function or structure charting methods	e.g., inherently formatted flow charts, such as, Warnier-Orr, action, HIPO
Non-procedural	e.g., organizational authority lines chart	e.g., entity-relationship chart, or bubble chart for logical data modeling

TABLE 4-1: The multiple uses of charts having hierarchical shapes.

chart—namely, the so-called "function" chart—which shows only the hierarchy. Thus, the structure chart illustrates a tree structure to which various details of the inputs/outputs and flow routing logic have been added. In fact, it is both a levels-of-control chart and a flow chart.

Moreover, confusion exists, since the term "structure chart" has also been used as a synonym for **all** levels-of-control charts, as well as for all hierarchy charts, including those used for multi-level decomposition. To summarize, the different senses of the term "structure chart" are represented by the ovals in Figure 4-4. The innermost oval represents the most restricted meaning, which is also the meaning adopted by the present book.

Single-Phase Charts Versus Multi-Phase Charts — Certain charting methods can be effectively used at different levels of detail, i.e., apply to both the analysis and design phases of system development, while others are more restricted in the level of detail to which they may be applied effectively. The former are categorized here as "multi-phase" methods, while the latter are called "single-phase" methods.

PROCEDURAL CHARTING SYMBOLS

Introduction

Basic Categories and Sub-Categories — Table 4-2 presents and briefly describes procedural diagramming symbols while grouping them into basic categories and levels of sub-categories. However, some obscure existing categories or sub-categories of symbols are not included in Table 4-2. Examples of these are the following:

—hierarchical relationship connecting lines, to show the tree structure in a chart with a hierarchical shape;

—flow scheduling symbols to show different time frames (see related publication on procedural diagramming).

Nomenclature — Table 4-2 also indicates a name for each specific symbol that exists, as well as other names used (synonyms).

Lists of Charting Methods Using Each Symbol — Lists of the methods which use each specific symbol can be found in Table 4-2. It should be noted that all procedural charting methods have symbols for opera-

<table>
<tr><th>SYMBOL CATEGORY</th><th colspan="4">SYMBOL SUB-CATEGORY</th><th>SYMBOL NAME</th></tr>
<tr><td>OPER-ATION</td><td colspan="4"></td><td>Operation</td></tr>
<tr><td>INPUT/ OUTPUT</td><td colspan="4"></td><td>Input/Output</td></tr>
<tr><td rowspan="17">FLOW</td><td rowspan="15">Flow Routing</td><td rowspan="5">Flow Routing Multi-Purpose
(used to symbolize many types of flow routes)</td><td rowspan="2">Flow Route Line</td><td rowspan="2"></td><td>Flowline (no arrowheads)</td></tr>
<tr><td>Flow Arrow</td></tr>
<tr><td rowspan="3">Connector
(used in conjunction with flow route lines)</td><td rowspan="2">Page Connector</td><td>On-page Connector</td></tr>
<tr><td>Off-page Connector</td></tr>
<tr><td>Flow Terminal Point</td><td>Flow Terminal Point</td></tr>
<tr><td rowspan="8">Basic Flow Routing</td><td>Sequential Routing</td><td></td><td></td></tr>
<tr><td rowspan="7">Branching</td><td rowspan="3">Parallel Branching</td><td>Parallel Branching</td></tr>
<tr><td>Parallel Branching, Terminal Point</td></tr>
<tr><td>"AND" (*) Boolean Gating Logic</td></tr>
<tr><td rowspan="4">Decision Branching</td><td>Decision Branching Point</td></tr>
<tr><td>Alternative Branching</td></tr>
<tr><td>"EXCLUSIVE OR" (⊕) Boolean Gating Logic</td></tr>
<tr><td>"INCLUSIVE OR" (O) Boolean Gating Logic</td></tr>
<tr><td rowspan="2">Current Contingency Flow Routing</td><td colspan="2" rowspan="2"></td><td>Redo Loop</td></tr>
<tr><td></td></tr>
<tr><td rowspan="2">Flow Scheduling</td><td colspan="3" rowspan="2"></td><td>Iteration</td></tr>
<tr><td></td></tr>
<tr><td>STORAGE</td><td colspan="4"></td><td>Store</td></tr>
<tr><td rowspan="2">ENVIRON-MENT INTER-FACING</td><td colspan="4" rowspan="2"></td><td>Environmental Entity</td></tr>
<tr><td>System Boundary</td></tr>
</table>

TABLE 4-2:
Procedural charting method symbol categories and sub-categories. *Thanks to Gabor Lorenz for devising this table.* [This table is also published in the related Ed Baylin book, Procedural Diagramming for System Development—From a More Scientific Viewpoint, to whose publisher the copyright belongs.]

OTHER NAMES SOMETIMES GIVEN TO THE SYMBOL	SYSTEM DETAILS INDICATED BY THE SYMBOL		CHARTING METHODS WHICH USE THE SYMBOL
	ALWAYS	SOMETIMES	
Functional Element, Function, Transform, Procedure, Activity, Process, Task	function	computerized or manual type, other type	All
	input or output	functional significance, informational or material type, medium	Several
Line, Connector, Connection Line	part or all of a flow route	telecommunication type of flow	Computer System Flowchart, Computer Program Flowchart
Connector, Arrow, Line, Connection Line	part or all of a flow route	telecommunication type of flow	Many
	part of a flow route		Many
	part of a flow route		Many
Terminal/Interrupt	beginning, ending, or interruption point in the system flow (implies a connection to a source or sink in the environment)		Computer Program Flowchart
Illustrated either by the conventions of the charting method or by Flow Routing Multi-Purpose Symbols			
	parallel flow branches		Action Diagram
	starting point or ending point of parallel flow branches		Computer Program Flowchart
"Parallel" Flow Branching Logic	"parallel" flow branching logic		Bubble Data Flow Diagram, Structure-Flow Chart
	source point of optional and/or alternative flow branches		Structure, Explosion-Flow, Structure-Flow Charts
Mutually Exclusive Flow Branching	mutually exclusive flow branches		Action Diagram
"Mutually Exclusive" Flow Branching Logic	"mutually exclusive" flow branching logic		Bubble Data Flow Diagram, Structure-Flow Chart
"AND/OR" Flow Branching Logic	"AND/OR" flow branching logic		Bubble Data Flow Diagram, Structure-Flow Chart
Repetition, Loop, Iteration	flow route taken when correcting error(s) by iteration		Structure Chart, Explosion-Flow Chart
Also illustrated by Flow Routing Multi-Purpose Symbols and by Basic Flow Routing Symbols			
Repetition, Loop, Do Loop	accumulation or depletion by repetition of operations	relative time frames	Structure, Explosion-Flow, Action, Structure-Flow Charts
Also illustrated by Many Flow Routing Symbols			
Reservoir, File, Depot, Storage Point, Accumulation Point	point in the system flow where one or more things can be deposited	input/output, accumulation/depletion, degree of permanency, organization/access method, medium	Several
Source, Sink	environmental entity		Many
	system boundary		Data Flow Diagram, Structure-Flow Chart

tions, whereas the presence of other types of symbols is more variable; e.g., structure charts do not have storage symbols, etc.

System Details Illustrated by Symbols

Each symbol illustrates a basic system detail corresponding to the category of that symbol. In certain charting methods, a symbol can represent additional details which do not correspond to that symbol's basic category. For example, a storage symbol in a computer system flowchart illustrates the medium used at the storage point indicated. As well, one symbol in a charting method can represent more than one category/sub-category of charting symbol at the same time; e.g., flow arrows in data flow diagrams indicate **both** the flow route and the input/output travelling along that route. These aspects of charting symbols are also included in Table 4-2.

A further complication arises with the use of the many charting methods whose conventions in representing given types of system details are not well established. Thus, different individuals may use a chart in different ways to represent the very same system detail.

Geometric Forms of Symbols

Table 4-2 does not demonstrate the actual geometric forms of the symbols. Rather, this is done in Figure 4-5. Although there are few examples in this latter figure to clearly illustrate the following point, it should be mentioned that the same geometric form can often represent two completely different system details in different charting methods; i.e., it can be one **type** of symbol in one chart and another type in another chart. This applies in particular to circular and square/rectangular shapes.

Figure 4-5 illustrates the symbols used in a number of so-called "third-generation" information system charting methods. In the related publication on procedural diagramming, most of these will be categorized as "single-phase" charting methods, since they each tend to be specialized for one particular stage of the system development process, sometimes involving only one particular type of system. Missing from this figure, although discussed in the related publication on procedural diagramming, are the following:

- —various third and fourth-generation diagramming methods used for hierarchical functional decomposition;
- —the SADT (Structured Analysis and Design Technique), another third-generation method;
- —this author's structure-flow chart.

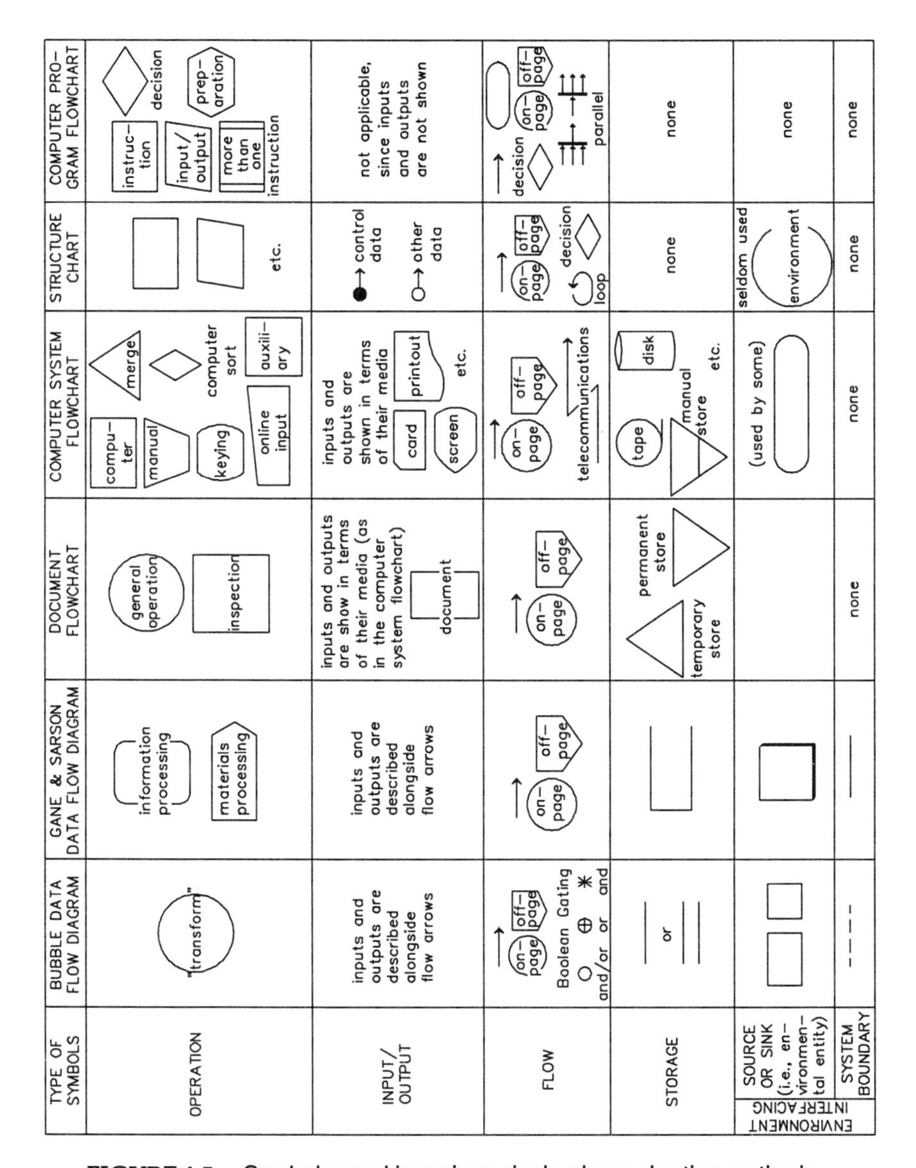

FIGURE 4-5: Symbols used in various single-phase charting methods. *The rows represent the different categories of symbols. The columns, which represent the different charting methods, are arranged from left to right in the approximate sequence in which the methods are used during the system development process.* [Auerbach©]

The latter methods are referred to in the related publication on procedural diagramming as "multi-phase" methods.

Consequences of Symbols Being Absent

It should be observed that the absence of symbols of a given type from a charting method does not necessarily mean that the method cannot represent a system detail of that type, since conventions associated with a charting method can sometimes represent a system detail without the need for a symbol. For instance, sequential flow routing symbols are not used in action charts, since it is understood that the flow is from top to bottom. Moreover, absence of a symbol from a given charting method may be often be offset by annotations in the chart; e.g., the lack of a source or sink symbol may be handled by a notation showing where the data originates or is sent.

FUNDAMENTALS OF EVALUATING PROCEDURAL DIAGRAMMING METHODS

The criteria for evaluating procedural charting methods described in this section, and developed in the related publication on procedural diagramming, are divided into two major groups; namely,

1) inherent illustrative capabilities, and
2) other use factors.

The first group of factors is divided into three major categories, consistent with the three ways of bridging different levels of system detail, as outlined in Chapter One. The other use factors are divided into major categories related to user-friendliness, computerizability, and usability of algorithms for functional decomposition.

Before continuing, it is suggested that the reader immediately refer to Table 4-2 and scan the headers of the columns and the first four rows. This will provide a preview of the individual methods discussed in the related publication on procedural diagramming and of some of the categorizations which are applied to them there.

In this section, the criteria described are particularly oriented towards understanding how suitable specific known diagramming methods are for particular stages of system development. Some of the criteria may also be used for comprehending the type of system for which the diagramming method is used. Thus, the subjects of computerizability and development of program code from diagrams are especially connected with computer information systems. In general, the criteria discussed below can also be used for understanding how procedural diagramming methods may be improved and their functions more easily communicated.

Major Groupings of Evaluation Criteria*

Various charting techniques are generally used in information systems work. Due to the characteristics of each form, and also because of accidents of history, most forms may be associated with a particular phase, or particular phases, in the system development process. Table 4-2 is a matrix that was created for evaluating various aspects of procedural charting. (Table 4-2 itself demonstrates the use of an action chart as a non-procedural, or **non**-action, diagramming method.) The evaluation factors are listed down the left hand side of Table 4-2, and are organized at various levels, with the two major groupings being given in the following sub-headers. The first major grouping of criteria is contained in Table 4-2a, while the second is in Table 4-2b.

Inherent Illustrative Capabilities (Table 4-2a)

One of the two major groupings is comprised of the factors that make it possible to control the amount of system detail illustrated by a charting method. These factors are the inherent capabilities of a charting method to represent, or exclude, various system details, and are the key to deciding which stage(s) of system development would be appropriate for the charting method. The basic rule of thumb is that the capability of leaving out certain system details is favourable for logical level perceptions of the system, while the capacity for including such features is needed for the physical level view. The illustrative capability factors, which are presented in Table 4-2a, are arranged in three sub-

---

* The evaluations criteria tables in this section are the copyrighted property, and reprinted with the permission, of the publisher of the related Ed Baylin book, Procedural Diagramming For System Development—From a More Scientific Viewpoint.

groupings. They correspond to the three ways of bridging levels of system detail illustrated in a chart.

Three Sub-Groupings of Illustrative Capabilities — One sub-group is comprised of *the capabilities of illustrating the various aspects of the umbrella of system elements as well as the capabilities of avoiding them.* These factors make it possible to control the extent of the umbrella spread for capturing system elements. They are grouped in Table 4-2a into the same five major categories as those of procedural charting symbols, which are the following:

1) function aspects (baseline and control functions, levels of controls);
2) input/output aspects (inputs/outputs, functional significances);
3) flow aspects (flow itself, specific types of flow routes);
4) storage aspects (storage points);
5) environment interfacing aspects (sources and sinks, system boundary).

Establishing an initial umbrella of system elements is necessary before the factors in the second and third sub-groups can be considered.

The second sub-group of evaluation factors concerns *the capabilities of illustrating system constraints as well as the capabilities of avoiding them.* These criteria make it possible to control the incorporation of constraints.

The third sub-group, which may be considered once constraints have been established, is comprised of *the capabilities of illustrating multi-level system elements as well as the capabilities of avoiding them.* These capabilities make it possible to explode (decompose) or deplode (concatenate) functions and inputs/outputs, thus enabling a multi-level system perspective to be captured on a single diagram. The latter factors must be considered in light of the fact that, while all diagramming methods are, in theory, "explodable," some are always used at approximately the same level of system perception, for reasons other than an inherent lack of either explodability or deplodability.

Thus, the amount of system detail which has to be, or may be, shown in a diagram is affected by three major groups of diagram attributes, namely, those enabling:

1) expansion or contraction of the umbrella of system elements;
2) incorporation or elimination of constraints;
3) explosion or deplosion of system elements.

Other Factors Affecting the Current Use of Charting Methods (Table4-2b)

The second major group of evaluation criteria, which is found in Table 4-2b, is comprised of miscellaneous charting method attributes, as well as external factors, both of which affect the use of charting methods. These criteria are arranged in three sub-groupings charting method attributes which

1) ensure achievement of more rigorous functional decomposition;
2) make for user-friendliness;
3) are related to the computerization of charting methods. This subgroup is comprised of (mainly inherent) capabilities enabling the computerization of the charting method, and of capacities provided by available computer software.

Most of the factors provided in Table 4-2b have an effect upon determining which charting methods are most appropriate for the various stages of system development. However, the computerization factors mainly affect the **type of system** which a diagramming method best suits; e.g., "ultimate decomposability" (translatability of a diagram into executable code) applies only to computerized information systems.

Important Further Evaluation Factors, Not Covered in Table 4-2

Before discussing the individual criteria indicated in the rows in Table 4-2, it is important to observe that certain essential considerations for evaluating procedural diagramming are hardly, or not at all, mentioned in this table, despite their potential importance in systems work, as well as in the development of any new charting methods. These considerations take us beyond what might be thought of as the "normal" frame of reference of evaluating charting methods (assuming that "normal" applies, since this fundamentally subject has hardly been addressed in any literature). The considerations in question are, however, important in understanding the contribution made by the structure-flow chart, and have to do with the use of the charting method for conceptual prototyping. Therefore, these factors will be introduced in conjunction with the structure-flow chart in the next chapter.

Explanation of the Evaluation Table

Use of Part "A" of the Evaluation Table

The use of Table 4-2a in evaluating the illustrative capabilities of charting methods can now be explained. Each row in Part "A" actually

EXPLANATION OF PART "A" OF THE TABLE

- A factor's effect on the particular system view taken (logical or physical) is nil ("0") or positive ("+") or synergistic ("*").
- Individual methods are ranked from 1 (strong) to 5 (weak). (NOTE: Many rankings are only intuitive.) Strength is unimportant or desirable or important, depending on the factor's effect on the system view adopted.
- The slash ("/") separates the effect or ranking of the capability of illustrating the system detail from that of the capability of avoiding the system detail.

	EFFECTS		RANKINGS													
	SYSTEM VIEW TAKEN		SINGLE-PHASE METHODS						MULTI-PHASE METHODS							
			SUITABILITY SPECTRUM--> LOGICAL--->--->PHYSICAL						HIERARCHICAL DECOMPOSITION METHODS						MISCELLANEOUS	
	LOGICAL	PHYSICAL	DFD	DOCUMENT	COMPUTER SYSTEM	FUNCTION	STRUCTURE	COMPUTER PROGRAM	HIERARCHY	IPO	WARNIER-ORR	ACTION	HOS	EXPLOSION-FLOW	SADT	STRUCTURE-FLOW
FLOW CHART	...	...	.X.	.X.	.X.	...	.X.	.X.	...	.X.	.X.	.X.	.X.	.X.	.X.	.X.
LEVELS-OF-CONTROL CHART	...	...	...	...	...	.X.	.X.	...	...	...	...	...	...	...	...	.X.
HIERARCHICALLY SHAPED CHART	...	...	...	...	...	.X.	.X.	...	.X.	...	.X.	.X.	.X.	.X.	...	.?.
CHART HAVING AN INHERENT BASIC FORMAT	...	...	...	...	...	.X.	.X.	...	.X.	.X.	.X.	.X.	.X.	.X.	.X.	.X.
INHERENT ILLUSTRATIVE CAPABILITIES																
CAPABILITIES OF ILLUSTRATING/AVOIDING THE VARIOUS ASPECTS OF THE UMBRELLA OF SYSTEM ELEMENTS																
Function Aspects																
functions	*/0	*/0	2/5	2/5	2/5	2/5	2/5	2/5	2/5	2/5	2/5	2/5	2/5	2/5	2/5	1/5
baseline functions	*/0	*/0	1/5	1/5	1/5	1/5	1/5	1/5	1/5	1/5	1/5	1/5	1/5	1/5	1/5	1/5
control functions	0/*	*/0	3/1	2/3	2/3	2/3	2/3	3/1	2/1	3/2	3/2	3/3	?/?	3/2	3/2	1/1
levels of controls																
external levels	0/*	*/0	4/1	4/1	4/1	1/5	1/5	4/1	5/1	4/1	4/1	4/1	4/1	4/1	4/1	1/1
internal levels	0/*	+/0	3/1	3/1	3/1	4/1	4/1	3/1	4/1	3/1	3/1	3/1	3/1	3/1	3/1	1/1
Input/Output Aspects																
inputs and outputs	*/+	*/0	1/?	1/5	1/5	5/1	2/?	5/1	5/1	1/5	4/1	1/1	1/5	2/?	1/5	1/?
functional significances																
significances of inputs	+/0	*/0	5/1	5/1	5/1	5/1	4/?	5/1	5/1	4/?	5/1	5/1	5/1	4/?	2/4	1/1
significances of outputs	*/0	*/0	5/1	5/1	5/1	5/1	4/?	5/1	5/1	4/?	5/1	5/1	5/1	4/?	5/1	1/1

TABLE 4-2a: Part "A" of procedural diagramming method evaluation criteria table. (*Page one of four for Part "A" of the table.*)

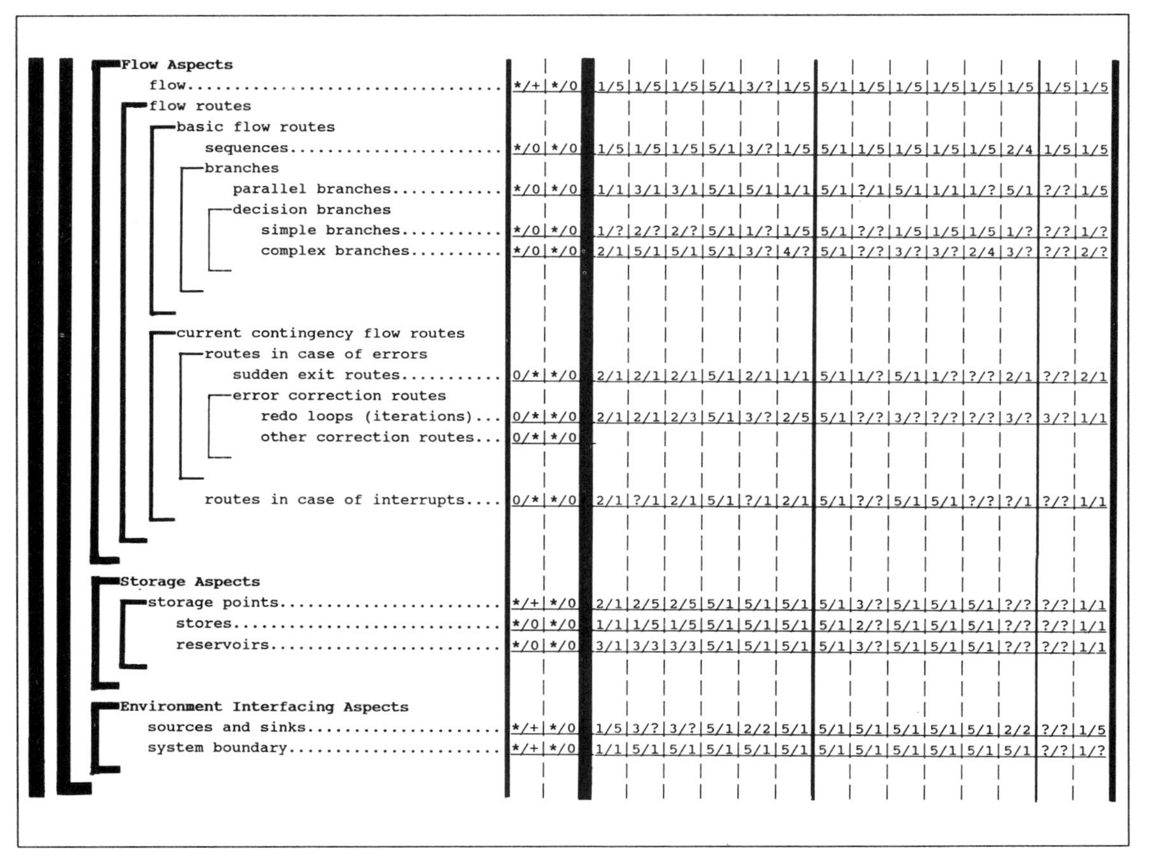

Flow Aspects																
flow	*/+	*/0	1/5	1/5	1/5	5/1	3/?	1/5	5/1	1/5	1/5	1/5	1/5	1/5	1/5	1/5
flow routes																
basic flow routes																
sequences	*/0	*/0	1/5	1/5	1/5	5/1	3/?	1/5	5/1	1/5	1/5	1/5	1/5	2/4	1/5	1/5
branches																
parallel branches	*/0	*/0	1/1	3/1	3/1	5/1	5/1	1/1	5/1	?/1	5/1	1/1	1/?	5/1	?/?	1/5
decision branches																
simple branches	*/0	*/0	1/?	2/?	2/?	5/1	1/?	1/5	5/1	?/?	1/5	1/5	1/5	1/?	?/?	1/?
complex branches	*/0	*/0	2/1	5/1	5/1	5/1	3/?	4/?	5/1	?/?	3/?	3/?	2/4	3/?	?/?	2/?
current contingency flow routes																
routes in case of errors																
sudden exit routes	0/*	*/0	2/1	2/1	2/1	5/1	2/1	1/1	5/1	1/?	5/1	1/?	?/?	2/1	?/?	2/1
error correction routes																
redo loops (iterations)	0/*	*/0	2/1	2/1	2/3	5/1	3/?	2/5	5/1	?/?	3/?	?/?	?/?	3/?	3/?	1/1
other correction routes	0/*	*/0														
routes in case of interrupts	0/*	*/0	2/1	?/1	2/1	5/1	?/1	2/1	5/1	?/?	5/1	5/1	?/?	?/1	?/?	1/1
Storage Aspects																
storage points	*/+	*/0	2/1	2/5	2/5	5/1	5/1	5/1	5/1	3/?	5/1	5/1	5/1	?/?	?/?	1/1
stores	*/0	*/0	1/1	1/5	1/5	5/1	5/1	5/1	5/1	2/?	5/1	5/1	5/1	?/?	?/?	1/1
reservoirs	*/0	*/0	3/1	3/3	3/3	5/1	5/1	5/1	5/1	3/?	5/1	5/1	5/1	?/?	?/?	1/1
Environment Interfacing Aspects																
sources and sinks	*/+	*/0	1/5	3/?	3/?	5/1	2/2	5/1	5/1	5/1	5/1	5/1	5/1	2/2	?/?	1/5
system boundary	*/+	*/0	1/1	5/1	5/1	5/1	5/1	5/1	5/1	5/1	5/1	5/1	5/1	5/1	?/?	1/?

TABLE 4-2a: Part "A" of procedural diagramming method evaluation criteria table. (*Page two of four for Part "A" of the table.*)

EXPLANATION OF PART "A" OF THE TABLE

- A factor's effect on the particular system view taken (logical or physical) is nil ("0") or positive ("+") or synergistic ("*").
- Individual methods are ranked from 1 (strong) to 5 (weak). (NOTE: Many rankings are only intuitive.) Strength is unimportant or desirable or important, depending on the factor's effect on the system view adopted.
- The slash ("/") separates the effect or ranking of the capability of illustrating the system detail from that of the capability of avoiding the system detail.

	EFFECTS		RANKINGS													
	SYSTEM VIEW TAKEN		SINGLE-PHASE METHODS						MULTI-PHASE METHODS							
			SUITABILITY SPECTRUM--> LOGICAL--->--->PHYSICAL						HIERARCHICAL DECOMPOSITION METHODS						MISCELLANEOUS	
	LOGICAL	PHYSICAL	DFD	DOCUMENT	COMPUTER SYSTEM	FUNCTION	STRUCTURE	COMPUTER PROGRAM	HIPO: HIERARCHY	HIPO: IPO	WARNIER-ORR	ACTION	HOS	EXPLOSION-FLOW	SADT	STRUCTURE-FLOW
FLOW CHART	...	...	.X.	.X.	.X.	...	.X.	.X.	...	.X.	.X.	.X.	.X.	.X.	.X.	.X.
LEVELS-OF-CONTROL CHART	...	...	...	...	...	.X.	.X.	...	...	...	...	...	...	...	...	.X.
HIERARCHICALLY SHAPED CHART	...	...	...	...	...	.X.	.X.	...	.X.	...	.X.	.X.	.X.	.X.	...	.?.
CHART HAVING AN INHERENT BASIC FORMAT	...	...	...	...	...	.X.	.X.	...	.X.	.X.	.X.	.X.	.X.	.X.	.X.	.X.
CAPABILITIES OF ILLUSTRATING/AVOIDING SYSTEM CONSTRAINTS																
Function Constraints																
types of functions																
computerized and manual types	0/*	+/0	5/1	2/4	1/5	5/1	5/1	5/1	5/1	5/1	5/1	5/1	5/1	5/1	5/1	5/1
other types	0/*	+/0	4/2	3/3	1/5	5/1	4/1	3/4	5/1	5/1	5/1	5/1	5/1	4/1	5/1	2/3
Input/Output Constraints																
material and data types	+/+	*/0	3/1	5/1	5/1	5/1	5/1	5/1	5/1	5/1	5/1	5/1	5/1	5/1	5/1	1/5
media	0/*	+/0	5/1	3/3	1/5	5/1	5/1	5/1	5/1	1/1	5/1	5/1	5/1	5/1	5/1	5/1
Flow Constraints																
flow scheduling algorithms																
algorithms for routes that branch	0/*	+/0	2/1	2/1	2/1	5/1	2/5	2/1	5/1	?/1	3/1	2/1	5/1	2/5	?/?	5/1
relative time frames	*/0	*/0	3/1	3/1	3/1	3/3	2/4	3/3	5/1	3/?	3/3	3/4	3/4	3/3	?/?	1/5
iterations to accumulate/deplete	+/0	*/0	3/1	3/1	3/1	5/1	1/?	2/5	5/1	?/?	2/5	1/5	1/5	1/?	?/?	1/1

TABLE 4-2a: Part "A" of procedural diagramming method evaluation criteria table. (*Page three of four for Part "A" of the table.*)

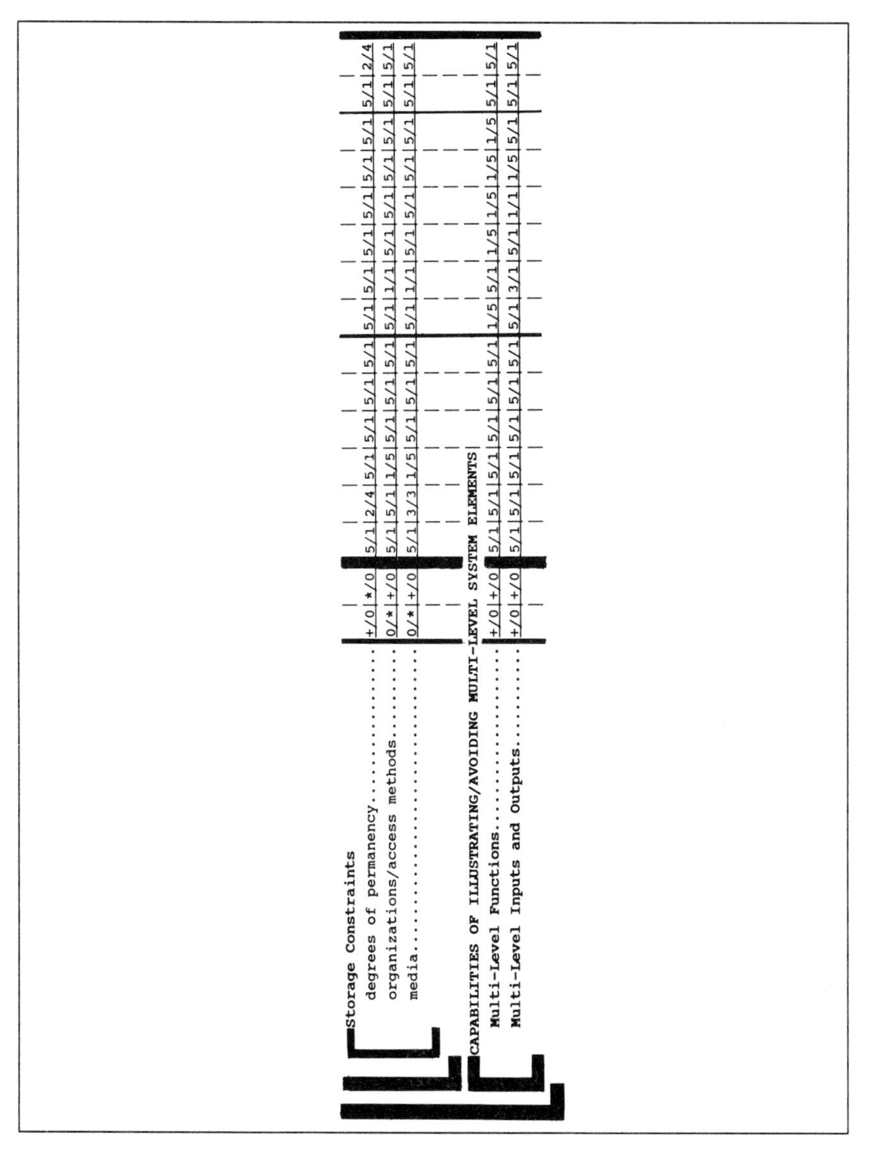

Storage Constraints																
degrees of permanency	+/0	*/0	5/1	2/4	5/1	5/1	5/1	5/1	5/1	5/1	5/1	5/1	5/1	5/1	5/1	2/4
organizations/access methods	0/*	+/0	5/1	5/1	1/5	5/1	5/1	5/1	5/1	1/1	5/1	5/1	5/1	5/1	5/1	5/1
media	0/*	+/0	5/1	3/3	1/5	5/1	5/1	5/1	5/1	1/1	5/1	5/1	5/1	5/1	5/1	5/1
CAPABILITIES OF ILLUSTRATING/AVOIDING MULTI-LEVEL SYSTEM ELEMENTS																
Multi-Level Functions	+/0	+/0	5/1	5/1	5/1	5/1	5/1	5/1	1/5	5/1	1/5	1/5	1/5	1/5	5/1	5/1
Multi-Level Inputs and Outputs	+/0	+/0	5/1	5/1	5/1	5/1	5/1	5/1	5/1	3/1	5/1	1/1	1/5	5/1	5/1	5/1

TABLE 4-2a: Part "A" of procedural diagramming method evaluation criteria table. (*Page four of four for Part "A" of the table.*)

EXPLANATION OF PART "B" OF THE TABLE

- A factor's effect on the particular system view taken (logical or physical) is nil ("0") or positive ("+").
- Individual methods are ranked from 1 (strong) to 5 (weak). (NOTE: Many rankings are only intuitive.) Strength is unimportant or desirable, depending on the factor's effect on the system view adopted.

	EFFECTS		RANKINGS													
	SYSTEM VIEW TAKEN		SINGLE-PHASE METHODS						MULTI-PHASE METHODS							
			SUITABILITY SPECTRUM--> LOGICAL--->--->PHYSICAL						HIERARCHICAL DECOMPOSITION METHODS						MISCEL-LANEOUS	
	LOGICAL	PHYSICAL	DFD	DOCUMENT	COMPUTER SYSTEM	FUNCTION	STRUCTURE	COMPUTER PROGRAM	H HIERARCHY	IPO	WARNIER-ORR	ACTION	HOS	EXPLO-SION FLOW	SADT	STRUC-TURE FLOW
FLOW CHART	...	...	.X.	.X.	.X.	...	.X.	.X.	...	.X.	.X.	.X.	.X.	.X.	.X.	.X.
LEVELS-OF-CONTROL CHART	...	...	...	...	...	.X.	.X.	...	...	...	...	...	...	...	...	.X.
HIERARCHICALLY SHAPED CHART	...	...	...	...	...	.X.	.X.	...	.X.	...	.X.	.X.	.X.	.X.	...	.?.
CHART HAVING AN INHERENT BASIC FORMAT	...	...	...	...	...	.X.	.X.	...	.X.	.X.	.X.	.X.	.X.	.X.	.X.	.X.
OTHER FACTORS AFFECTING THE CURRENT USE OF PROCEDURAL CHARTING METHODS																
CHARTING METHOD ATTRIBUTES WHICH ENSURE ACHIEVEMENT OF MORE RIGOROUS FUNCTIONAL DECOMPOSITION																
Requirement that Mathematical Axioms be Used in Functional Decomposition	+	+	5	5	5	5	5	5	5	5	5	5	1	5	5	5
Requirement that Functions be Comprised of Operations Directly Necessary for Achieving Objectives																
only directly necessary operations	+	+	5	5	5	5	5	5	5	5	5	5	5	5	5	1
all directly necessary operations	+	+	5	5	5	5	5	5	5	5	5	5	5	5	5	1
CHARTING METHOD ATTRIBUTES WHICH MAKE FOR USER-FRIENDLINESS																
Fewness of Symbols	+	0	2	2	5	1	3	3	1	3	2	3	1	3	1	5
Clarity of Symbols	+	+	1	1	2	1	2	2	1	2	3	2	1	2	1	3
Problem-Relatedness of Terminology	+	+	2	1	1	?	3	2	?	1	3	1	3	?	1	3

TABLE 4-2b: Part "B" of procedural diagramming method evaluation criteria table. (*Page one of two for Part "B" of the table.*)

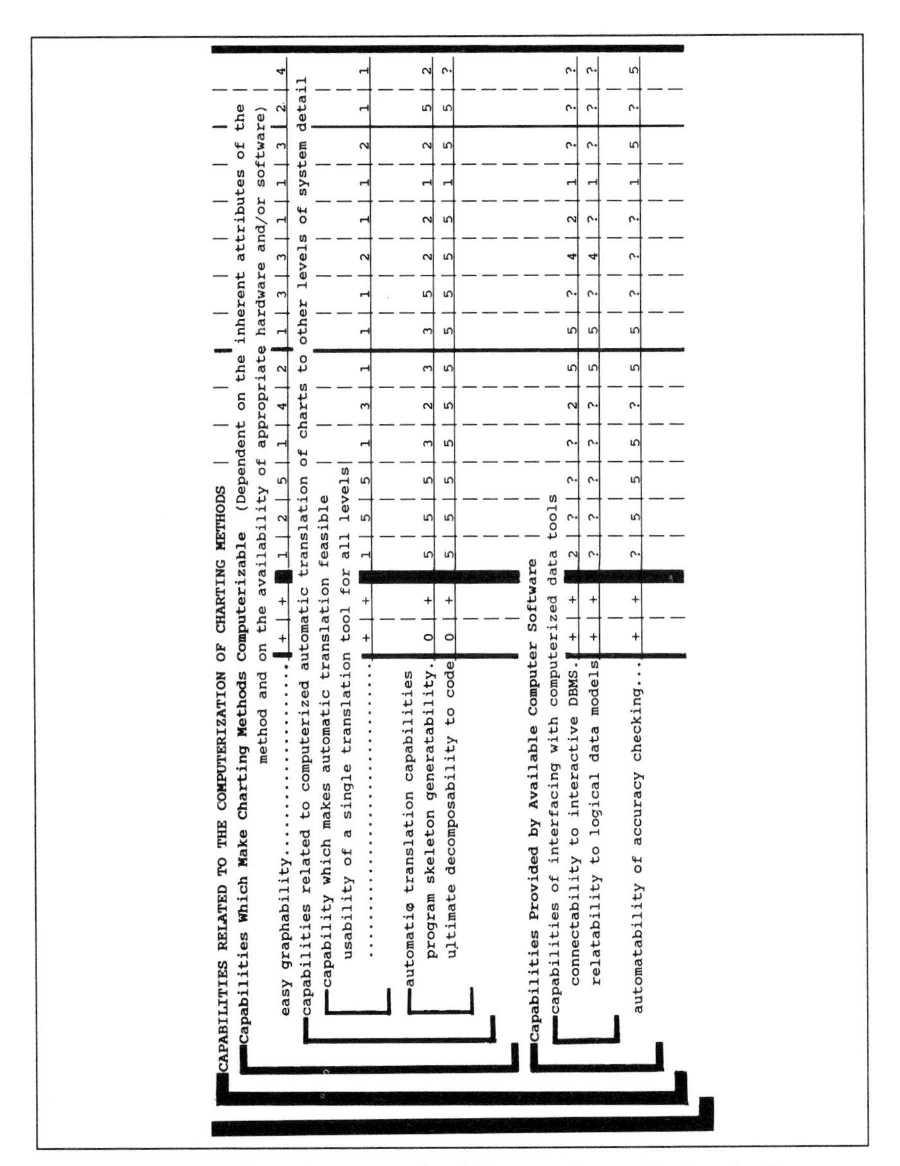

CAPABILITIES RELATED TO THE COMPUTERIZATION OF CHARTING METHODS	1	2	3	4	5	6	7	8	9	10	11	12	13	14	15	16
Capabilities Which Make Charting Methods Computerizable (Dependent on the inherent attributes of the method and on the availability of appropriate hardware and/or software)																
easy graphability	+	+	1	2	5	1	4	2	1	3	3	1	1	3	2	4
capabilities related to computerized automatic translation of charts to other levels of system detail																
capability which makes automatic translation feasible																
usability of a single translation tool for all levels	+	+	1	5	5	1	3	1	1	1	2	1	1	2	1	1
automatic translation capabilities																
program skeleton generatability	0	+	5	5	5	3	2	3	3	5	2	2	1	2	5	2
ultimate decomposability to code	0	+	5	5	5	5	5	5	5	5	5	5	1	5	5	?
Capabilities Provided by Available Computer Software																
capabilities of interfacing with computerized data tools																
connectability to interactive DBMS	+	+	2	?	?	?	2	5	5	?	4	2	1	?	?	?
relatability to logical data models	+	+	?	?	?	?	?	5	5	?	4	?	1	?	?	?
automatability of accuracy checking	+	+	?	5	5	5	?	5	5	?	?	?	1	5	?	5

TABLE 4-2b: Part "B" of procedural diagramming method evaluation criteria table. (*Page two of two for Part "B" of the table.*)

includes two evaluation factors: the capability to illustrate the system-feature named in that row, and also the capacity to avoid illustrating it. The slashes ("/"), which occur throughout the body of this part of the table, are merely used to separate the entries in each box, as will be explained below. There are two kinds of entries in the table cells: effects and rankings. The effects involve the relationships of the evaluation factors to the logical and physical perceptions of the system, while the rankings represent the potency of the diagramming methods in relation to the factors.

Effects of Evaluation Factors on Logical and Physical Perceptions of Systems — The entries occurring in the two leftmost columns are arithmetic symbols (i.e., 0, +, *). Each such entry represents the effect that one of the two criteria included in that row has on the system view (logical or physical) indicated in that column. Specifically, an arithmetic symbol to the left of the slash in one of the cells is the effect of the capability of **illustrating** the system detail named in that row, while a similar entry to the right of the slash is the effect of the capacity to **avoid** the system detail named in that row.

The effects are simply rules of thumb. They can be explained as follows:

- The **plus** sign to the left of a slash means that the capability of illustrating the system detail in that row is desirable for the view in question (logical or physical), while the **zero** means that this capability is unimportant for that view.
- The **plus** sign to the right of a slash means that the capacity to avoid the system detail in that row is desirable for the view in question, while the **zero** means that this capacity is unimportant.
- The **asterisk** (multiplication sign) to the left of a slash means a synergistic effect is achieved by including the system detail. Specifically, not only does it mean the same as the plus sign in relation to the particular factor to which it applies, but, in addition, it indicates that the capacity to illustrate the system detail in that row is desirable because it provides a way of including and/or avoiding other undesirable system details, associated with other factors in the table. Roughly speaking, it may be said that the effects of these criteria are "multiplied" (the asterisk is the symbol for multiplication in most programming languages). For example, the capacity to illustrate parallel branches helps avoid forcing operations into a serial (sequential) arrangement when they are logically in parallel.
- The **asterisk** to the right of a slash means that a synergistic effect, in the sense just explained, is obtained by being able to exclude the system detail.

Rankings of Charting Methods in Relation to the Evaluation Factors — The entries in the columns corresponding to the various charting methods are integers ranging from 1 to 5, and question marks. Each numeric entry represents a measure of how well the charting method in that column can illustrate or avoid the system feature named in that row. (A question mark indicates an uncertain ranking.) Specifically, an entry to the left of the slash in one of the cells is the ranking of that charting method's capability of illustrating the system detail in that row, while a similar entry to the right of the slash is the ranking of the method's capacity to avoid the system feature in that row.

Evaluating the Suitability of Charting Methods for Particular Stages of System Development — The evaluations are obtained by relating the rankings to the effects. The rankings were arrived at subjectively by rough guesswork. In fact, three of the charting methods referenced in the column headers, namely, DFDs, document flowcharts, and IPO flowcharts, actually each refer to two varieties of the named charting method, with each variation having somewhat different characteristics. Since the different variations are not represented by different columns, a single ranking is applied to each pair of variations of these charting methods.

A charting method is evaluated in each row with respect to each of the two factors in that row (the capability to illustrate, and the capability to avoid, the system detail in question), and in relation to the factor's effect on each of the two system views. In other words, for every row of the table, there are four distinctive evaluations of each diagramming method. For example, one such evaluation may be determined as follows:

1) *A charting method is selected for evaluation.* Let us take the DFD.
2) *A level of perception is chosen,* e.g., to evaluate the data flow diagram as a physical viewing device.
3) *A particular evaluation criterion, is chosen,* e.g., the capability of avoiding the illustration of sudden exit flow routes in case of errors.
4) *The row in the table corresponding to the above criterion is located.* In this example, this row occurs in the table under current contingency flow routes, within the flow aspects sub-section of the capabilities of showing/avoiding the various aspects of the umbrella of system elements.
5) *Continuing along the same row in the table, the appropriate effects column is located.* In this example, this would be the physical system view column, which happens to be the second column to the right of the factors. This places us at a particular cell.
6) *The entry on the appropriate side of the slash in this cell is determined.* In this case, the entry in question is on the righthand side of the slash,

since this side corresponds to the capacity to avoid illustrating the system detail in question. Looking at this point in the table, it may be seen that the entry is a zero. This means that the capability of avoiding the illustration of sudden exit flow routes is irrelevant to the ability of any charting method to perceive the system at a physical (detailed) level.

7) *Continuing still further to the right along this same row in the table, the appropriate charting method column is located.* In this example, this would be the data flow diagram column. This places us at another cell in the table.
8) *The entry on the appropriate side of the slash in this cell is determined.* In this case, the entry in question is again on the righthand side of the slash, for the same reason as in point #6. Looking at this location in the table, it may be seen that the entry is the digit one (1). This means that the data flow diagram is maximally potent in its capability to avoid showing sudden exit flow routes.
9) *The ranking just determined is now related to the effect found in point #6:*
   —If the effect is positive, or, synergistic ("ultra-positive," represented by an asterisk), and the ranking is strong, then the evaluation is favourable. On the other hand, a weak ranking, combined with the same positive effect, results in an undesirable evaluation.
   —If the effect is nil, then, whether the ranking is strong (1 or close to 1), or weak (5 or close to 5), the factor is irrelevant. This is the case in the present example.

It is important to emphasize that a charting method's ranking must be related to the factor's effect on each system view, in order for the evaluation to be determined. For example, if a diagramming method with great potency (i.e., a ranking of 1), is related to a nil effect on the logical view, this method is considered irrelevant as a logical viewing technique despite its great potency. On the other hand, if a ranking of 1 is combined with a positive effect on the physical view, it means that this diagramming method is excellent for physical viewing.

Use of Part "B" of the Evaluation Table

Table 4-2b works somewhat differently than does Table 4-2a. Each row in this part of the table contains only a single factor, which is exactly as worded. Thus, no slashes occur in the various boxes. In other words, the effects and rankings are straight interpretations of the factors as worded, without any distinctions being made between the capacity to include or exclude the feature described in each row. Moreover, ultra-positive (synergistic) effects are absent.

PART TWO—
THE NEW CHARTING METHOD

CHAPTER FIVE

BASICS OF THE STRUCTURE-FLOW CHARTING METHOD*

Having completed Chapters Three and Four, which dealt with underlying concepts of functional modeling, the reader will now be prepared for Chapters Five and Six. These two chapters are in actuality a continuation of one another. Chapter Five discusses most of the introductory concepts, and Chapter Six provides an example case. It should be noted that the case does not cover all the concepts from Chapter Five, while, conversely, it develops a number of new concepts not introduced in Chapter Five, since their presentation is best accomplished within the framework of a case. Certain elements of the structure-flow chart, including certain ideas introduced only in connection with the case, are

* The materials in the first section of this chapter are based on an article by Ed Baylin, "The Structure-Flow Charting Method," Cybernetics and Systems: An International Journal, published in Washington, D.C. by Hemisphere Publishing Corp©. It should be observed that the copyright on the **diagrams** does **not** belong to Hemisphere.

not discussed until later chapters, pending further discussion of the concepts of functional modeling.

INTRODUCTION

One of the most important subjects in information systems work is the conceptual structuring of systems, which uses diagrams as its basic exposition tool. With present methods, modeling is such that each individual system tends to be treated as a unique occurrence, instead of in terms of prototypes. Moreover, different diagramming methods are used for different levels and types of systems, as well as for different phases of system development.

The structure-flow chart not only provides a method for conceptual prototyping, but also a method of handling all levels and types of systems in all phases of system development. The structure-flow chart might appear to be overly sophisticated, as it captures an inordinate amount of information. However, the amount of information given can be limited at the discretion of the diagrammer, thereby allowing the method to be used simply as a data flow diagram which has been specially formatted for conceptual prototyping. At the very least, an appreciation of the structure-flow charting method enables better usage of existing information system charting techniques, and helps to organize thinking about complex problems. While the mastering of this method is no easy task, it can handle the true complexity of the problems to which it is applied, thus rendering overly simplistic those methods which rely mainly upon intuition to understand a problem.

ADVANTAGES AND DISADVANTAGES OF THE METHOD

Current Methods in Perspective

First, even the most experienced readers should be familiar with the procedural charting concepts covered in Chapter Four, which contains many original and essential perspectives, including the following:

- the poorly understood concept of flow as opposed to levels-of-control charts;

- the concept of "formatted" charts, based on various inherent or special page alignments;
- criteria for evaluating charting methods.

Second, it would be desirable if the material in the related publication about procedural diagramming were familiar to the reader. This publication reviews and compares the majority of current flow and levels-of-control charting methods, and introduces the structure-flow chart and its use in conceptual prototyping.

Evaluation Criteria

To refresh the memory of the reader, the conclusions (summarized in Chapter Four) regarding the stage of system development for which a given diagramming method is most appropriate are based on factors such as the following:

- A charting method with few symbols is better for analysis (logical system viewing), as users can understand it more easily while using it more conceptually.
- A charting method which does not force the showing of controls is better for analysis than for design, as most indirect system operations are omitted from the umbrella of system elements in order to avoid too much detail during analysis. Conversely, a charting method which allows showing of controls is needed to show details during design.
- A charting method which forces the specification of constraints, such as whether an operation is performed manually or by computer, is not very useful for obtaining a logical picture of the system, and is thus not applicable during analysis.
- A charting method which cannot show various types of flow routing logic (parallel branching, alternative paths, etc.) during flow is of doubtful value, especially for complex systems.
- A charting method which cannot demonstrate unexpected interrupts has doubtful utility in the analysis and design of complex systems, since operations of these are frequently attended by unexpected interrupts from the system environment.

Relatively Unique Features of the Structure-Flow Chart

The following sub-headers organize the discussion of the relatively unique features of the structure-flow chart:

Forced Clear Functional Thinking

Clear functional thinking is forced in the structure-flow charting method. This results from the following:

—page layouts of functions which are formatted to demonstrate functional class/time-orientation, and provide rules allowing the use of the same essential model of relationships when system elements are seen from the viewpoint of a subsystem rather than from that of a parent system;

—algorithmic definitions of functions, which ensure that each function is an objective-defined function (Chapter Three), including sophisticated specification of the functional significances of both inputs and outputs, when these are shown;

—clear depiction of **both** levels-of-control, when shown, **and** system flow, including the showing of levels-of-control relationships, when demonstrated, in a context of three-dimensional understanding.

One of the benefits of forcing clear functional thinking is that it results in lucid partitioning into subsystems, e.g., with the functional cohesion method. The greatest benefit of clear functional thinking, however, is that it allows one to model systems in terms of conceptual prototypes, as opposed to treating each system modeled by a diagram as a unique occurrence. Again, "prototype" here does not refer to *physical* prototyping, e.g., the quick mounting of a computer system which approximates what the final computer system will be. Instead, the reference is to *conceptual* prototyping, in which an approximate diagrammatic picture of the system is quickly obtained. These prototypes, applicable to all types and levels of system, are extremely useful for tasks such as:

—evaluation of software packages, since a checklist of functions to look for in a system is effectively provided by the conceptual prototype;

—learning about new systems;

—organizing the writing of case books on systems applications.

With respect to the latter, the conceptual prototyping method has been used by the author in a related publication, about conceptual prototyping of business systems, to create a number of prototypes of typical business systems, all of which are decomposed into the same functional elements, down to several levels of detail. Thus, the set of data flow diagrams or action charts for one system can be used for another system simply by replacing certain of the textual details. This approach allows accurate multi-level system specification to be done with great accuracy by a beginner in systems analysis. This is similar to what has been achieved with respect to computer programming using fourth-generation languages (rather than the more procedural and less user-friendly second or third-generation ones).

Sophisticated Arsenal of Diagramming Symbols

In the structure-flow chart, levels of control or functional significances of inputs and outputs can be shown in ways which are both much clearer and more precise than in any other charting method (currently known to this author). Should inclusion of such detail be found to be confusing, the structure-flow charting method gives the diagrammer considerable discretion in calling up different parts of the arsenal of available symbols, depending upon the audience and stage of system development for which the chart is designed. Thus, although the chart has the unique ability to serve as both a levels-of-control and a flow chart, the levels-of-control aspects may be omitted entirely, or in part. As well, not all the functional significances of inputs and outputs need be shown.

The Structure-Flow Chart's Practicality

The following are reasons why practitioners might want to use the structure-flow chart rather than the data flow diagram:

- Data flow diagrams do not require incorporation of a conceptual prototype, and, in any case, cannot incorporate this nearly as well as the structure-flow chart, which is designed specifically with conceptual prototyping in mind. The advantages of conceptual prototyping are comparable to the advantages of drawing a picture on a connect-the-dots-diagram as opposed to a blank sheet of paper. The former ensures the correct scale of the picture and the accurate portrayal of the object, while the blank sheet of paper may be used in any way. Not only is rigor ensured by the structure-flow chart, but, in most cases, the picture can also be drawn more quickly.
- Notwithstanding the page spatial arrangement conventions, the DFD is no easier to read than a **simplified** structure-flow chart. To this end, the reader may refer to the comparison of the two in the accounts payable system example in the appropriate related publication on procedural diagramming.
- The structure-flow charting method simplifies **thinking**, as well as diagramming. To illustrate by analogy, a word processor is considerably more difficult to learn to use than a typewriter, but a word processor is both more efficient and more effective.

- The structure-flow chart may be employed in design stages of system development, as well as in logical system viewing, since it contains levels-of-control features as well as those which describe the main flow line of the system. Translating data flow diagrams used in analysis into levels-of-control charts used in design is unnecessary, since levels-of-control features can simply be added to the same basic diagrams.

Technical Limitations of the Structure-Flow Charting Method

The structure-flow charting method has certain limitations, as follow:

- Since it cannot show certain types of physical details, and because it is limited to its own inherent format, this method may require certain other charting methods to be used in conjunction with it. For example, computer systems flowcharts showing media and the operations instruments may be helpful, along with various forms of specially formatted flow charts, such as those indicating flow between and among organizational entities (the "organization-based" flow chart, covered in the related publication on procedural diagramming). These supplementary diagrammatic forms may or may not be desirable in the exposition of a system.
- The method is cumbersome for hierarchical functional decomposition. Although multi-level functional decomposition of the wide-shafted arrow—the operations symbols of the structure-flow chart—is possible (with various Boolean logic symbols as in the Yourdon DFD, along with arrows for flows internal to the wide-shafted arrow), the hierarchical functional decomposition methods* are generally better for doing hierarchical functional decomposition. This is true especially if the more advanced fourth-generation methods, e.g., HOS, are used to decompose the interior of the wide-shafted arrow. These latest functional decomposition methods further develop existing well-known paradigms, and in no way conflict with the structure-flow charting method. Rather, they enhance the structure-flow chart, by supplementing it. The role of the structure-flow charting method is to first provide the basic framework in which to do hierarchical functional decomposition, by providing the wideshafted arrows to be hierarchically decomposed.

---

* See related book, authored by Ed Baylin, *Procedural Diagramming for System Development—From a More Scientific Viewpoint.*

ESSENTIAL ASPECTS OF THE STRUCTURE-FLOW CHART

Functional Analysis Features

Meaningful Black Box Aspect

The structure-flow chart uses the wideshafted arrow, the symbol for an objective-defined function, as its operations symbol. Although the function symbol looks like nothing more than a black box having direction and component steps called "flow functions," it is in fact much more because of the variety of other symbols and conventions associated with it.

First, the wideshafted arrow's contents consist of one, and only one, objective-defined function, which is a group of operations defined in a rigorous way. Although they may not be fully understood by the reader at this point, the restrictions on the contents of each wideshafted arrow in a structure-flow chart have some interesting consequences. For example, deplosion (or implosion, or concatenation, or going from the micro to the macro level of detail), is limited in certain cases. Although objective-defined functions can sometimes be combined into a more macro-level, two wideshafted arrows can never be reduced to one when two objective-defined functions have a sequential relationship but belong in different functional classes or at different control levels within the same functional class. For instance, an initiative (feedforward) coordination objective-defined function and the baseline objective-defined function to which it relates cannot be imploded into a single wideshafted arrow, since, being an adaptation function, the initiative function is not in the same direct line of flow.

Second, a non-meaningful black box, such as the rectangle, without the types of symbols and conventions associated with the wideshafted arrow, cannot be used to clearly identify the functional significances of the inputs and outputs; for example, some outputs may correspond to objectives, others to feedback, others to "symbiotic" relationships of a function to other functions, and so on. Similarly inputs can be stimuli, transactional supplies which are the operands or which play the immediate causal role in functional execution, guidelines (controls) to be used in task execution, or agents and instruments who and which execute the functional operations. These arrows and symbols have been developed to describe the functional significances of inputs and outputs of the function.

Layout to Capture Differences in Functional Class, Time-Orientation, and Level

That each wideshafted arrow's contents are restricted to the operations of a single objective-defined function is not an end in itself, as merely picking out one function at a time in a system will not help analyze the interrelationships of the various functions. The wideshafted arrows are assembled into charts in which their relative spatial relationships are used in ways to capture a variety of understandings. Briefly, one or more base (or "baseline") functions is/are selected to carry out the system's purpose directly. The other wideshafted arrows point at various angles towards, or away from, this base system wideshafted arrow, or towards or away from the non-base wideshafted arrows themselves. Each angle and relationship to other wideshafted arrows indicates something about the meaning and significance of the function within the system. Figure 5-1 provides a simple model for showing the relative spatial arrangements of wideshafted arrows on the page, as based on functional classes and time-orientations. The conventions adopted are the following:

- Current adaptation functions are either in exactly the same direction as or at right angles to the baseline. Those functions which
  —initiate the baseline are in the same alignment as the baseline;

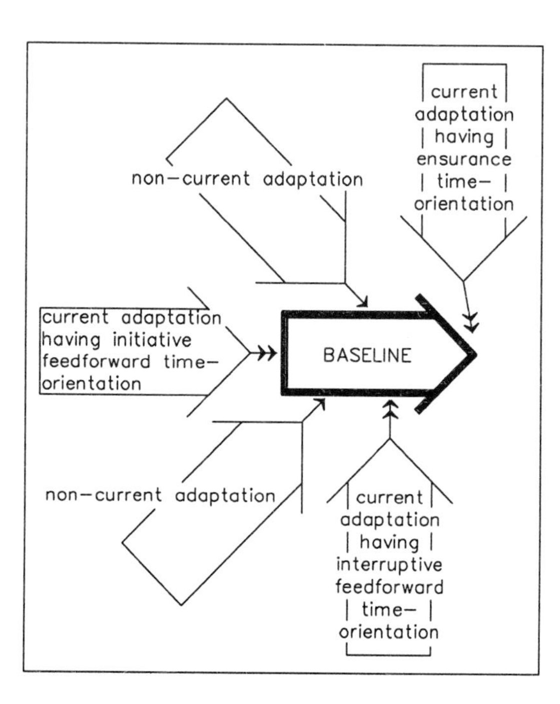

FIGURE 5-1: Page layout conventions for functional class/time-orientation.

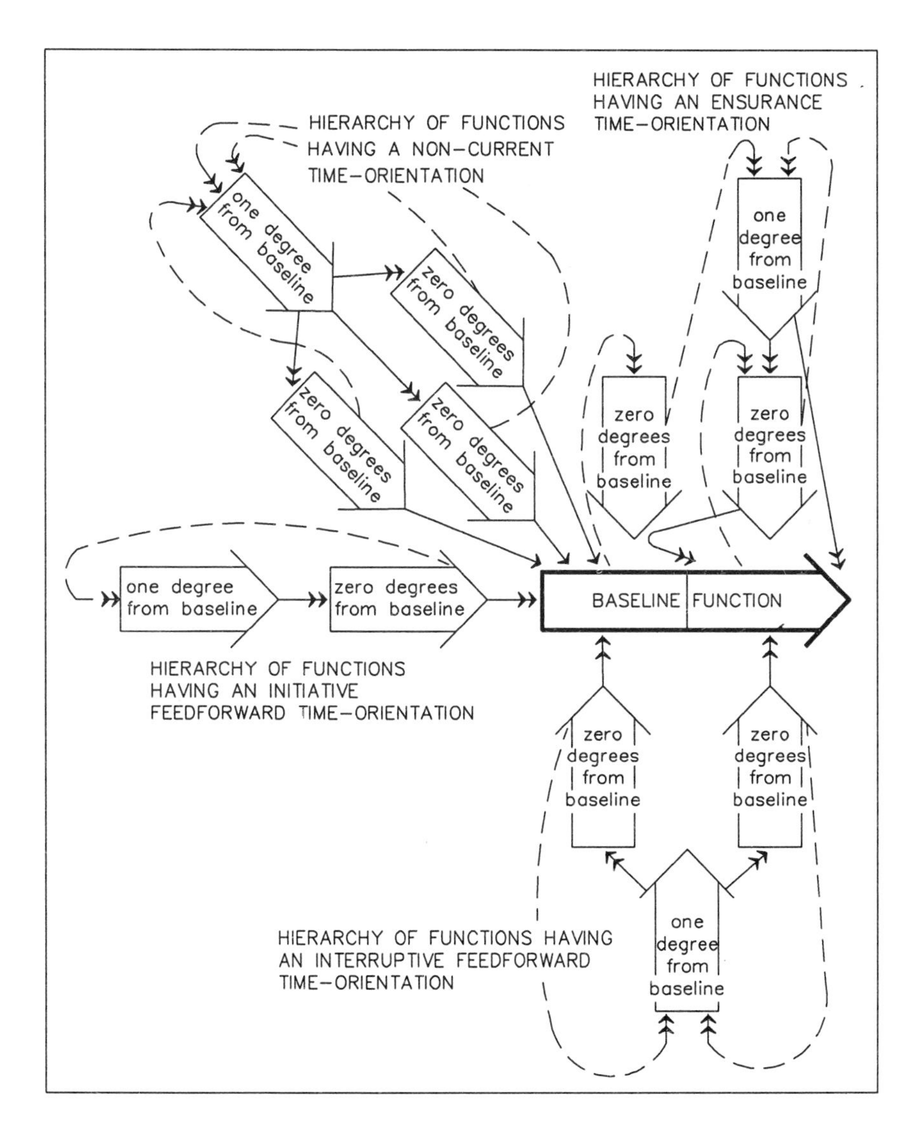

FIGURE 5-2: External levels of control in the structure-flow chart. *These are levels of adaptation functions of increasingly general scope.*

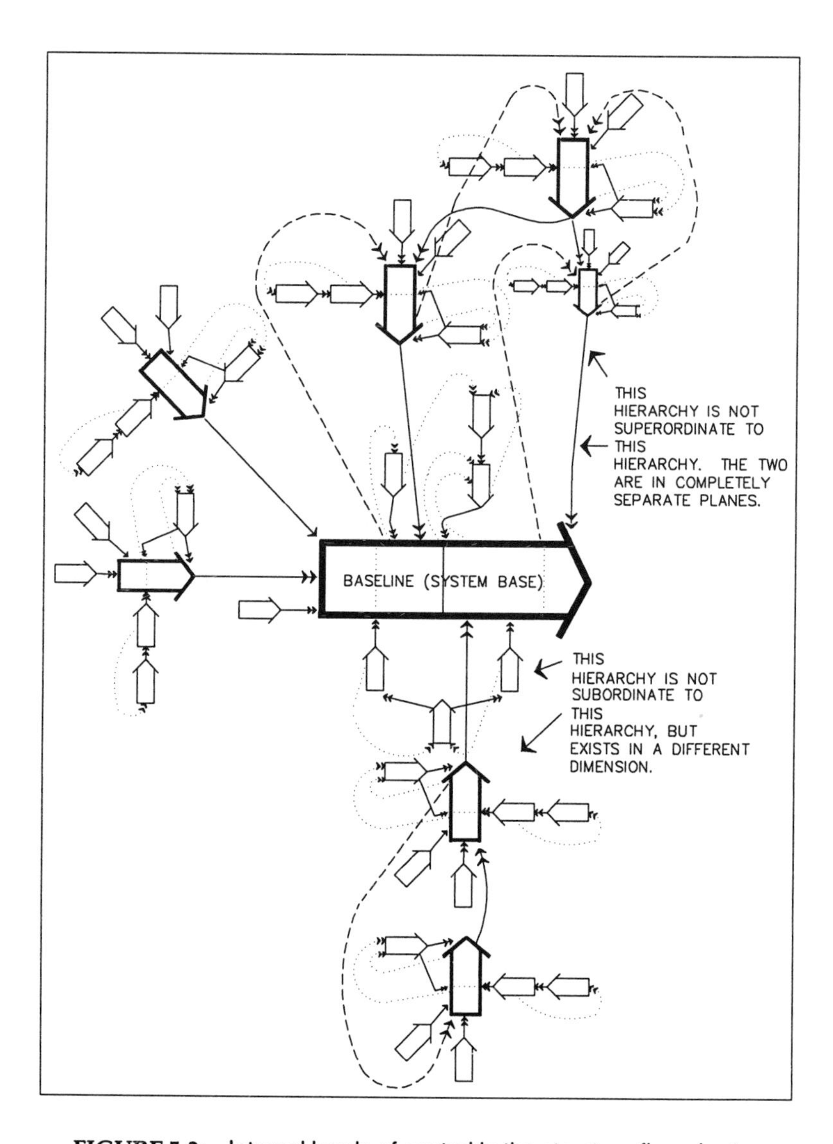

FIGURE 5-3: Internal levels of control in the structure-flow chart. *This diagram is best understood if the more thinnly drawn, smaller objective-defined functions and their connecting lines are imagined in a dimension pointing into the page. These represent the local, internal adaptation functions surrounding either the baseline or other, more primary adaptation functions.*

—interrupt the baseline—e.g., to cancel, slow down, speed up, prevent, etc.—point into the underside of the baseline;
—ensure effective completion of the baseline point into the baseline from above, with decision-making ensurance adaptation operations occurring at the end of the baseline step(s) to which they apply.

- Non-current adaptation operations point towards the baseline at a slant.

With the preceding rules, different types of control levels may be captured, as follows:

- As in Figure 5-2, different hierarchies of wideshafted arrows may be created in each of the different dimensions. The are "external" levels of control (see below).
- As in Figure 5-3, adaptation operations, or localized parts of the baseline, may be surrounded by a more local level of functions, known as "internal" adaptation functions. Each surrounded function thus forms the base of a new system. The relationships of these local, internal functions to their base function are similar to the relationships of the more primary adaptation functions to the **system** base (baseline) function.

Three-Dimensional Space for Showing Levels of Control

Unlike regular levels-of-control charts, no common apex for all the different hierarchies is present in the chart. Thus, a triangular representation of system levels of control is inadequate. To achieve a common apex for the primary hierarchies of adaptation functions, a

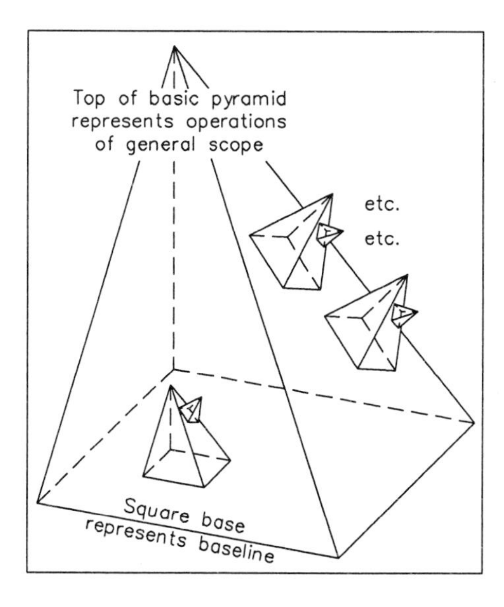

FIGURE 5-4: Pyramids affixed to pyramids and different types of levels. *Internal adaptation operations of local scope are symbolized by pyramids affixed to the sides and bases of other, more primary pyramids. The higher the parts of the triangular pyramid walls, the more general in scope the adaptation operations are within that pyramid in which they occur.*

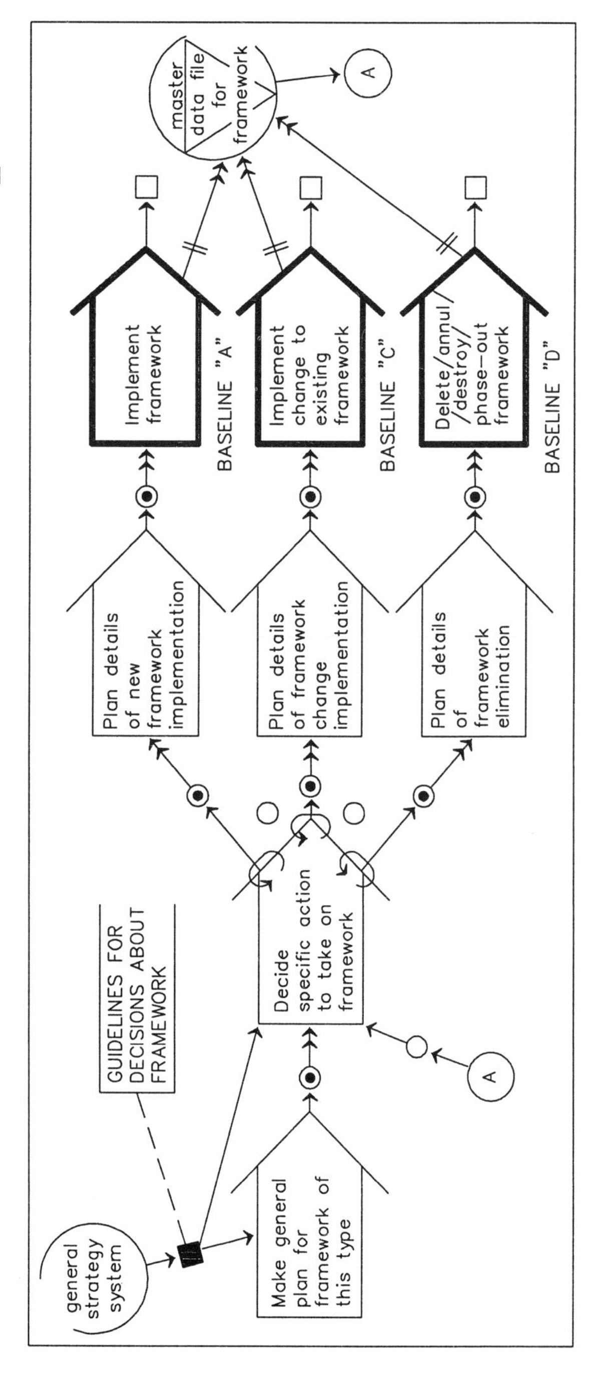

FIGURE 5-5: Prototype structure-flow chart of framework a/c/d system.

three-dimensional structure, such as a three or four-sided pyramid, can be used. In such a pyramid, the base represents the system baseline, and each of the triangular walls symbolizes one of the hierarchies of adaptation functions. To capture the local internal levels of adaptation functions demonstrated in Figure 5-3, smaller pyramids can adhere to the base or to the walls of more primary pyramids, as in Figure 5-4. This hologram type of effect is needed to capture the system levels-of-control paradigm.

Because of the restriction in contents of each wideshafted arrow to only those operations in the same class and control level which directly achieve an objective, adding extra control levels in the structure-flow chart requires expanding the umbrella of functional elements, and cannot be done by explosion during system development. This idea—

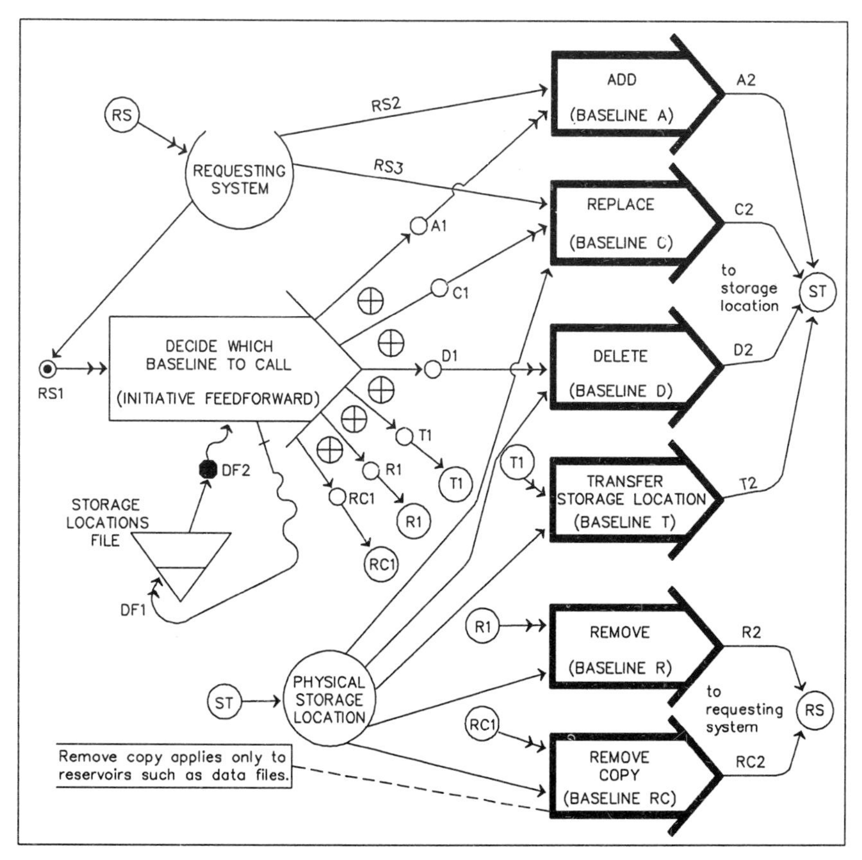

FIGURE 5-6: Prototype structure-flow chart of reservoir operating system.

which will be explained in the third section of this chapter—makes the structure-flow chart different from any of other charting methods presently known to the author.

To conclude, the structure-flow chart can add levels-of-control details to an underlying flowchart, and forces its user to discern the difficulties of capturing all the dimensions of levels of control in the mere two dimensions of a page. Also, in contrast to the structure chart (variety of the levels-of-control charts), which has neither of the preceding abilities, the structure-flow chart can capture "and/or" and parallel flow gating logic in the levels-of-control parts of the chart. These various issues will be further discussed later in this chapter.

Prototype Modeling Applications

The structure-flow chart allows for the derivation of a few prototypes, which provide various general-purpose standards applicable to many

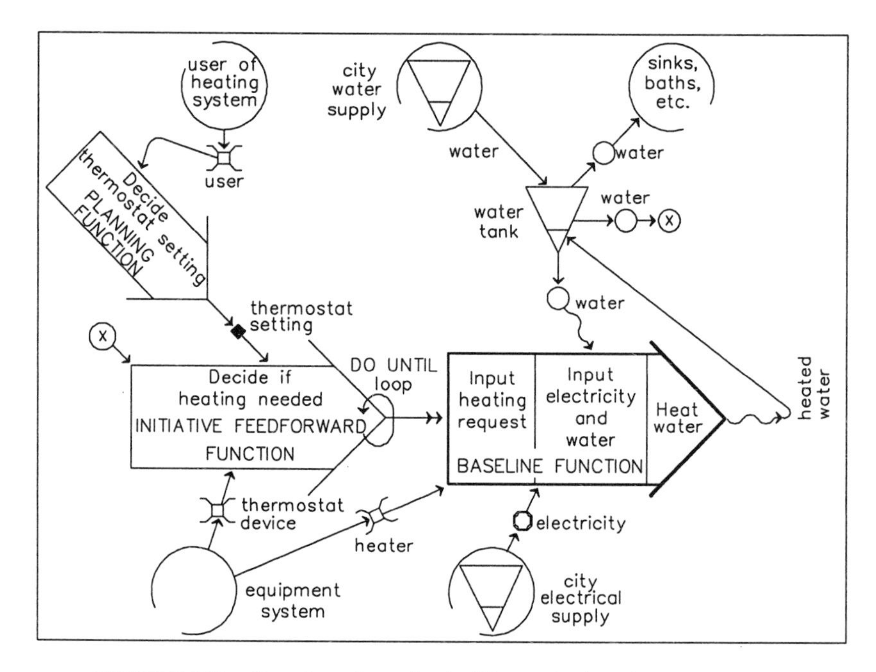

FIGURE 5-7: Structure-flow chart of the water heating system. *The shape of this chart is similar to those of the baseline subsystems of the business (both product and accounting ones) in the functional model of the business, and also resembles the morphology of the structure-flow chart of the business organization as a whole.*

systems. Experimentation with the structure-flow chart on various business systems leads to the following conclusions regarding their basic morphologies:

- The morphologies of the various **baseline product and accounting** subsystems are all essentially similar, resembling that of the organization as a whole. (See Chapter Two, the functional business organization model.) A third example is the supply acquisition system, to be developed in the next chapter, and the sales orders subsystem of the business.

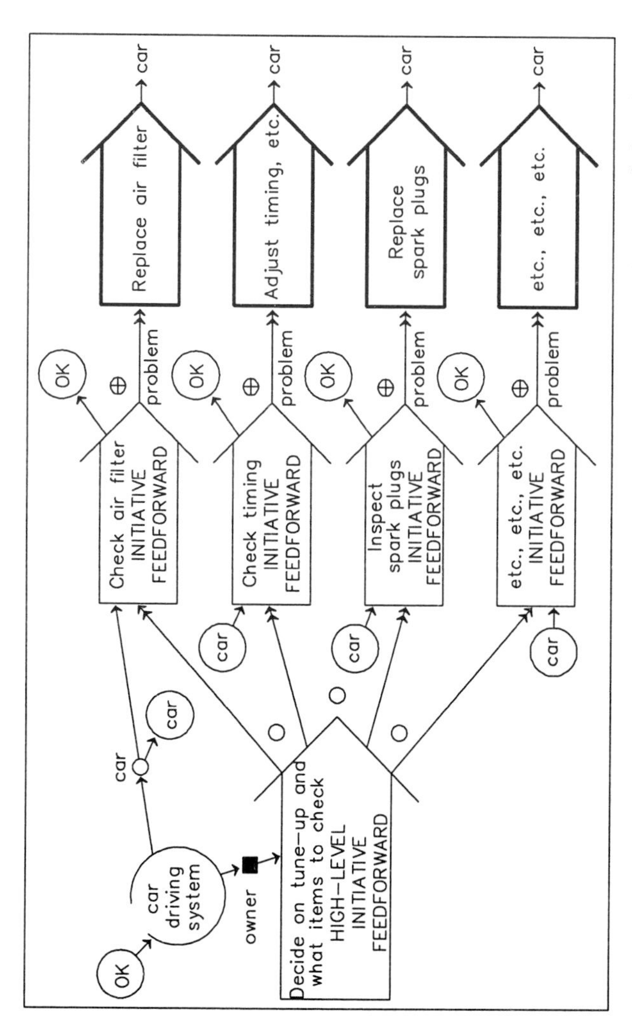

FIGURE 5-8: Structure-flow chart of the car tune-up system. *This is similar to the shape of the framework system prototype.*

- The morphologies of the **framework and buffer** subsystems are similar, as seen in Figures 5-5 and 5-6. The general long-term planning subsystem of the organization also has this prototype.
- The morphologies of the **operational control level information systems** are similar, tending to resemble the shape of Figure 5-9. Actually, these are subsystems of the baseline systems of the business organization, as will be demonstrated below, through deriving the supply acquisition information subsystem from the supply acquisition parent system. It might be observed that, insofar as the framework systems have many levels in their tree structure, a single set of consecutive levels of the tree looks somewhat like an **operational control level information system**. However, the information system is essentially different from even this, since the different wideshafted arrows stringing out from left to right represent the many **baselines** of the information system; whereas, in a framework system, these represent different **levels** of initiative coordination functions leading into the last wideshafted arrow, which forms the baseline.

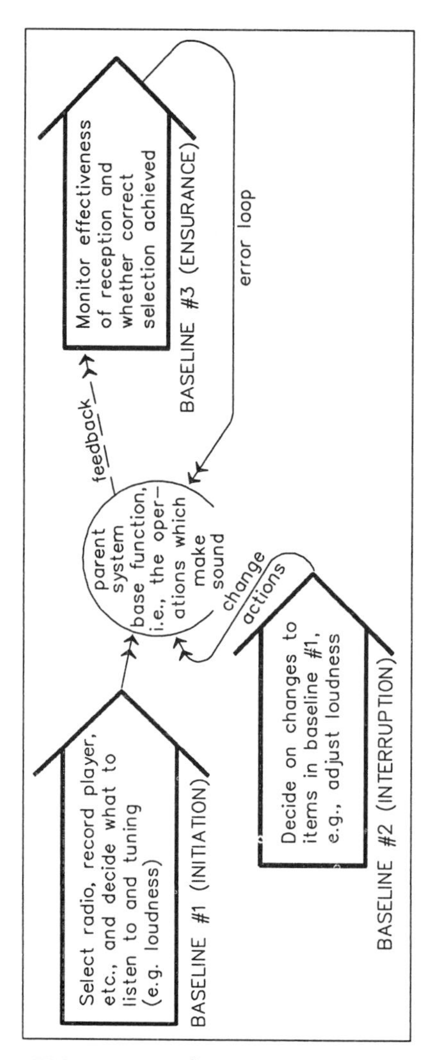

FIGURE 5-9: Structure-flow chart of the sound reproduction control system. *This is similar to the morphology of the normal operational control level information system.*

So far, the discussion of prototypes has centered on organizations and their various subsystems. The same kinds of prototypes are present in non-business systems, for example, mechanical, biological, and man-machine systems. Mechanical systems include house heating, car electricity regulation, air conditioning, food blending, sound reproduction, coffee making, bread toasting, and microwave oven heating. Other types of systems include body heating in mammals, water treatment, cake cooking, gas fill-up, and car tune-up. Figures 5-7, 5-8 and 5-9 present a few of these systems, and indicate to which prototype they bear the closest resemblance. To repeat, the different basic prototypes are those labelled as **framework systems, reservoir operating systems, baseline systems, and operational control level information systems.**

INPUTS AND OUTPUTS OF THE WIDESHAFTED ARROW

INTRODUCTION

The Small Arrows of Entry and Exit

Significance of Entry or Exit Angles

The small arrows leaving the wideshafted arrow represent outputs of the objective-defined function. Those outputs (Figure 5-10) which leave at angles of less than ninety degrees from the direction of the wideshafted arrow symbolize objective or future-oriented feedback, or symbiotic output. Each is distinguished by unique symbols. The objective arrow may also fold back in a somewhat reverse direction soon after exit from the wideshafted arrow.

In contrast to these outputs, some small arrows, either of non-continuous line type or with bars near the start of their shafts, leave the wideshafted arrow either at right angles or at an angle more than ninety degrees from the direction of the wideshafted arrow. Of these,

- those which slope backward against the direction of the wideshafted arrows are:
  - —feedback affecting the current run, if their shaft lines are non-continuous;

—outputs extracted from the function by other functions which actively seek something from this function (but without the "cooperation" of this function), if their shafts are continuous and they have a double bar across the tail part of the shaft;

- those which exit perpendicular to the wideshafted arrow with a double bar on their tail signify offshoots, of no particular functional significance to the objective-defined function.

The small arrows entering the wideshafted arrow can enter at any point along the line of flow of the objective-defined function. Once in, the inputs stay inside the wideshafted arrow, and need not be repeated i.e., need not be shown as entering the wideshafted arrow at some later point along the line of flow. Thus, if more than one stage of the flow represented by sequentially related functional elements within the wideshafted arrow requires the same input, the input need only be shown entering the earliest functional element in the sequence, from which it is assumed to be transmitted to any subsequent functional elements needing it.

As in Figure 5-11, the entry angle may be less than ninety degrees from the direction of the wideshafted arrow. In such cases, the input involved is something which promotes further execution of the objective-defined function. On the other hand, if the small arrow is at an entry angle less than ninety degrees from the **contrary** direction of the wideshafted arrow, it could only represent opposition, e.g., a stimulus (transaction) which retards, suspends or cancels execution of the function at the point of entry along the line of flow. Finally, if the entry is at right angles, this means that the significance of the input is not defined in the chart. The latter could, for instance, be used to avoid having to show a complex or rather unstructured (non-repetitive) relationship of a current adaptation function to the function receiving the input. In other words, no commitment is made in the diagram as to its functional significance.

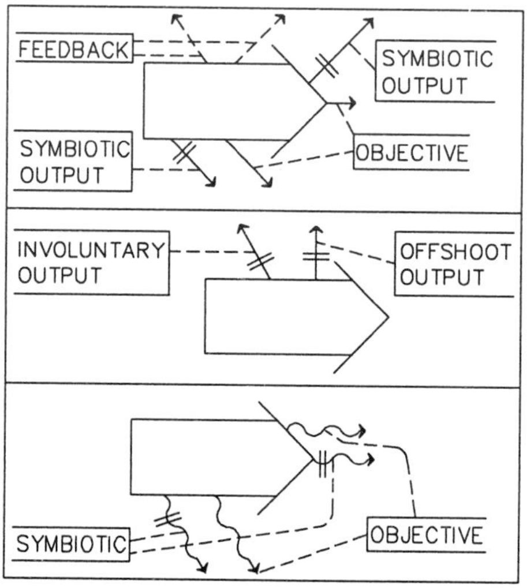

FIGURE 5-10: Features of small arrows leaving the wideshafted arrow.

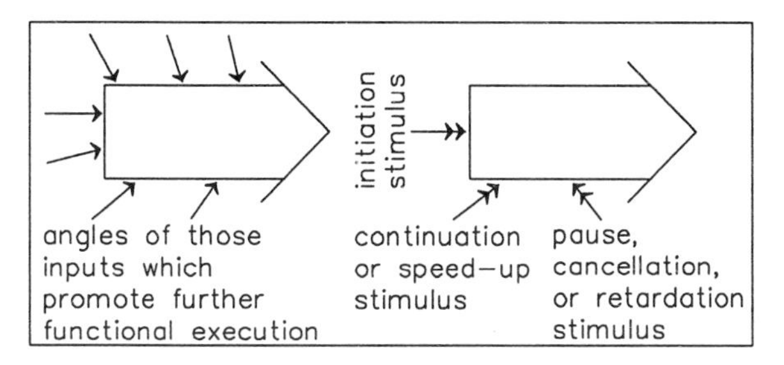

FIGURE 5-11:
Features of the small arrows entering the wideshafted arrow.

Direction At Points Other Than Entry or Exit

The small arrows may split in several directions after exit from the wideshafted arrow, or converge from multiple directions before entry to it. Figures 5-12a and b show one of two things, or a hybrid thereof. First, the same output may go to two or more places. Second, the output is made up of two or more simultaneously produced items. Similarly, two or more simultaneously acquired inputs may be represented, as in Figure 5-12c.

In addition, two or more arrows may exit at the same point, as in Figure 5-12d. This indicates that the same output may have more than one significance, e.g., it may be both an objective and a symbiotic output.

Finally, any arrow may take any direction after its initial exit trajectory. Although this may be a rather fine point to have included as a convention in the structure-flow chart, an arrow which folds back **before** the input/output type symbol attached to its tip represents something which goes out of the system to bring something back into a later system function. Otherwise, what is represented is simply a change in direction. The difference between the two ideas is graphed in Figure 5-13.

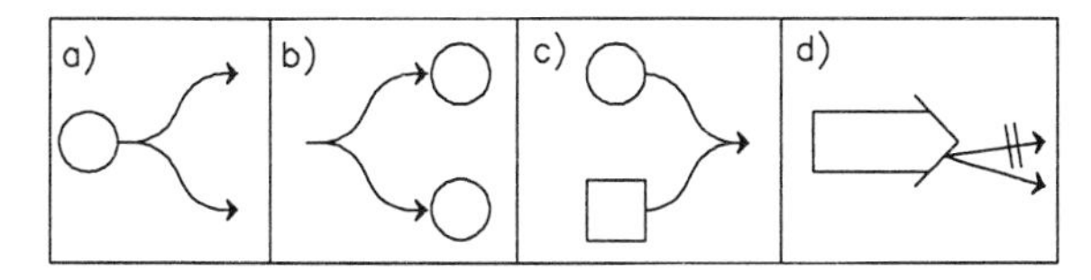

FIGURE 5-12: Branching/merging of flow.
a) single output sent in two different directions;
b) two simultaneous outputs;
c) simultaneous inputs;
d) output sent in two directions, each with different significance, one a symbiotic output, the other an objective output.

Arrow Heads

The arrow head symbol may be explained in relation to transactions, defined as specific events which generate, or stimulate, a series of ensuing events. For example, a customer placing an order, a request that a certain piece of work be done, the calling of a subroutine by a computer program, etc. all generate further events. The double-headed arrow is used here to represent the stimulus associated with a transaction.

The double head on an arrow (see Figure 5-14a) often implies activation, although it may also mean re-activation (continuation after pause), interruption, slowing down, speeding up, cancellation, etc., depending upon the angle of the input/output arrow shaft in entering the wideshafted arrow, and upon the point in the flow at which the effect occurs. The double-headed arrow tip is connected with current control relationships affecting an objective-define function, whether control of the base function affected by the stimulus input originates from within the system or from the system environment.

In contrast, the single-headed arrow represents one of the following:

—a connection in the diagram to an output symbol, e.g., to a small circle representing transactional input;

—a link in the diagram to an on-page or off-page connector, to avoid crossing of lines in the diagram, or to connect to another page of the diagram, respectively;

—an input received without associated stimulus being involved.

The first two above uses are at the discretion of the diagrammer, as the direction of flow would be understood without the arrow head at these points.

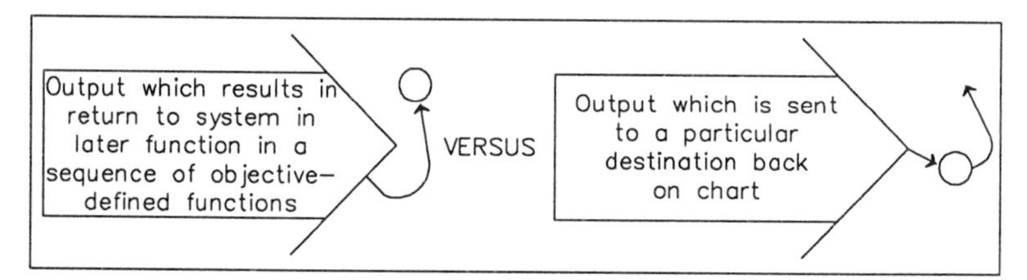

FIGURE 5-13: Meaning of the hook in the objective output small arrow. *The output on the left results in something being brought back into the system in a later objective-defined function. The crook in the arrow on the right is simply for directional purposes.*

Input/Output Symbols on Tips and Tails of Arrows

The reader should review Figures 5-14, 5-15, and 5-16 while reading the following.

Transactional Input

A transaction results in events which act upon something and often transform or transmit that which is operated upon (i.e., the operand). In fact, the operand may be thought of as that which continues to drive the operation after the stimulus has begun the chain of events. For instance, the manufacturing process transforms raw materials into finished products. The transformed inputs, along with the causal input which starts the process, are called "transactional inputs," and are each

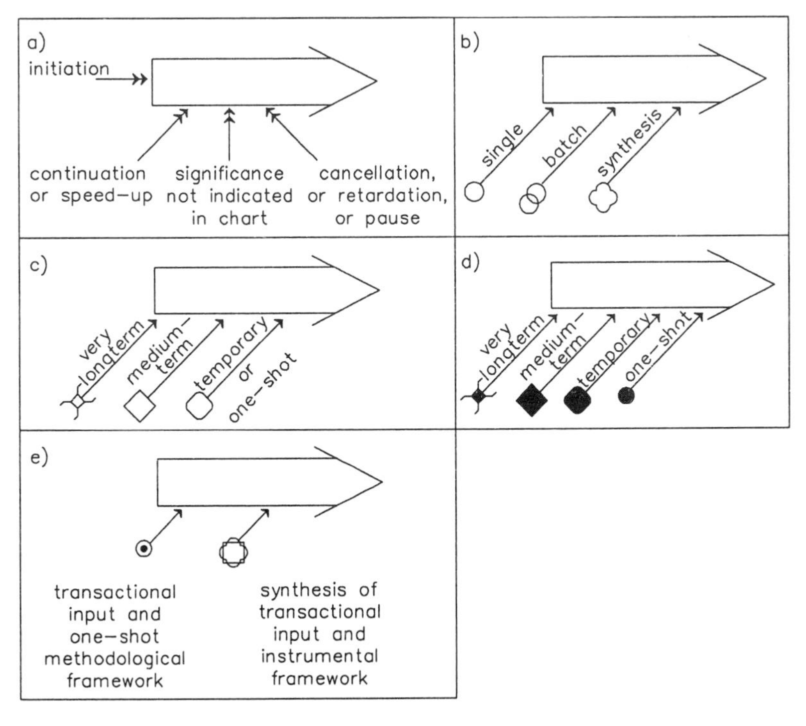

FIGURE 5-14: Examples of the various types of inputs:
a) stimulus inputs;
b) transactional inputs;
c) instrumental framework inputs;
d) methodological framework inputs;
e) hybrids of various types of inputs.

represented by a small circle (see upper left-hand box in Figure 5-15). The causal type of transactional input is differentiated in the structure-flow chart's symbols from the operand type by its entering the function via a double-headed arrow as opposed to a single-headed arrow.

Transactional inputs of the operand type are often transformed or transmitted by the objective-function which inputs them, or by a later function in the chain of events started by the transaction. (In the case of a loop, the later function may be a future run of the inputting function itself.) However, some transactional inputs just serve as operands

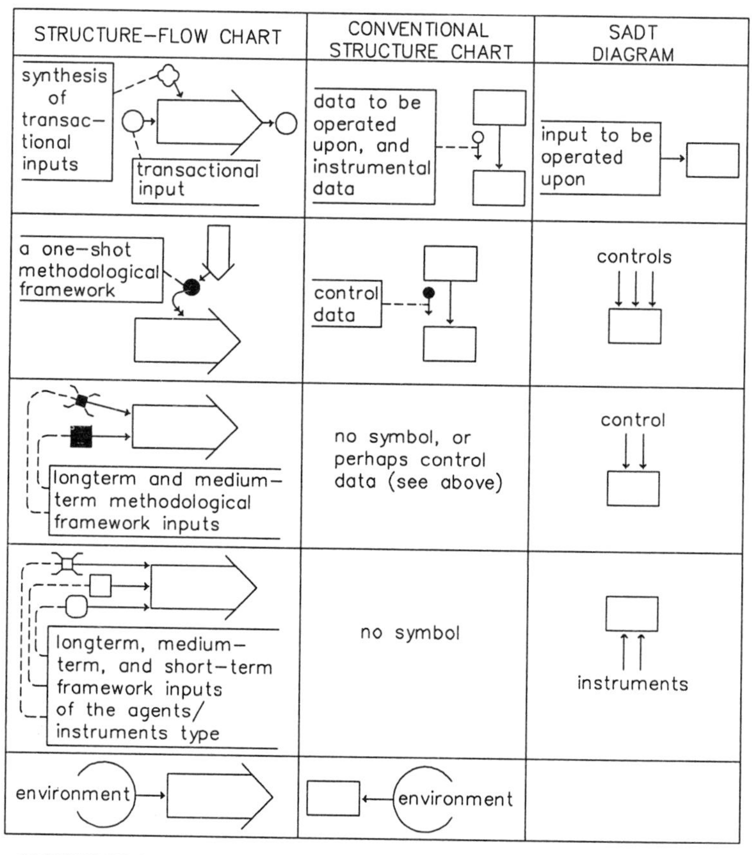

FIGURE 5-15:
Comparison of inputs in three methods which show significance.
In the structure-flow charting method, the symbols affixed to the tails of input arrows may also be attached to the heads of output arrows. However, the symbol chosen for an output depends upon the use made of the output by its destination.

or inputs without being transformed or transmitted; for example, data from which a report is produced may neither be transformed nor transmitted, as it is still intact in the file from which it originated after the report has been produced. Similarly, a customer order is a transactional input to the order entry function, although the customer order is itself not transformed by the order entry process.

Transactional input may be "aggregated," that is, either "batched" or "synthesized." A batch is constituted by a grouping of two or more transactional inputs, while a synthesis is the merging of many transactional inputs to form something which represents them collectively. For example, a total of sales orders is a synthesis of the data about individual sales orders. Depending upon the perspective of the viewer, any single transactional input item may also be seen as a synthesis and/or as a batch. Thus, a customer order for two or more products may be seen as a batch of orders, one order per product, or as a single order. Also, a single order may be seen as a synthesis of order header, order total, and order line item data. These various ideas are captured in the structure-flow charting symbols presented in Figure 5-14b.

Finally, it should be noted that loose usage of the word "transaction" in information systems work often confuses data which represents the transaction with the transaction itself. Rather, such data should be thought of as transactional input.

Frameworks and Framework Input

Framework input tends less than operand transactional input to be transformed or transmitted by the operations which use it. However, one must be careful not to define the difference between transactional and framework input by what are just frequently accompanying characteristics of transactional inputs. For example, transactional inputs, unlike framework ones, are often transformed or transmitted by the operations which use them. However, in some instances, an instrumental framework may not only be depreciated by being used, but may be totally transformed (or consumed), as in the case of fuel being used as an instrument to transport a passenger between locations. Conversely, as has been discussed, instances exist in which operand transactional input is neither transformed nor transmitted. Therefore, the issue of transformation or transmission is separate from the issue of the defining differences between transactional and framework input.

The types of framework inputs are as follows:

- *Methodological* framework input (also called "guidelines" or "control data") is any data used to indicate **how** a particular run of the supplied function is to be achieved, or, in the case of a more long-term methodological framework, how more than one run is to be achieved. Transactions are often accompanied by methodological framework data. For instance, such data may give the source of supply to obtain inputs, or the destination to which a functional output is to be delivered, the quantity and/or quality of the product to be engendered, i.e., the alternative to be selected in this particular run. Thus, as in Figure 5-16, the order entry function in a business sends methodological frameworks for handling of the order to the business baseline. This includes the quantity of each item to be produced, the shipping and billing destinations, special manufacturing and shipping instructions to be followed, etc.

- Framework inputs of the *agent/instrument* variety are the persons or machines who or which carry out functional operations. Some of the framework inputs of this second type are in the nature of data, e.g., an item of data which makes a code more readable to human operators, such as a customer name corresponding to a customer code.
- A framework may be an *environmental entity,* as represented by unclosed circles in the structure-flow chart, e.g., the supplier, the production subsystem, the data environment, etc. This is an external framework, as opposed to the methodological and instrumental framework types.

Methodological framework symbols are similar to those for instrumental ones, the essential difference being that the methodological framework symbols are solid-filled. The degree of circularity of the methodological and instrumental framework input symbols indicates the degree to which the framework input item is stable, i.e., used on a long-term basis. Thus, a highly stable framework item is represented by a symbol which looks a little like a square with its sides pushed in. By pushing out the four depressed sides of this symbol, a square is obtained, which represents a medium stability framework. Pushing out the sides of the square renders a relatively temporary framework, which looks like a square with arched sides.

Finally, in the case of a **methodological** framework, pushing out the sides even further, into a circle, obtains the symbol for a one-shot framework. In the case of a methodological framework, this results in the same symbol as that used in the structure chart for control data. The circle, however, is not used to represent a one-shot **instrumental** frame-

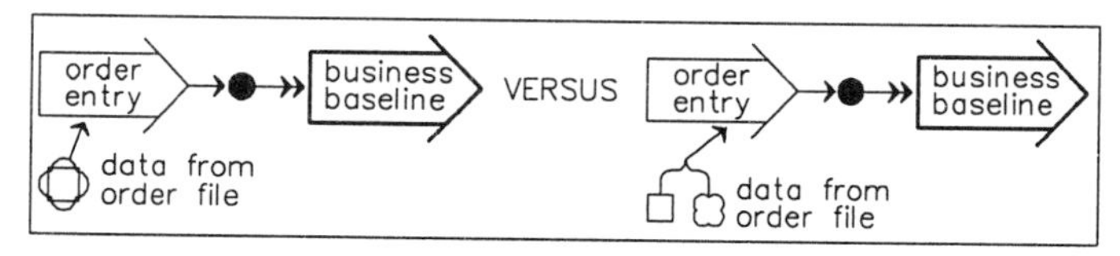

FIGURE 5-16: Alternative representations a hybrid input type.

work, as this results in the same symbol as that for transactional input, which is a different idea from framework input. Instead, the temporary instrumental framework symbol is also used for a one-shot instrumental framework.

Confusion between Transactional and Framework Inputs

Transactional and framework inputs are by definition different from one another, as framework inputs, unlike transactional ones, are neither operands nor causal forces in the functions which use them. However, transactional and framework inputs are often confused with one another. This occurs for the following reasons:

- **First**, in the case of the operand type of transactional input, what is an operand and what is not may be unclear, especially where a framework is of a one-shot durability. When building a wood shelf, the difference in roles played by the saw and the hammer (the instrumental frameworks), on the one hand, and the wood and the nails (the operand type of transactional inputs), on the other, provides a clear example of what is and what is not an operand. However, other examples exist which may lead to confusion in making this distinction. For instance, a sales forecast and a current inventory available quantity are input to a function which decides whether to replenish stock, where they are compared against one another. Since it is only the difference between the two which is of interest in making the purchasing decision, which ones of these, if any, are operands? While, in the mathematical equation, both are operands, only the inventory available quantity is an operand in the purchasing function, as the sales forecast data has merely been established in the purchasing system in order to provide a basis of comparison (i.e., a methodological framework) in making a decision. This is true even if the sales forecast is re-evaluated before every new round of making stock replenishment decisions.

  The difficulty of identifying an operand can be exemplified by an analogy to a metal joint linking the leg of a table to the table top. The joint here is an operand, since it is combined with other operands to achieve the product of the operation. If the joint were to be an instrumental framework, it would neither be combined nor synthesized together with the other ingredients needed to compose the product. In contrast, a piece of wire used to temporarily hold the joint together while the metal joint piece is being attached is an instrumental framework, even if the wire is "consumed" by the operation, i.e., discarded after serving its purpose. This is because the wire never becomes part of the operation's product.

- **Second**, even though framework inputs may not be operated upon by operations which use them as framework inputs, transactional and framework input are often confused because frameworks having a durability beyond the current run may be transformed by specialized framework functions. When a framework is input to be transformed by such a function, it is represented as transactional input, whereas it is represented as a framework on the output end of the framework function. Thus, that which may be seen as framework input by one function in a system may be seen as transactional input by another function, depending upon the use made of the input item by its immediate destination. In deciding how to classify an input/output item, only the use made by the input/output by its immediate destination is considered. Where the exact use by the immediate destination is not known or is multiple—e.g., when the item is passed to a reservoir whence it is sent in multiple directions—no particular input/output symbol should be employed; e.g., see item #10 in later figures of the supply acquisition system (Chapter Six).
- **Third**, confusion between transactional and framework input may happen because a single entity may actually combine both framework and transactional input ideas; that is, input/output symbols can be of a hybrid nature. Thus, Figure 5-16 combines a synthesis of transactional inputs symbol with a medium stability instrumental framework one. This, for instance, could be used to represent the data in the customer open order file, in which the customer code is an instrumental framework input, while the collective details of a backorder are the transactional input synthesis. The hybrid idea may also be represented as simultaneous inputs/outputs, as also seen in Figure 5-16.

Despite the ease of confusion of the ideas of transactional and framework input, and despite the possibility of some grey areas between the two, it would seem to this author that inputs to an operation can usually be classified into either one or the other.

FURTHER DETAILS ON OUTPUT SIGNIFICANCES

Symbiotic, Offshoot, and Competitive Outputs

Symbiotic Output

The nature of symbiotic outputs, as opposed to that of objectives, will now be explored. Figure 5-17a shows part of the functional breakdown of an order entry system, consisting of baseline and validating functions. This diagram also demonstrates part of the functional breakdown of a purchasing system, in which only the baseline and initiative coordination functions appear. A bilateral exchange of data between the decision-making functions of the two systems is shown in the figure. **On the one hand**, order entry receives data from the purchasing control function on what inventory is available to fill customer orders. **On the other hand**, purchasing receives data from order entry on the amount of available stock committed to new customer orders. Actually, the exchange between the two systems passes via a reservoir, namely, the inventory master file, although this is not indicated in the diagram. If the two systems are viewed as having a cooperative relationship with one another, then, insofar as the outputs in question are not the objectives of the control functions producing them, these are understood to be symbiotic outputs. The symbiotic outputs do not have to flow in both directions, although the present example works this way.

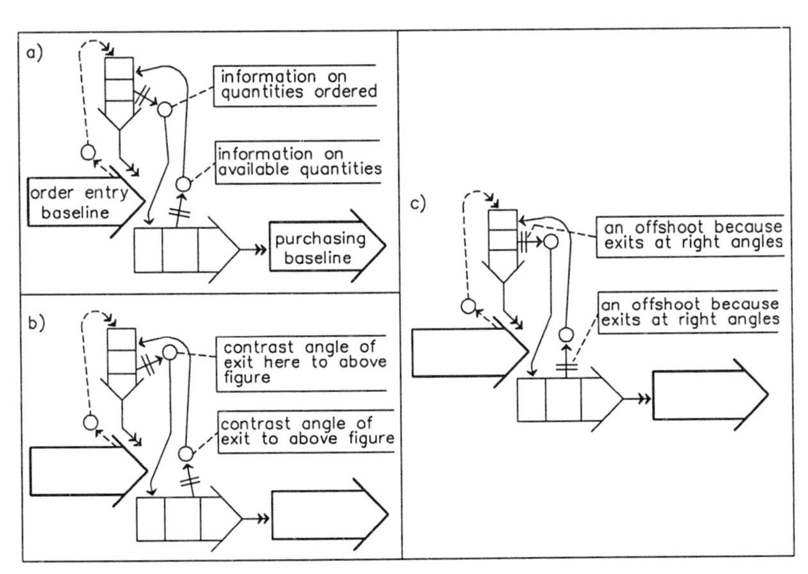

FIGURE 5-17: Symbiotic, competitive, and off-shoot relationships. *The various exchanges between different functions represented by an output arrow with a double bar on its tail:*

a) symbiotic relationships;
b) competitive relationships;
c) off-shoot relationships, between functions.

Competitive and Off-Shoot Output

If the two systems are viewed as being in competition with one another, the outputs would be shown as in Figure 5-17b, where data from one system would be extracted by the other system, e.g., by espionage or by a raid. In cases where the systems are seen neither in a cooperative relationship, nor in a competitive one, the outputs would be diagrammed as in Figure 5-17c, as offshoot outputs, having no particular significance to the sending system.

Symbiotic Outputs in Information Systems

The various types of non-objective output mentioned above can be produced by any class of system function. However, symbiotic outputs are usually associated with informational (decision-making) functions. The latter not only make decisions needed to plan, and coordinate and validate operations within their scope of authority, but also tell other functions, in other authority chains in the levels-of-control chart, what has been decided.

In information systems work, it is common to see the symbiotic outputs of a system erroneously referred to as "system objectives." For instance, the information system for accounts receivable in a business monitors whether the receipt is correct and whether the bank deposit has occurred correctly. Following valid performance of these operations being determined by the information system, general ledger postings may be made by the information system. The G/L (general ledger) interface is a **symbiotic output**, not an **objective output**, of the accounts receivable control system. The objective of this system is not to provide any and all information about cash receipts for any purpose whatsoever, but to authorize deposit of payment and to confirm successful completion of the cash receipts activity. The fact that the cash receipts information system may in some companies *organize* the G/L data before transmitting it—e.g., by producing a batch total to accompany the details—does not imply that this data is a system **objective**. It simply implies that the cash receipts information system is being "courteous" to the general accounting system, to use a humanizing term.

More Examples

Here are some examples of offshoot, symbiotic, and competitive, output types. The objective of a driver is to control the operation of driving himself and his passenger to work. In doing so, the driver takes cognizance of his geographical location, and communicates this to his passenger. Depending upon its nature, this communication could be one of the three non-objective output types, as in the follow instances:

- At one point, the driver points out to his passenger a good place to go shopping. This communication to the passenger would be an offshoot output if it has nothing either directly or indirectly to do with the objective of controlling the drive to work.
- However, maintaining a pleasant relationship with the passenger may have an effect on **future** successful accomplishments of trips with this same passenger, from whom the driver receives a fee. If pointing out a good place to go shopping contributes to this end, it would be classified as a symbiotic output.
- On the other hand, if the passenger is very offensive, the driver may wish to discourage him from taking future rides. Therefore, the passenger must extract information about the shopping location from the annoyed driver. The latter is an extracted, or competitive output. Like symbiotic output, this has an effect on future rides to work.

Outputs Returning Via a Loop

Reservoir Along the Line of Flow

One variety of interaction occurs via a reservoir along the line of flow of an objective-defined function. In this case, the output of one step is placed in a reservoir, the contents of which are built up gradually, by an iterative process, or by a single iteration. The contents of the reservoir are then drawn down, iteratively or not, by a later flow function in the same run of the objective-defined function. An output returning as an input in this manner is symbolized by a squiggly line for the small arrow shaft, as in Figure 5-18a. An example of this is the build-up of a temporary transaction file along the line of flow of a batch processing computer information system.

Run-to-Run Recursive Interactions

A second variety of interaction involving a loop occurs between different runs of the same objective-defined function, wherein the output of one run becomes the **transactional** input of the next run. This is recursive interaction. For instance, a total may be accumulated from run to run, as in Figure 5-18b. Also, as in Figure 5-18c, the output of one run may be operated upon by the next run in a recursive process which gradually refines the output over many runs.

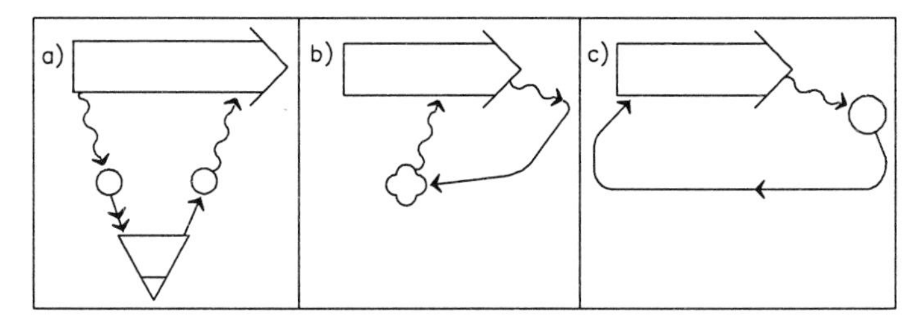

FIGURE 5-18: Relationships between the function and itself.

a) *squiggly arrows showing relationship within same run, in which the output of one step of the objective-defined function is collected in a reservoir, from which it returns in a subsequent step of the same functional run;*

b) *run-to-run relationship shown by squiggly arrow, in which the output of one run of an objective-defined function returns as an input in a later run. In this example, a total is being accumulated from run to run.*

c) *run-to-run relationship shown by squiggly arrow, intended to show a recursive relationship between different runs of the same objective-defined function, in which the objective output of one run becomes the transactional input to the next run.*

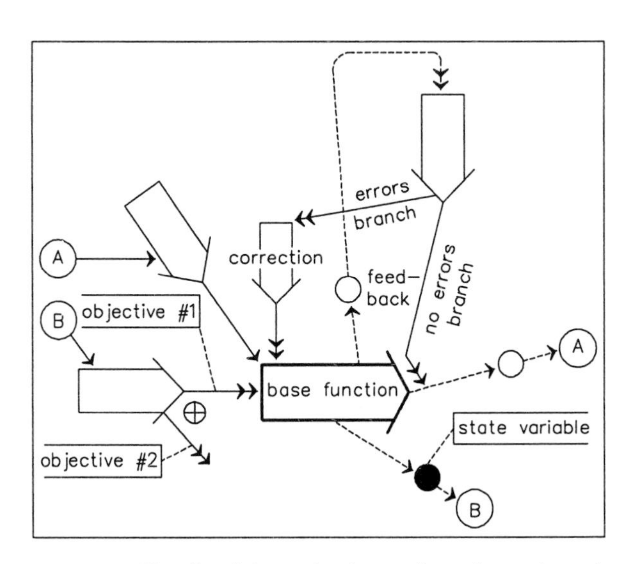

FIGURE 5-19: Feedback loops having various time-orientations. *A feedback loop is begun by each of the dashed arrows leaving the base wideshafted arrow. Those linked by on-page connectors A and B are future-oriented feedback cycles, while the other loop affects the current run.*

Feedback Loop

A feedback loop, originated from a function, returns to the originating function, perhaps after having passed via one or more decision-making functions which may change the course of action in the originating function. The output starting the loop consists of data about what has transpired during functional operation. This data may be picked up by the decision-making function, which then initiates action to affect the behaviour of the originating function. The decision-making function may be part of either a current run ensurance feedback loop, or part of a loop affecting future runs of the originating function.

In an ensurance loop, of which one is shown in Figure 5-19, the data is handled by a validating function, which checks the effectiveness of functional execution to the point of feedback occurrence in the base function originating the feedback loop. If no errors have occurred, the output of the validating function is the completion or continuation stimulus. Otherwise, a return loop is needed, leading back, perhaps via another function to perform corrections, to the originating base function.

Also seen in Figure 5-19 are two future-oriented feedback loops. One, linked by on-page connector A, passes via planning operations which establish long-term frameworks for future operation of the originating base function. The second loop, linked by on-page connector B, affects the initiation of a future run of the originating function. A specific case of the type of logic portrayed by on-page connector B occurs where an objective-defined function has a number of different mutually exclusive objectives, where the objective achieved in one run may determine the objective to be achieved in the next run, as in a round robin. In this, each objective takes its turn to be achieved in a particular, pre-defined sequence, based on the value of a so-called "state variable," to reflect the state of the system as of the previous run. This is reset each run, and input to the next run. This state variable, a one-shot methodological framework, is seen in Figure 5-19.

The reader should not fail to note that the recursive relationships in Figures 5-18b and c are **not** future-oriented feedback ones. However, in these figures, if the item shown returning to the originating function were a methodological framework instead of transactional input, this would be a future-oriented feedback relationship rather than a recursive one.

BASIC FUNCTIONAL INTERACTIONS IN THE CHART

INTRODUCTION

The following shows how various kinds of "first-order" interactions may be charted using the structure-flow chart. These interactions are the ones which occur

- **directly** between baseline and adaptation operations;
- or, between adaptation operations and those framework adaptation operations (called the "basic" framework operations) which are in direct contact with the baseline operations;
- or, between decision-making functions and the decision-implementing functions which they externally control, provided that the decision-implementing functions are those which are in direct contact with either the baseline or the basic framework operations.

Stated in another way, "first order" interactions exclude secondary mutual relationships among the adaptation objective-defined functions, in which they enable, direct, and support **one another**, as opposed to either the first-order (basic) framework or the baseline operations. The excluded types of relationship are to be discussed in Chapter Seven, in connection with the subject of levels of control in the structure-flow chart. In still other words, using terms which will be explained later, first-order interactions are interactions between baseline or basic framework operations and adaptation operations which are "zero degrees removed" from the "operating core" (baseline and basic framework operations). Figures 5-20a and 5-20b represent the first-order interactions of the classes of objective-defined functions.

Following are a number of illustrations, involving current adaptation operations, of the different time-orientations of first-order interaction

<u>Example #1—Initiative Coordination (Figure 5-21a)</u>

1. **Situation Input Step**: Obtain information needed to determine whether purchases are needed to replenish inventory levels, and detect shortage situations.
2. **Decision Step**: Decide to purchase particular inventory supplies in particular ways.
3. **Decision Communication Step**: Generate and send purchase orders for given quantities of particular items from specific vendors.

Example #2—Interruptive Coordination (Figure 5-21b)

1. **Situation Input Step**: Obtain information needed to determine whether the customer has requested modification of a late customer order not yet delivered, and detect situations in which change is needed.
2. **Decision Step**: Decide what revisions to make to the customer order, and what action to take to pacify the discontented customer.
3. **Decision Communication Step**: Instruct a procuration function to handle the customer; or, change the customer order.

Example #3—Interruptive Procuration (Figure 5-21b)

1. **Situation Receipt Step**: Receive information on required changes to customer order (see example #2 preceding) from coordination operations.
2. **Processing Step**: Prepare steps for pacifying customer.
3. **Correction Implementation Step**: Pacify customer.

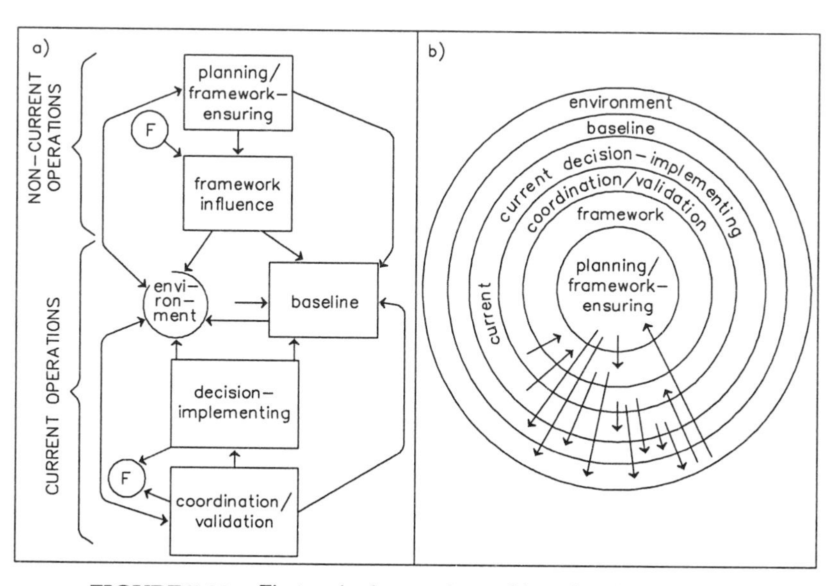

FIGURE 5-20: First-order interactions of functional classes.
a) chart format #1; Note: On-page connector "F" includes the basic framework operations in the system "operating core," along with the baseline ones;
b) chart format #2, using circles within circles.

Example #4—Validation (Figure 5-21c)

1. **Situation Input Step**: Receive information needed to detect error in previous step, or, alternatively, to detect lack of error in previous step, and determine situation.
2. **Decision Step**: Decide what to do about situation.
3. **Decision Communication Step**: Return execution to the next step of the base function being validated, if no error, or instruct the appropriate correction function to correct error in previous step.

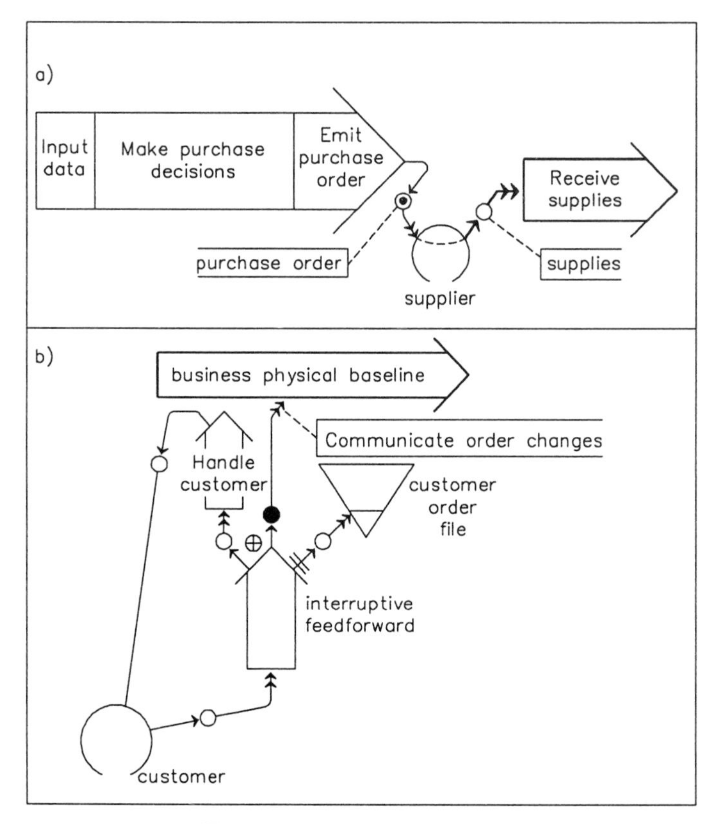

FIGURE 5-21: The different time-orientations of interaction.
a) example of initiative (feedforward) interaction;
b) example of interruptive (feedforward) interaction;
c) facing page
d) facing page

Example #5—Ensurance Correction (Figure 5-21c)

1. **Situation Receipt Step**: Receive information on error in previous step (see example #4 preceding), from validation operations.
2. **Processing Step**: Prepare corrective action.
3. **Correction Implementation Step**: Implement correction.

Example #6—Interruptive Coordination (Figure 5-21d)

1. **Situation Input Step**: Obtain information needed to detect the need to cancel the customer order.
2. **Decision Step**: Decide what to do about cancellation needed.
3. **Decision Communication Step**: Stop the processing of the customer order, if the decision is to cancel.

In examples two and four (above), the mere communication of the decision by the decision communication step of the decision-making function is insufficient to achieve the desired effect. Therefore, implementation requires action by current decision-**implementing** functions.

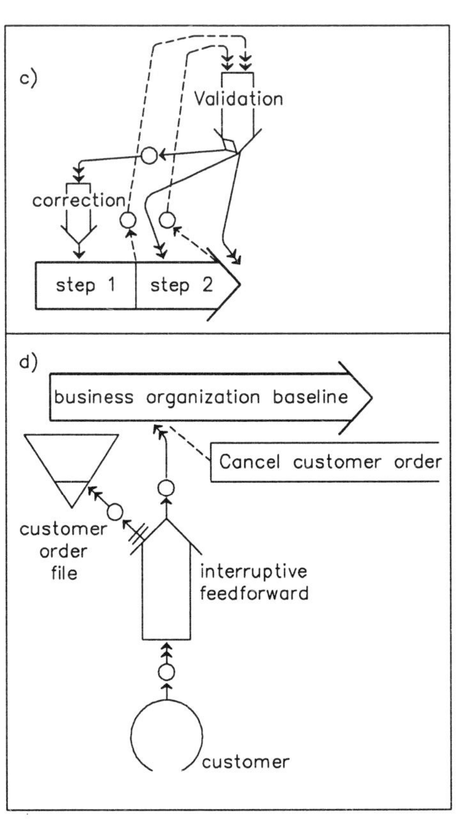

FIGURE 5-21 c and d:
c) example of ensurance time-orientation;
d) example of interruptive time-orientation leading to cancellation.

CLASSIFICATION OF FIRST-ORDER INTERACTIONS

The following scheme may be used to classify patterns of interactions between objective-defined functions in different classes. The subject of patterns of interaction is further discussed in the supplement to Chapter Eleven (the complete comprehension of which may also require that the reader has completed Chapter Eleven).

Initiative and Speed-up Feedforward Interaction

One important variety of first-order interaction is that which initiates or speeds up operations in a base objective-defined function being controlled. This involves a coordination function and its affected base function. As in Figure 5-22, the coordination function transmits a stimulus, perhaps along with transactional input, to the base objective-defined function.

Initiative feedforward interaction may also coordinate the arrival of supplies needed in the subsequent running of the base function. If the supplies are assumed to be accumulated in reservoirs, the coordination function, as in Figure 5-22, contacts the reservoir control functions to bring about the input of either transactional or framework supplies. Similarly, changes in supply arrival schedules may be required when speed-up feedforward is applied. This would be diagrammed as in Figure 5-23a. Transactional inputs (supplies) obtained from other functions or systems may not be accumulated in reservoirs. In this case, the supply-providing functions are contacted directly, rather than by contacting the reservoir control operations. Figure 5-23b shows the obtaining of supplies on this basis from external sources.

As noted in the comments in Figures 5-22 and 5-23, contacting the reservoir operations and/or external entities to signal that supplies are needed is, for purposes of simplification, usually not shown in the structure-flow chart.

Feedback Followed by Continuation or Rectification

Feedback interaction, a second type of first-order interaction, begins when information is sent to a validation function concerning what has happened in the preceding step of the base objective-defined function being coordinated/validated. This information is in the form of transactional input. The arrival of the transactional input at the input flow function of the validation function is accompanied by a stimulus which initiates its execution.

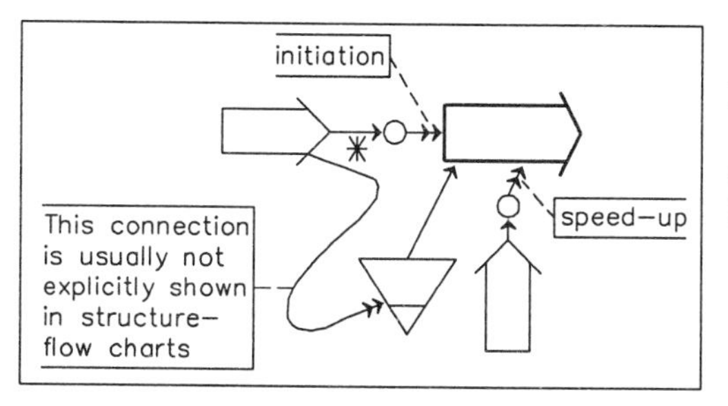

FIGURE 5-22: Initiation and speed-up interaction.

If no problem is found by the validation function with the execution of the part of the base objective-defined function which it validates, execution of the next step, or successful completion of the base function (if there is no next step), is initiated by the validation function. The stimulus sent to the base function when no problem has been discovered may be accompanied by guidelines (one-shot methodological frameworks) as to how the next step is to be executed.

If, on the other hand, problems are detected by the validation function, the stimulus may either be used to cause repetition of the operations which executed in error (iteration to redo), or it may be fed forward to a decision-implementing (correction and/or procuration) objective-defined function, instead of to the base function. These decision-implementing current function operations, demonstrated in Figures 5-24a and 5-24b, are initiated by the stimuli in question. Each stimulus may be accompanied by guidelines and/or by information in the form of transactional input, transmitted from the validation to the

FIGURE 5-23: Interaction to obtain supplies.
This shows a facet of initiative or speed-up interaction. Examples are given of the following:
a) interaction to speed-up arrival of supplies from internal reservoir;
b) interaction to initiate and obtain supplies from external source.

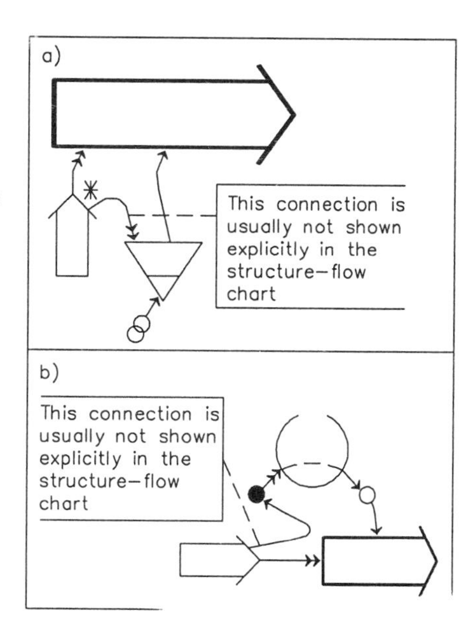

decision-implementing objective-defined functions, providing the latter with what they need to know to solve the problem in question.

Different Ways of Implementing Corrections

Implementation of corrections may simply be achieved by the validation objective-defined function stimulating rerun of the operations which executed with error. Alternatively, decision-implementing functions may be invoked, as in Figures 5-24.

First, in Figure 5-24a, mutually exclusive options are represented for the correction-implementing function. These assume that this function can act in different ways, depending upon the circumstances. These options are as follows:

- Restart execution of the step, but not at its beginning. When restarting, provide the rerun segment of operations with guidelines, as the case may be.
- Correct the error within the correction function itself, and then go to the end of the corrected step of the base function for a check that all is now O.K. This check is done via the feedback from the base function.
- Take either one of the above two paths. This is represented by the non-specific control arrow at the right-angles to the base objective-defined function.

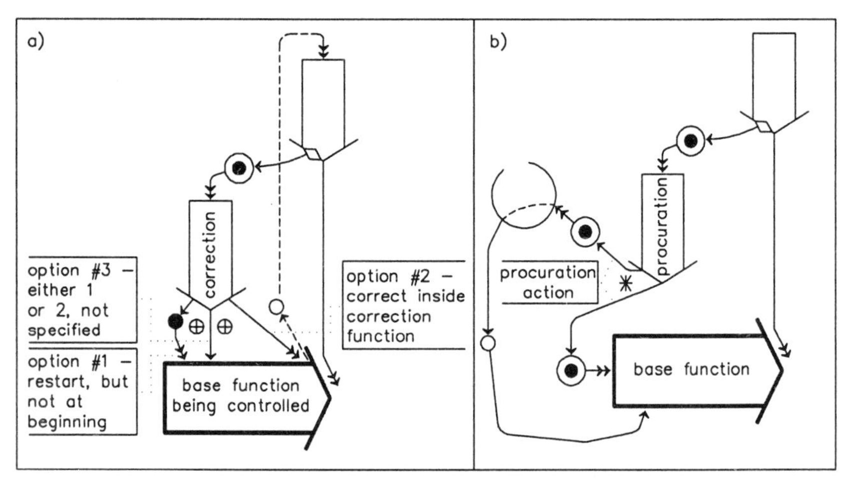

FIGURE 5-24: Interaction to ensure completion of the base function.
a) different ways in which decision-implementing current operations can correct the situation;
b) going outside the system to procure rectification.

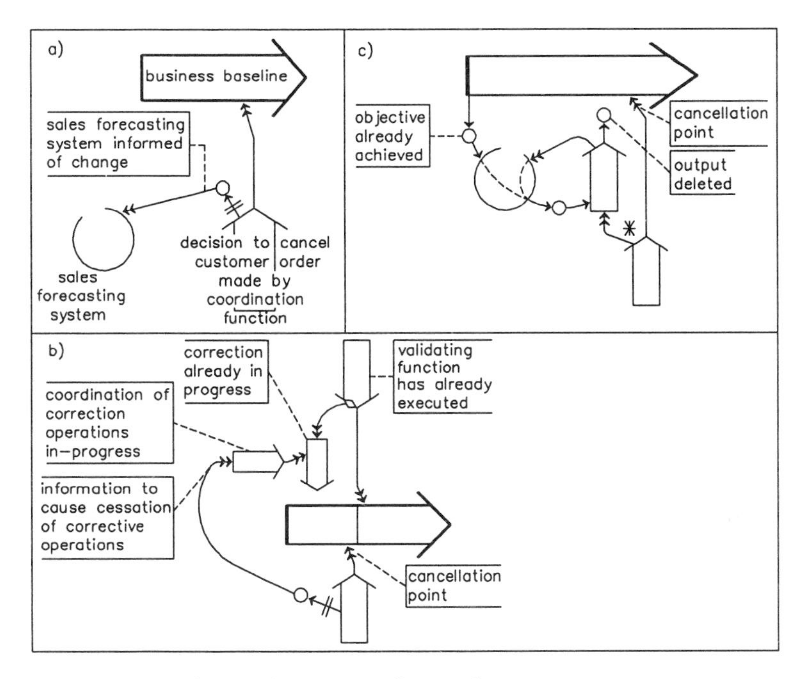

FIGURE 5-25: Interaction to cancel operations.

a) ***Example #1:*** *A customer order is cancelled or its delivery date delayed. However, the sales "actuals" figures have already been updated in another system based on this order. The sales forecasting information system has to be informed of the change.*

b) ***Example #2:*** *At the time of cancellation, some of the feedback of the base function has already been made use of by other objective-defined functions. For example, correction of an error in previous operation of the base function has already begun. Further operation of the correction function must be stopped. To this end, information is transmitted by the cancelling coordination function to other cancelling coordination functions. The latter are internal functions, local to the function whose operations are being cancelled.*

c) ***Example #3:*** *Some of the objectives of the base objective-defined function have already been achieved at time of cancellation, and delivery of these outputs made elsewhere inside or outside the system. Return of these outputs may then be obtained by procuration operations. This involves the role of a decision-implementing function in a feedforward relationship; that is, special action is needed, over and above sending information to another system or function.*

Second, in Figure 5-24b, two parallel objectives are shown for the procuration function. The latter decision-implementing function causes something to happen in the system environment, and, in parallel, initiates a rerun of the base function step monitored for problems by the validation function. For instance, invalid input from the system environment could be handled by the procuration operations, thus allowing the rerun to complete successfully.

Cancellation, Slow-down, and Speed-Up Feedforward

Cancellation or slow-down (see Figures 5-25), a third type of first-order interaction, involves stopping or slowing down the operations in a particular run of the base function being controlled. This is achieved by transmitting a stimulus, possibly accompanied by guidelines, from a coordination function to a base function being controlled by it. However, more may be involved than simply stopping or slowing down operations in the base function, since effects outside the base function caused by execution of the base function to this point may have to be undone. These effects may have occurred in other systems, which may have to be informed; or they may have taken place in other functions of the system in which the base function being controlled operates. In the latter case, the other operations in question have to be informed; for example, reservoir management operations have to be informed that scheduled deliveries of input have to be withheld. Similar considerations apply to speed-up feedforward (see Figure 5-26).

Step Insertion Feedforward Interaction

In a fourth type of first-order interaction, feedforward may involve the insertion of steps prior to executing any step, including the first one, in a base objective-defined function. This may involve changes to guidelines and/or transactional inputs and/or framework inputs used in the next step. The need for such change may be detected by planning or

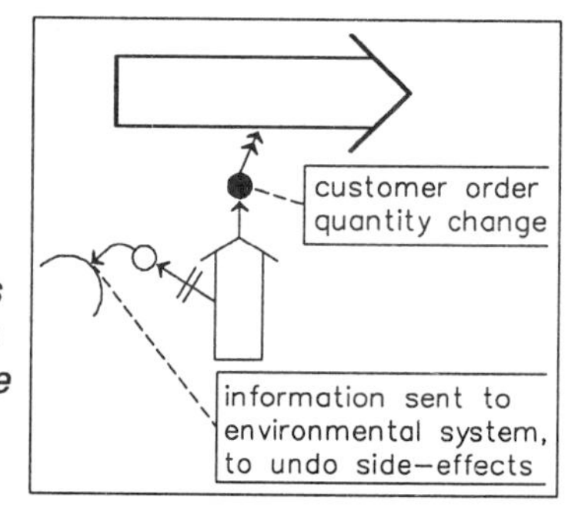

FIGURE 5-26:
Example of undoing side-effects in speed-up interaction.
This example shows speed-up in the form of moving up of the required order delivery date and/or the increase in order quantity. This example also shows side-effects to be revised, and is the reverse of the above example, Figure 5-25a.

coordination operations. The change may be made in the following ways:

By planning via the combined efforts of coordination and framework operations: Where planning operations detect a needed modification to frameworks in advance of the next step, as in Figure 5-27a, coordination operations postpone execution of the next step until the framework changes needed in subsequent steps have been prepared and implemented.

By coordination in conjunction with the prevention or procuration operations: Here it is not a matter of informing some other function, as in #1 above, but of taking advance action which would not otherwise be taken, even if some other function were informed. Thus, as in Figure 5-27b, transactional inputs coming from outside the system will be delayed unless special assistance in delivering these is provided to the external system originating these; i.e., since it is not simply a matter of informing the external system that inputs will be needed in the next step, decision-implementing functions are needed to carry out the decision.

Environmental Equilibrium Interaction

A fifth type of first-order interaction may be represented with the structure-flow chart by a wideshafted arrow **not** pointing towards some other wideshafted arrow. This includes both baseline and adaptation functions.

Environmental equilibrium interactions often lead to other, sometimes unexpected, multiple-order environment-linked interactions. For example, a company whose baseline operations produce consumer kitchen products sells a line of toasters with a guarantee. Certain toasters sold come back to the company because they have factory defects. The company response is first handled by initiative coordination operations, following which baseline operations may be initiated to produce a replacement toaster for a dissatisfied customer. All these operations form part of an environment-linked interaction chain.

Baseline involvement — In an open system, the baseline functions may interact with the environment in such a way that equilibrium is restored. A stimulus from the environment applied directly to the baseline function, or to a feedforward function initiating baseline execution, leads to execution of the baseline, which eventually restores equilibrium with the environment by delivering requested outputs to the environment. To illustrate, a customer order stimulates execution of the business baseline, which eventually delivers goods to the customer.

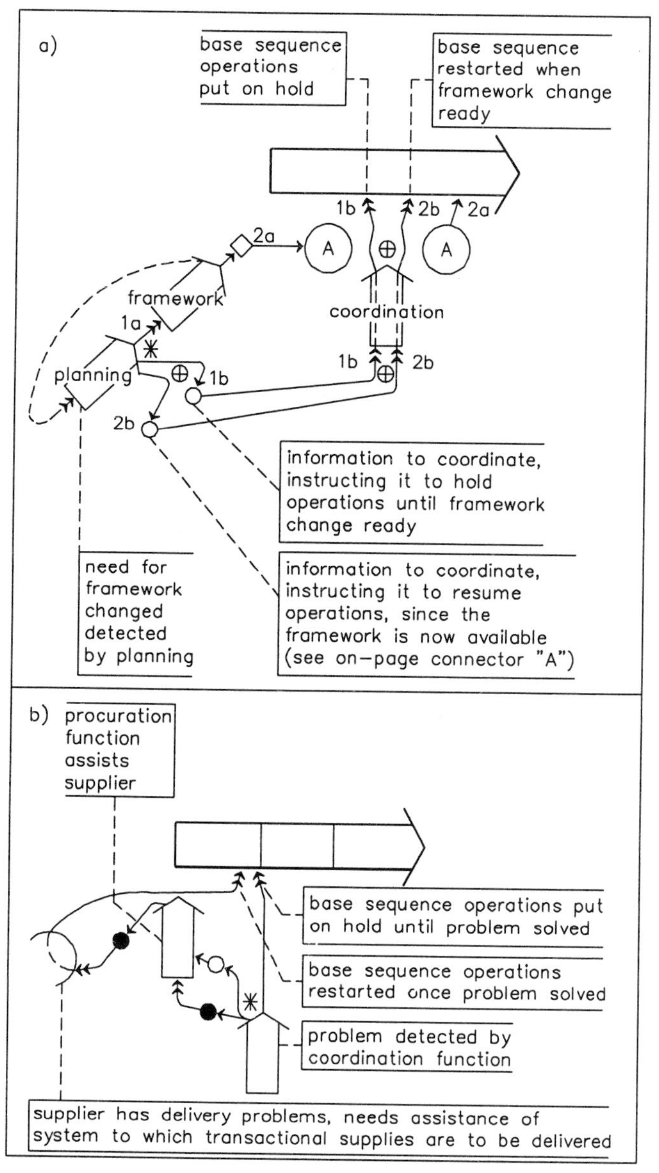

FIGURE 5-27: Step insertion feedforward interaction examples.

a) *1a and 1b are first output in parallel from the planning function. 1b leads to action 1b by the coordination function, to put operations on hold. 2a and 2b are produced after placing the operations on hold. 2b is the continuation message eventually passed along to the base function. 2a is the framework to be used in the second half of the base function's run.*

b) *Step insertion feedforward involving only current operations.*

In this way, equilibrium between the business and the customer may be restored.

Adaptation functions pointing away from baseline — So far, as structure-flow charts have been explained, all wideshafted arrows, except those representing baseline functions, point towards other ones. This includes framework influence and current procuration operations, which need not point out of the system, depending upon their usage in maintaining equilibrium with the system environment, as is to be described. In contrast, Figure 5-28a demonstrates certain functions which point out of the system, rather than towards another objective-defined function. This type of representation applies to those planning and framework functions which maintain a certain general "equilibrium" with the system environment, but which do not interact in either a direct or determinate way with other objective-defined functions of the system.

Examples of Environmental Equilibrium Interaction

For example, operations defending the borderline of the system against invasion by other systems are framework upkeep operations pointing out the system at a slant to the baseline. Here, as in Figure 5-28b, the attack is pushed back by framework upkeep operations, and by a long-lasting barricade erected by the framework add/change/delete operations.

Another illustration is accounts payable and receivable operations in a business organization. Based on one view of such a system (see Chapter Two), accounting operations are seen as maintaining a certain energy interchange equilibrium with the system environment. Here, the framework of system relationships with environmental entities must be kept up, first by compensating other systems which provide this system with supplies (inputs), and second, by obtaining similar compensation from other systems, where this system is a supplier of these other systems. This view of the business organization regards payables and receivables as framework upkeep functions (while another view, in which profit is the business mission, sees them as baseline functions). The framework view is illustrated in Figure 5-28c.

The model of accounting operations just mentioned views these operations as framework upkeep ones having a **current** time-orientation. Although these accounting operations do not directly affect the baseline or any other system functions, the signal for their execution is generated on each run of the baseline; that is, they are in regularized interaction with the baseline operations. Because of their being executed in connection with every run of the baseline, they are also current operations, pointing out of the system **at right angles** to the baseline operations.

Link Forward

This discussion of the structure-flow chart will be continued in Chapter Six with examples of the supply acquisition system.

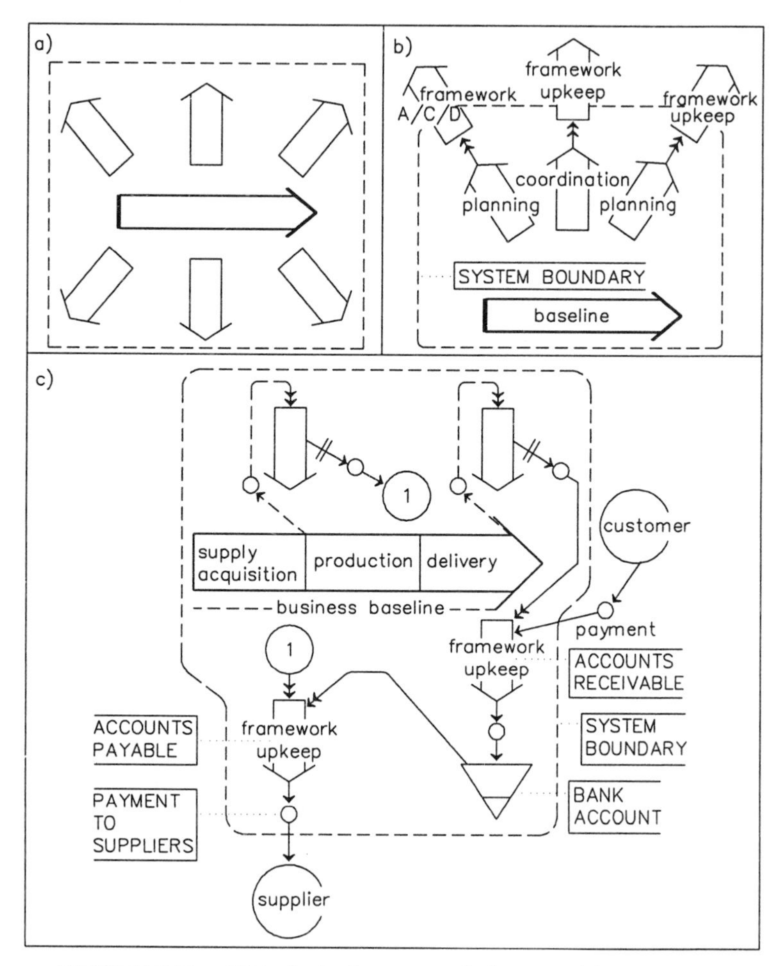

FIGURE 5-28: Wideshafted arrows pointing out of the system.
a) different directions in which pointing may occur;
b) repulsion of attack from outside the system;
c) environmental equilibrium interaction via framework upkeep operations.

CHAPTER SIX

STRUCTURE-FLOW CHARTING OF THE SUPPLY ACQUISITION SYSTEM

Levels of System Development to be Covered

The model of the business organization sets the stage for developing the supply acquisition information subsystem, the system selected here to demonstrate system development using the structure-flow chart. A number of finer points and grey areas not covered in Chapter Five will be discussed as this case unfolds. Hopefully, their explanation will be clearer within the framework of a case. Moreover, taking advantage of the context of a case, this chapter touches upon certain new ideas, which will be discussed in greater length in later chapters. With respect to this, the reader is warned that certain ideas may seemingly "pop up out of nowhere" as the case is discussed.

The supply acquisition information subsystem is developed using a top-down methodology going from preliminary analysis to the middle or even later levels of this strategic level of system development. With each step, more and more choices are incorporated as constraints into the model of the supply acquisition information subsystem.

In the first step, the most salient general features of the supply acquisition system are developed. Step two further extends the umbrella of informational functions and data files in the system, including both internal data files, and files maintained in buffer subsystems of the business organization. Step three views the details provided by step two from the perspective of the supply acquisition **information** subsystem only. This step also discusses which operations may be computerized, thereby beginning the logical design (analysis/design) of the new subsystem.

It should be noted that it is mainly the current functions of the supply acquisition information subsystem which are discussed in this case; i.e., it is mainly the operational control level information subsystem that is dealt with. Two major non-current areas, however, exist in the information subsystem, namely, the long-term planning of supply acquisition facilities and methods, and the long-term material requirements planning, based on sales forecasting data. While the basics of the latter, the MRP activity, are included within the umbrella of functional elements at a more detailed level of system presentation, the planning of warehouse facilities and methods is hardly mentioned. This oversight is merely because the discussion of warehouse planning would probably require a chapter of its own if it were to be discussed.

Concurrently with strategic analysis, the broader system environment must be studied to determine how it affects system development. This concurrent study is introduced toward the end of the second section of this chapter. The third section of this chapter then takes the development process into the level of detail which interfaces analysis and design (called the "analysis/design" stage). Here, each of the internal information sub-subsystems of the supply acquisition system is developed in more detail, including the addition of certain non-current functions, of new control levels, and of scheduling constraints.

Demonstration of the System Development Process Concepts

By the end of the development of the supply acquisition information subsystem, the reader notices that certain dependencies in the model result from constraints selected in the model of the business organization. In other words, the analysis level, like the design one, actually relies on certain hypotheses, since there seems to be no such thing as a conceptual model free of **environmental** (i.e., logical) constraints. The best that any analyst or designer can do is either to suggest that certain constraints be removed from the previous level of system (in whose context the system under consideration is a subsystem), or, given the established constraints, develop a model as free as possible of any **further** constraints.

Detail within the supply acquisition system is added in the sense of expanding the umbrella of system elements and incorporation of physical constraints. However, little detail is added in the sense of explosion. That is, the analyst's perception of functional elements is similar throughout the development, moving from preliminary analysis to analysis/design (the area between analysis and design), since it is only once into the strictly design level that explosion with the structure-flow chart becomes needed and appropriate.

FIRST LEVEL ANALYSIS OF THE SUPPLY ACQUISITION SYSTEM

Main Line System Functions

Figure 6-1 is a high-level representation of the supply acquisition system using the structure-flow chart. Based on the model of a manufacturer, the supply acquisition system is a physical-base subsystem of the business organization, because its base activities process physically tangible supplies, as represented by #17 in the figure. The supply acquisition system objectives are also physical, as identified by the output labelled #23 in Figure 6-1. Number 23 represents the raw material and components parts made available to the next subsystem of the business, be this next subsystem the warehouse or the production subsystem.

The baseline function is not the only one of the supply acquisition system needed to fulfill the system essence-identity (mission plus boundaries). Various other functions, classified as "adaptation" ones, are of course needed. The latter include not only decision-making functions, but also decision-implementing ones needed to implement decisions when mere communication of the decision is insufficient to cause decision implementation. Figure 6-1 shows only one of these so-called "adaptation" functions, namely, a decision-making function which initiates the system baseline operations, and which together with the baseline forms the "main line" group of objective-defined functions of the system.

The "initiative (feedforward) coordination" function, as it is called, decides whether, and in what way, to issue new orders for purchase of physical supplies from suppliers. These physical supplies, such as raw materials and component parts in a manufacturer, are ordered from suppliers by some form of communications medium, such as a purchase order document. The initiative function emits purchase orders, and forwards them to suppliers. In response to such purchase orders, suppliers are expected to forward supplies to the input operations of the supply acquisition baseline function. The initiation of the baseline passes from the initiative function, through the supplier environment, back to the baseline. Thus, the system initiation occurs indirectly, via the environment.

SYMBOLS USED IN THE STRUCTURE-FLOW CHART

Operations and Environmental Entity Symbols

Wideshafted Arrow

The single adaptation wideshafted arrow in Figure 6-1 is aligned in the same direction as the baseline function's wideshafted arrow, and to the

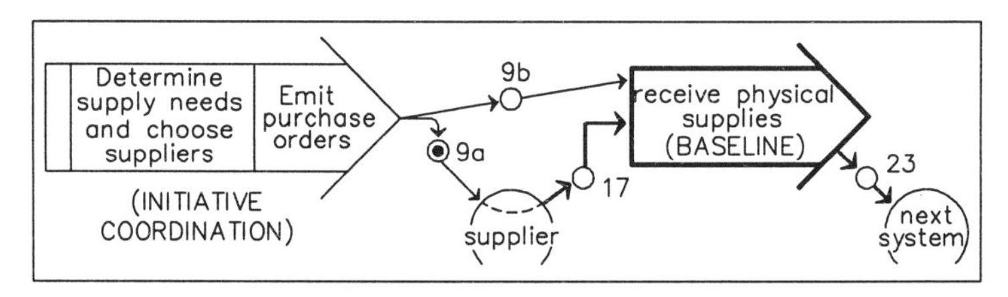

FIGURE 6-1:
Simplistic representation of the supply acquisition system.
The numbering scheme used to identify inputs and outputs is the same as that to be used on the more detailed versions of diagrams of this system. The three items identified in this very general figure are as follows:

- 9a: purchase order sent to supplier;
- 9b: purchase order receiving copy sent to baseline;
- 17: physical supplies from supplier delivered by the receiving operations.
- 23: physical supplies transmitted by the receiving operations to the next system, be it the warehouse or production.

left of the baseline, both arrows going from left to right. This type of alignment shows a sequential relationship of the two objective-defined functions. As well, it immediately indicates that the adaptation function, if the wideshafted arrow on the left is one, has an initiative role. However, if this were a two-baseline system (see later in this chapter), an objective-defined function in this left position in the chart might be another baseline function, rather than an adaptation one. To avoid ambiguity in this figure, as well as in all figures in this book, a baseline function is highlighted by the heavier trace line used for its wideshafted arrow.

Unclosed (Open) Large Circle

An open circle represents an environmental entity. In Figure 6-1, one of the unclosed circles represents the supplier. This environmental entity is not only external to the supply acquisition system, but is also external to the business organization. The second unclosed circle in the diagram represents the system to which the physical supplies are delivered after being received by the supply acquisition system. This destination subsystem of the business will be either the supplies warehouse or the production system (see the model of the business organization in Chapter Two). No choice is made at this level of the analysis as to what the next system is to be. However, this next system would be known if the unclosed circle had an inverted triangle, the reservoir symbol, enclosed within its boundaries as this would indicate a buffer system such as the supplies warehouse or a data file. No designation of a buffer system is in Figure 6-1.

Symbols on Tips/Tails of Small Arrows

Small Circle

Four small circles in Figure 6-1, referenced by numbers 9a, 9b, 17, and 23, represent "transactional" input, as follows:

- Number 17 represents the physical supplies, e.g., raw materials and component parts coming from suppliers and entering the baseline.
- Number 23 represents these supplies in their passage to the next system, which, while not transformed by the supply acquisition system itself, may eventually be changed into a finished product by the later production system. This transformation may involve a synthesis, or simply an assembly, of component parts and raw materials of various kinds to form a new product.

- Number 9b represents the receiving copy of the purchase order, which will generate the receiving activity when physical supplies are received.
- The larger of the two concentric circles forming #9a stands for transactional supplies. What is being referred to is the purchase order notification going to the supplier, in the form of a document, a phone call, a combination of the preceding, or any other medium which performs the function. This notification indicates which products to supply, in what quantities, when and where to deliver supplies, how to package the delivered items, what customized changes are to be made to the supplied items for this delivery only, etc. This data, symbolized as transactional input, is so considered because it is the immediate causal input leading to execution of the supplier system's initiative coordination function.

Darkened Circle

The smaller, darkened circle within the transactional input circle in #9a also represents data. This book's specialized term for this type of data is "one-shot methodological framework" input data. Briefly, this may be referred to in more familiar terms as "guidelines," or, as the computer literature would sometimes call it, "control data." This data is integrated with the transactional input symbol in #9a, since it is also supplied in the purchase notification going to the supplier. Unlike the transactional input data in the purchase notification, the guideline data is not itself either an operand or a causal input in relation to its receiving function. Rather, its role is to indicate to the supplier **how** to carry out only the supplier's **order receiving** operations themselves. For example, this data may indicate that an order confirmation document should be sent to the customer, so that any faulty communications can be detected at an early stage of order processing.

Thus, while the transactional input data on the purchase notification is the immediate cause leading the supplier to create a workorder for the supplier's own baseline, some data on this notification tells the supplier how to conduct order receiving operations. Since these guidelines may apply to only a single run of the supplier's baseline, i.e., may be different on every purchase order, they are called "one-shot," although some of the data represented as being of the one-shot methodological framework variety may, in fact, be considerably more long-term. Nonetheless, all frameworks are assumed to be one-shot at this high level of analysis, since this leaves the viewpoint flexible, and avoids incorporating scheduling constraints into the model.

Connector Circles

No example of an on-page connector circle is present in Figure 6-1. The reader may, however, refer to Figure 6-2 below for examples. The on-page connector symbol is, like that for transactional input, a circle. However, the size of the on-page connector circle is larger, corresponding to the size of this symbol on the computer system flowcharting template, from which it has been copied. Unlike transactional input circles, on-page connector circles are labelled **within** the circle in the structure-flow chart.

The Small Arrows Themselves

Arrow Head

The two types of arrow heads used in the structure-flow chart are, first, the single-headed version, and second (not exemplified until Figure 6-2), the double-headed variation.

One would think that the arrival of the purchase notification at the supplier would be synonymous with a stimulus to activate the supplier operations leading to filling of the purchase order. This is quite realistic, and in fact, the arrow connecting #9a to the supplier is actually double-headed in all diagrams subsequent to Figure 6-1. However, at this high level of analysis, it is not a good idea to commit oneself to this scheduling constraint. For instance, the purchase order's use at the supplier company could be activated by something other than by the arrival of the purchase order. A number of purchase orders from different sources might be left in storage in the supplier's incoming mail stack for a week until the supplier finds time to consider any one of the newly received purchase orders. The single-headed arrow leaves open this possibility. Thus, although in the cases of #9a to the supplier, #17 to the baseline, and #23 to the next system, the single-headed arrow is presently used. These will all be converted to double-headed arrow format in subsequent diagrams.

Angle of Entry or Exit of Arrows

In Figure 6-1, all connective arrows which enter wideshafted arrows contribute to further execution of the function represented by the wide-shafted arrow. This is known, since the angle of the entry trajectory into the wideshafted arrow is at a slant in the general direction of the wideshafted arrow. Similarly, all connective arrows in Figure 6-1 exiting wideshafted arrows are headed in the same essential direction as the wideshafted arrows. That is, the outputs in question each represent functional objectives ("objective output"), or future-oriented feedback, or symbiotic output. In the case of the present diagram, all exit arrows represent **objective** output.

The objective output in question is constituted by #9a and #9b, in the initiative coordination function, and by #23, in the baseline function. Number 23 is not only a functional objective, but is also a system, or final objective, since it is produced by a baseline function. In contrast, #9a and #9b form an intermediate objective within the system, albeit an end objective of the **initiative coordination** function.

Finally, the crook in the arrow which emits #9a from the initiative function has a particular meaning. As seen, the arrow folds back in somewhat the reverse direction prior to reaching #9a. This means that #9a is an output whose purpose is to produce an effect whereby an input is created for a subsequent system function. Number 9a is the purchase order document, sent to the supplier with the intent that the supplier remit physical supplies to the subsequent baseline function. The link between the initial effect on the supplier and the sending of supplies is indicated by a non-continuous line through the unclosed circle representing the supplier.

Thickness of Small Arrow Shaft

A trace (thick) line is used to represent material (physical) as opposed to data (information) flow.

FURTHER DEVELOPMENT OF THE SUPPLY ACQUISITION SYSTEM

ANALYSIS: LEVEL TWO

New Features of the System

The preliminary analysis involved looking at the main flow line (main line) of the physically based supply acquisition subsystem of the business. That which follows first sketches some of the other supply acquisition system current control functions, as well as the major data files needed for communication between functions **within** the supply acquisition system. Then the major external data interfaces are added.

Figure 6-2 (to be read along with Table 6-1, which decodes the reference numbers for input and outputs), contains some new items. These include the following:

- the open purchase orders data file (used for communication between the system coordination/validation functions);
- greater detail representing constraints (namely, the double-headed arrow and identification of the next system to which the supply acquisition system sends its output of physical supplies);
- two further current decision-making functions classified as interruptive feedforward and ensurance types, according to their time-orientations. For sake of brevity, when referring to the feedforward function in the following, the term "coordinate" is often applied. This term may be used for coordination functions occurring in both initiative and interruptive feedforward time-orientations. As will be pointed out in Chapter Seven, "coordinate" is often used in regular structure charts instead of the more scientifically correct noun "controller."

Open Purchase Orders File

Each record in the open purchase orders file contains details such as the following: purchase order number, supplier code, requested delivery date, delivery mode, packaging, delivery destination, items ordered, item quantities, item prices agreed to, etc. In other words, what is essentially contained in the records of this file is the data needed in the purchase order, including data needed to cross-reference other data not carried in this file.

For the moment, it is assumed that the open purchase orders file contains one record associated with each purchase transaction. However, it is quite likely that at least two record types will be needed, one for the purchase line items, and one for more general data on the purchase order. These different record types would be carried in two separate files, cross-referenced to one another via the purchase order number.

Initiative Coordination Function

Data to the Open Purchase Orders File

Number 10 represents the purchase order record to be added to the open purchase orders file. This has no particular input/output symbol on the arrow tip, since the data from the record will be dispersed from the open purchase orders file in a couple of different directions, in relation to each of which the data would use a different input/output

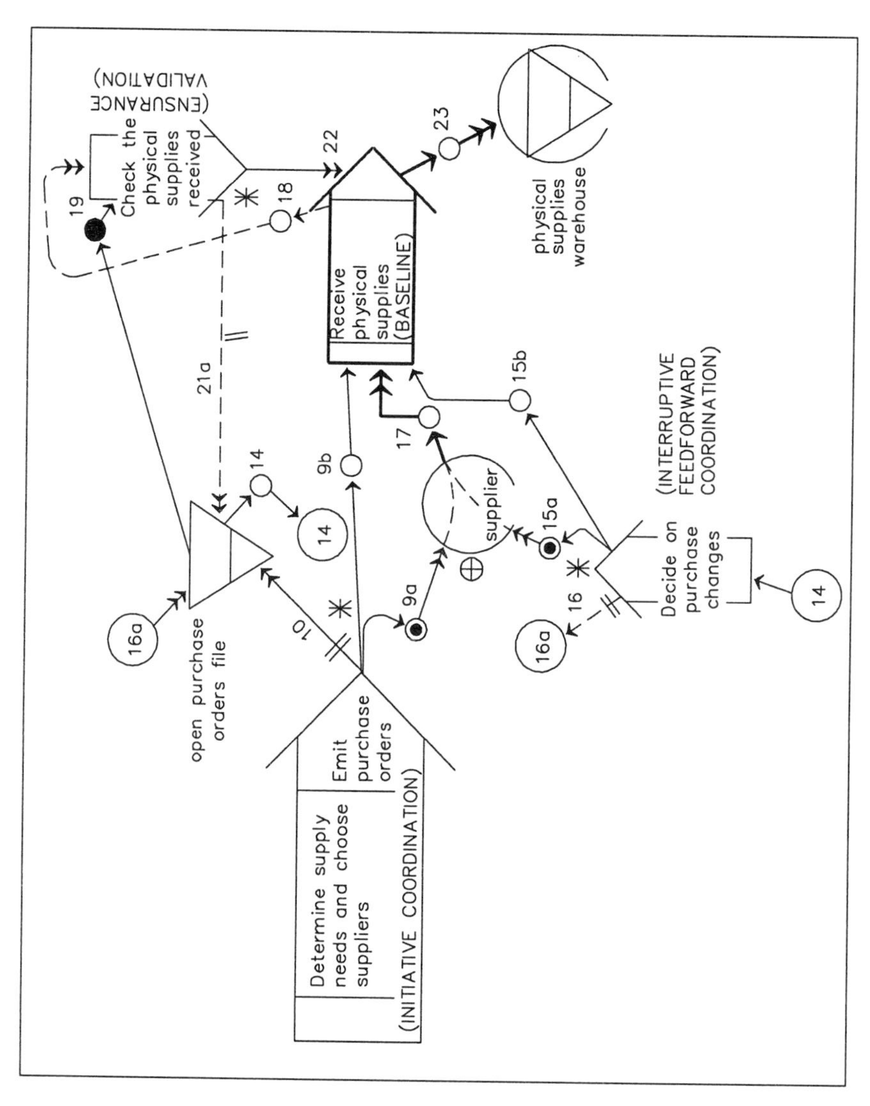

FIGURE 6-2: Level #2 of detail of the supply acquisition system. *The same numbering scheme for inputs and outputs is used in this diagram as in the other figures of this same system. These numbers are decoded in Table 6-1, opposite this page.*

NO.	DESCRIPTION OF ITEM
1	sales forecast data
2	bill-of-materials (product structure data)
3	material requirements plan data
4	total quantities of supplies on-hand, on-order from suppliers, and committed to customer orders
5	data used to guide supply needs determination, e.g., safety stock levels, reorder lot sizes, lead times, purchase prices, etc.
6	data from the supply inventory master file used to make printout readable, e.g., inventory item description, units, etc.
7	data from supplier master file used to make printouts readable, e.g., supplier name
8	data about supplier used to guide supplier selection, e.g., rating
9a	purchase order notification to supplier
9b	receiving copy of purchase order to baseline
10	purchase order record to open purchase order files
11	purchase (on-order) quantity, to be added to on-order total in supply inventory master file record
12	miscellaneous data leading to change or cancellation of an open purchase order
13	data from the supplier leading to the change or cancellation of an open purchase order
14	data from open purchase orders file used to print reports
15a	purchase order change or cancellation notification to supplier
15b	purchase order change or cancellation notification to baseline
16	change or cancellation data used to update purchase orders and inventory files
17	physical transactional supplies, e.g., raw materials, components
18	data on supply receipt fed back to review the receipt
19	data from open purchase order file used to verify receipts
21	data about the receipt used to update the various files
22	the "okay" to accumulate the goods (supplies) in the warehouse
23	physical transactional supplies forwarded for accumulation in the warehouse
24	receiving report sent to accounts payable

TABLE: 6-1 Decode of numbers for inputs and outputs. *These apply to the various figures of the supply acquisition system in this chapter*

symbol. Omitting the particular symbol associated with #10 leaves open the door to specifying more than one functional significance upon exit from the file.

It may not be too subtle a point to say that the purchase order data might be added to the open purchase orders file **as a result of** data transmitted from the initiative coordination function. The operations which, in fact, update the file with the new purchase data might be contained within a set of file update functions implied by the existence of the inverted triangle symbol. Since these reservoir operating functions are not explicitly indicated in this diagram, a special structure-flow chart of the reservoir operating system would be needed to demonstrate this detail.

Both numbers 9 and 10 reference essentially the same data items, as is indicated by the common origin of the arrows leading to numbers 9 and 10. The difference in the flow line symbols for 9 and 10 has nothing to do with the different media used, such as paper for the purchase order and magnetic medium for the file. The output arrow with the double bar on its tail leading from the initiative coordination function to #10, along with the angle of its trajectory in relation to the wideshafted arrow, indicate that this output is of the symbiotic type. In other words, it is produced by the initiative function only to maintain its cooperative relationship with other functions, in this or other systems. Specifically, this data is useful to the other two current decision-making functions, which obtain it via the open purchase orders file.

Symbiotic output #10 and the objective represented by #9 are produced in the same functional run. A Boolean algebraic symbol, the asterisk ("*" for AND) may be inserted between the two output arrow shafts, near their point of departure, signifying that both are necessarily produced in the same run. Such symbols may be used in the structure-flow chart whenever it is both feasible and desirable to be explicit about the relationship between neighbouring arrows going either out of or into the wideshafted arrow.

Double-Headed Arrow

Number 10, connected to the inverted triangle symbol by a double-headed arrow, indicates that the file update operations to add a record to the file are initiated by the arrival of #10 into the reservoir operating system. As explained, this type of logic representation embodies a constraint into the model.

Interruptive Coordination Function

The interruptive coordination function ("interruptive coordinate," for short) serves to change or cancel or otherwise intervene in the supplier's processing of the outstanding (open) purchase orders. This intervention may occur at any point between the issuance of the purchase order and the delivery of physical supplies in response to this order. Number 15, the objective of this function, is a purchase order change/cancellation notification in some form. This is represented in the same way as #9, for the same reasons.

The relationship of the interruptive coordinate to the open purchase orders file occurs in a bi-directional sense. In the one direction, #14, from the file to the function, represents the records in the file considered for change or cancellation; while, in the reverse direction, #16 symbolizes the data on the purchase order change or cancellation going from the function to the file. Number 16 can be explained in the same way as #10, with the exception that #16 is data leading to file record changes or cancellations, as opposed to file record additions.

Validation Function

Role of Ensurance Function

The validating function (or ensurance function, as it may be referred to below) is concerned with validating the effectiveness—but not the efficiency—of the physical supply receipt operations. In contrast, efficiency improvements can be handled on a feedforward basis, in future runs of the baseline. Notwithstanding, better efficiency may become a concern of a validating function when it invokes further operations, either for purposes of correction or to continue the base function.

If the baseline functioning has not been effective, the ensurance function may invoke a decision-implementing current function, such as a corrective one, to remedy the situation before the current run can be considered complete. Alternatively, in case of error, the ensurance function may simply cause the operations to be rerun. But if the run has been found by the validating function to have been successfully completed, it signals successful completion, as indicated by #22 in the diagram. Number 22 is a stimulus causing the outputs of the current baseline run to be forwarded to the next step, i.e., to the next system in the present case. Similarly, any corrective action needed will involve a stimulus being sent by the validating function to the decision-implementing current adaptation function. For simplicity, the latter is not shown in the diagram.

Input/Output Types Associated With Validating Function

Some details of the workings of the validation function are shown in the diagram. The report of the results of the baseline run, namely, the feedback item #18, is categorized as transactional input, since it constitutes the immediate cause of execution of the validation function which receives it. The ensurance time-orientation of the feedback item #18 is known, because the angle of the output arrow slants back against the direction of the wideshafted arrow. The non-continuous (dashed) flow line used for feedback avoids crossing over other flow lines in the structure-flow chart diagrams. The feedback output, one of those output types whose role is to bring something back into the current run of the function, brings back either the completion authorization stimulus or remedial actions. The latter complete the deviation-reducing feedback loop involved in any ensurance time-orientation.

Symbiotic Outputs

The ensurance function and the open purchase orders file are reciprocally coupled. Number 19 represents purchase order records going from the file to the function, while #21 indicates a symbiotic output going from the validation function to the file. In the one direction, #19 involves the passage of purchase order records from the file to the function, where they are used to validate the receipts of physical supplies. Thus, the purchase order data, now in the role of a methodological framework, is compared to the feedback data to ensure that what has been ordered is what has been received. In the other direction, if the receipts are acceptable, #21 is data sent back to the file to update the corresponding purchase order record to reflect the receipts. Once the purchase order has been filled (or at least filled to a point after which no further backorders or partial receipts are anticipated or desired), the purchase order record may be flagged for eventual deletion from the file. Otherwise, the open purchase order will be, in certain cases, updated with acceptable partial receipts, with the balance being registered as being on backorder with the supplier.

Further Constraints Incorporated into the Model

Some of the detail in Figure 6-2 represents incorporation of constraints. **First**, numbers 17 and 23 are connected to their destinations with a double-headed arrow, meaning that they invoke operations in their destination systems. Thus, the arrival of supplies, #17, provides the stimulus which sets the receiving operations into motion, etc. **Second**, the next system, namely, the warehouse, is made explicit in Figure 6-2.

The warehouse system is represented by both an unclosed circle and an inverted triangle, meaning an environmental buffer system. In the warehouse case, the reservoir represented by the inverted triangle represents an accumulation of physical transactional supplies; whereas, in the open purchase orders file case, the inverted triangle means an accumulation of data transactional supplies. Since the same modeling is used for both informational and physical functions, the same symbols may be used.

ANALYSIS: LEVEL TWO CONTINUED

This sub-section further develops the details of the supply acquisition, as in Figure 6-3. With respect to how further detail is added, this is done mainly by expanding the umbrella of system elements, and somewhat by incorporating constraints. No detail is added by explosion.

Added here are some of the external data files with which the control functions in the system interact, namely, the supply inventory and supplier master data files. The files in question are not strictly internal to the supply acquisition system; i.e., they belong in buffer subsystems of the business, since they are used in, and/or updated as a result of, data received from more than one of the business subsystems (see the functional model of the business organization in Chapter Two). The supply acquisition system has its own particular two-way interaction with these data files. The various inputs and outputs involved in interfacing these files with the supply acquisition system are explained in the following.

Data From the Supply Inventory Master File

To make the diagram somewhat simpler at this level of detail, the input/output types for the data input from these files to the supply acquisition control functions are not yet shown in Figure 6-3. This data is, however, categorized in later diagrams where it is referenced by numbers 4, 5, and 6. These items—numbers 4, 5, and 6 in Table 6-1—refer to supply inventory quantity totals, and methodological and instrumental frameworks for determining supply needs. The framework items will be described in the third section of this chapter, while #4 is described in the following paragraph, since it is an important transactional input item.

Number 4 represents the statistics for supply inventory quantities on-hand, committed to customer orders already received, and on-order from suppliers but not yet received. From the supply inventory master file, this data is transmitted to the flow functions of both the initiative and interruptive coordinates. The symbol used for #4 is for a **synthesis** of transactional supplies, because the various totals represent cumulatives of various business transactions, such as customer order filling, purchasing of supplies, etc. The transactional input designation of this data results from the on-order, committed-to-customer-orders, and on-hand totals being operated upon in order to detect inventory situations leading to further execution of the functions which receive this data. These totals are compared against standards by the initiative function to determine supply needs, and by the interruptive function to detect situations wherein changes or cancellations to open orders might be needed. As well, the data represented by #4 plays a one-shot methodological framework role in the evaluation of requests received by the interruptive function for changes or cancellations to open purchase orders. The totals represented by #4 will eventually be modified as a result of purchasing actions (see items 11 and 16, next).

Data To the Supply Inventory Master File

Numbers 11, 16, and 21 represent future-oriented feedback data outputs from the initiative, interruptive, and ensurance functions, respectively. The arrows leading to these output items all have double bars on their tails, indicating a symbiotic type output as well as a future-oriented feedback one. This dual role is explained in the following:

- **First**, data item #11, from the initiative function, represents the order quantities which are added to the on-order totals in the supply inventory master file. This is future-oriented feedback, since a loop connects the initiative function with itself. Increases in quantities on order will diminish needs for new orders in the next runs of the initiative function. As well, #11 is a symbiotic output, as the on-order quantity is of interest to the interruptive function as well as the initiative one. The interruptive function takes the on-order quantity into account in its decision making related to changing or cancelling open orders.
- **Second**, data item #16 represents adjustments to quantities on-order. These adjustments are recorded in both the open purchase order and the supply inventory master files (as indicated by on-page connectors 16a and 16b, respectively). This is both a future-oriented feedback and a symbiotic type of output, for the same reasons as apply to item #11.
- **Third**, #21 represents data about what has been received. This is transmitted in three different directions, as indicated by on-page connectors 21a, 21b, and 21c. Number 21a updates the open purchase order file with the quantity of the receipt; #21b updates the

supplier master file (see below); and #21c is used to decrease the on-order total and increase the on-hand total in the appropriate records of the supply inventory master file. An interesting question about #21 has to do with why it is both future-oriented feedback and symbiotic output. As for the symbiotic output role, this is partially explained as were the same roles for items #11 and #16. The

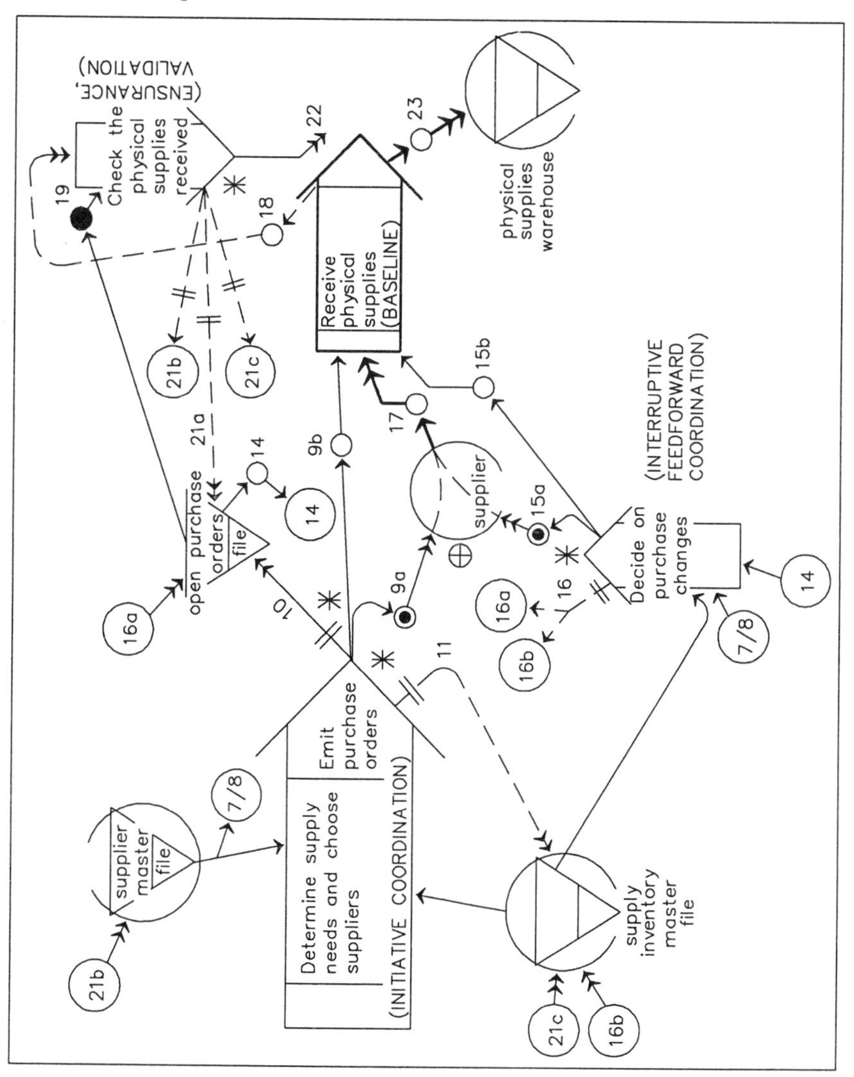

FIGURE 6-3: Level #3 of detail of the supply acquisition system. *Again, the same numbering scheme is used for inputs and outputs as in other diagrams of this same system*

future-oriented feedback role may, however, be questioned, as no apparent loop exists between different runs of the ensurance function, at least not a loop via the supply inventory master file. In fact, however, there is another reason for the future-oriented feedback role, which is that the ensurance function is sending future-oriented feedback **on behalf of the physical baseline function**. This concept was explained in Chapter Three. A future-oriented feedback output of the system baseline function, passed via the ensurance function, is emitted only after successful receipt has occurred in the supply acquisition system baseline.

Data From the Supplier Master File

To keep the diagram as simple as possible, the output type symbols for data items from the supplier master file are not yet demonstrated in Figure 6-3. They are methodological and instrumental frameworks, which will be described later in this chapter.

Data To the Supplier Master File

Insofar as it travels via on-page connector 21b, #21 represents future-oriented feedback data, transmitted on behalf of the baseline system function, going from the ensurance function to the supplier master file. This is quantities data, used to update the supplier history totals in the supplier master file.

ANALYSIS: LEVEL THREE

Separation of Informational Operations

In the next step of analysis, the supply acquisition system is divided by logical cohesion into two subsystems, namely, the information subsystem and the physical subsystem. Next, at the logical design (analysis/design) level, the information subsystem is itself further divided, into computerized and non-computerized subsystems, as well as into various subsystems identified along functional lines. The information subsystem contains the data collection, decision-making, and decision communication operations of the supply acquisition system. The physical subsystem, on the other hand, contains the baseline operations and those adaptation operations which implement decisions which are not informational. The latter, along with the baseline operations, are "physical" operations, in the sense that they handle physically tangible transactional supplies, i.e., raw materials and component parts. As well, certain framework operations of the supply acquisition system are physically based. These, to be mentioned below, in the sub-section on concurrent study of the information subsystem environment, include operations to acquire machinery needed in the supply acquisition system.

Conversion of Functional Significance

The logical cohesion method is used to separate the supply acquisition system into information and physical subsystems. The information subsystem is circled by a non-continuous line in Figure 6-4. Laid out in its own structure-flow chart, the information subsystem takes on a different morphology, as in Figure 6-5. The jump from Figure 6-4 to Figure 6-5 is obtained by a process called "the conversion of functional significance." This changes the functional class of certain functions, whenever a function plays a role in the subsystem belonging in a class which is different from the role it plays in the parent system. Thus, the three current decision-making functions of the supply acquisition system change to baseline ones, thereby giving the subsystem three different baseline functions. These are sequentially related to one another via environmental entities, of which the supplier environment relates the first and second baselines. The supplier environment then connects to the baseline of the **parent** system. The latter is signified by an **environmental entity** symbol (open circle, or horseshoe) in the information subsystem. The physical baseline, i.e., the environmental entity, in turn connects to the third baseline function of the information subsystem, namely, the former validation of the parent system now turned baseline along with its initiative and interruptive decision-making colleagues.

The three baselines, all of which point from west to east (as required for baseline function representation), form part of a sequence. The second baseline is an optional step in the sequence, and is placed below and between the other two baselines, but in the same direction. While this placement of the optional baseline is used here to indicate that it constitutes an optional step, it is also possible in certain systems that one baseline could be placed below another because it is parallel to the other, or an alternative to the other. However, in the latter cases the two baselines would be one directly under the other.

The existence of different baselines in the supply acquisition information subsystem demonstrates the following important ideas:

- A system may have more than one baseline.
- The various baselines may be related to one another in serial, mutually exclusive, or parallel fashion.

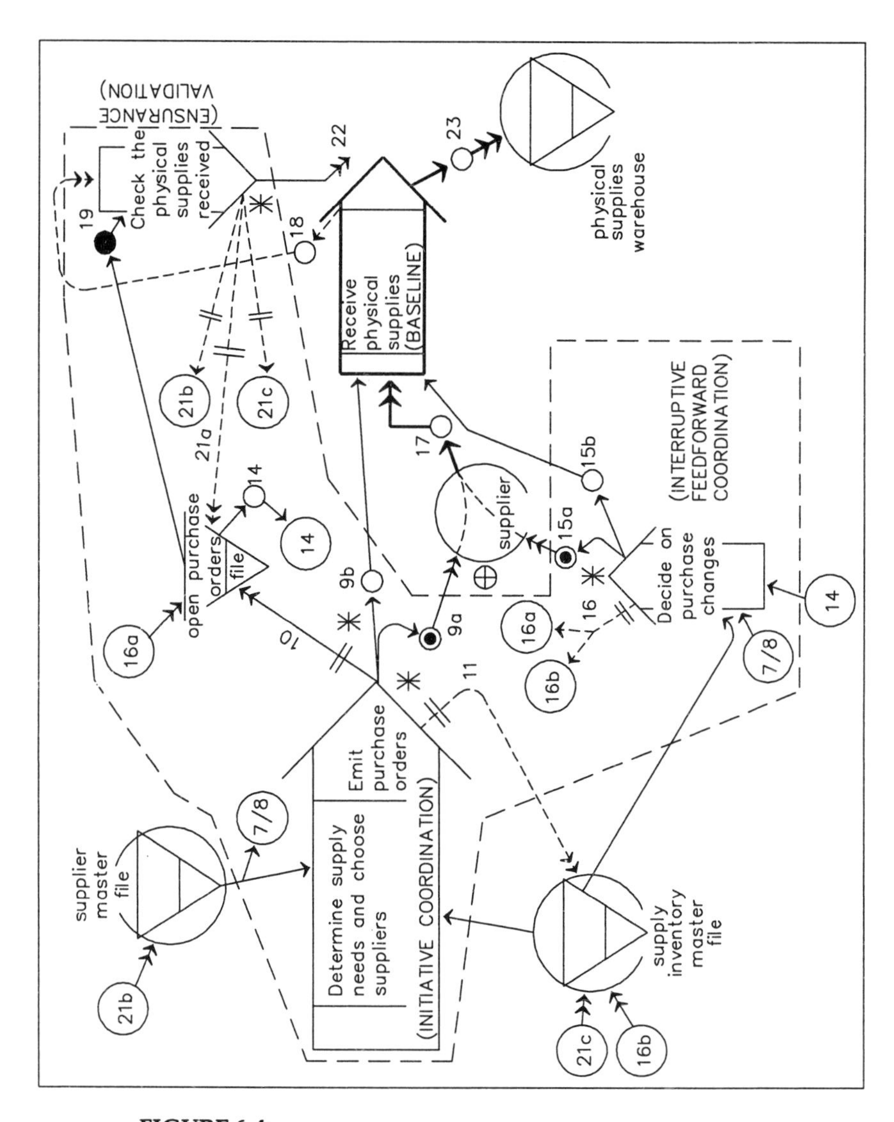

FIGURE 6-4:
Identification of the supply acquisition information subsystem.
The information subsystem is identified on the figure by the contents of the broken line which encircles various objective-defined functions and other elements.

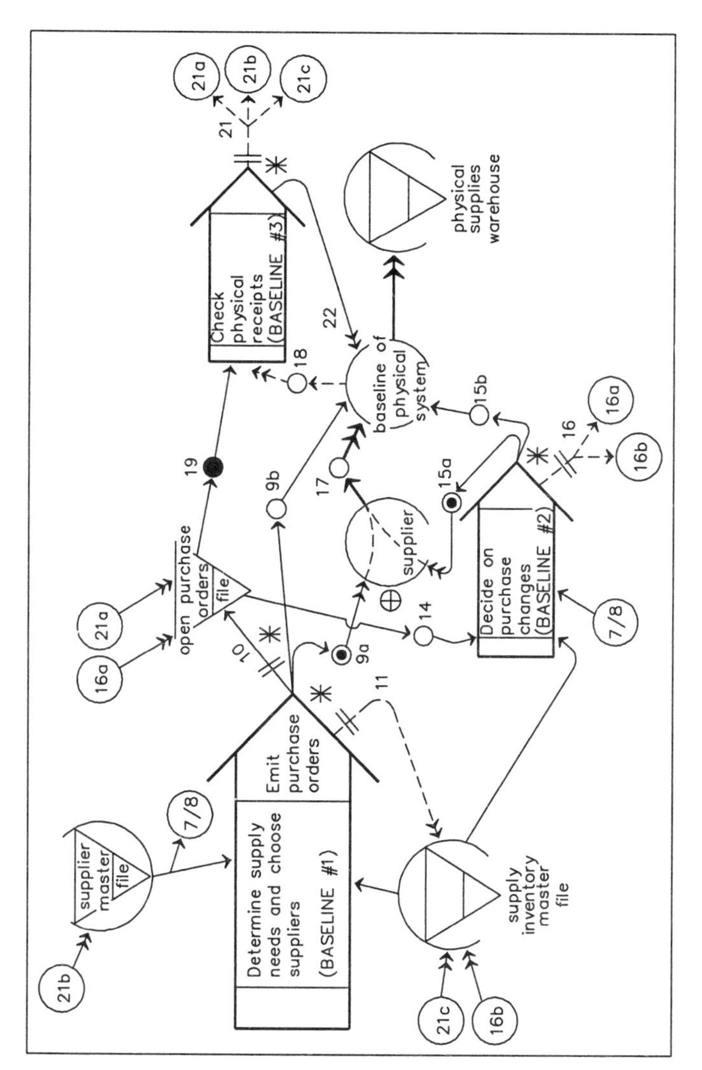

FIGURE 6-5: The information subsystem from its own viewpoint. *The supply acquisition information subsystem is portrayed here in the structure-flow chart from its own viewpoint. Of note is the ninety degree change in orientation of the wideshafted arrows for baselines two and three, as compared to their orientations in the structure-flow chart of the supply acquisition system as a whole. This information subsystem has three objectives, represented by numbers 9, 15, and 22. These objectives have a sequential dependence. Thus, the baseline functions are sequentially related, although they are not in the same* ***direct*** *line of flow. The step which produces #15 is optional in any given run of this sequence.*

- Even a baseline sequence may contain indirectly as well as directly related steps. The directly related steps are the flow functions within each objective-defined function, i.e., within the wideshafted arrow. The indirectly related steps are each represented by a different wideshafted arrow, where one wideshafted arrow precedes the other in the flow.

System Objectives When There is More than One Baseline

With more than one baseline function, a system must have more than one system (final) objective. This is because the objective(s) of each baseline function are **system** objectives, regardless of the fact that the achievement of one system objective is a step on the way to the fulfillment of another final objective. In the present example, the system has three mutually contingent objective sets, as follows:

- to issue new purchase orders, or to emit decisions not to purchase;
- to issue purchase change/cancellation notices, or to emit a decision not to change or cancel an outstanding purchase order;
- to emit a decision to accept receipts of supplies, or to emit a decision to correct such receipts.

So far, only the positive decisions listed are recorded in the diagrams, although further addition of detail would reveal the other above alternative in each case; e.g., the decision **not** to purchase.

The idea of **three** sets of sequentially related system, or final, objectives may trouble the reader. The reader should observe that the supply acquisition information subsystem is usually known in industry as the "purchasing and receiving" one. If only the last of the three sets of final objectives were to be considered to fulfill the subsystem objective, this subsystem would perhaps have to be called the "receiving" information one. The latter descriptor obviously leaves something lacking, although "receiving" alone would adequately describe the objective of the parent, physical-base system. The information subsystem's mission, stated generally, is "to control the supply acquisition operations," or, stated more specifically, "to make new purchase orders, change open purchase orders, and control receipts against purchase orders."

Computerization

Identifying Computer Subsystems

The next step is to identify further constraints under which the information subsystem is to operate. For example, different alternatives for computerization of parts of the supply acquisition information

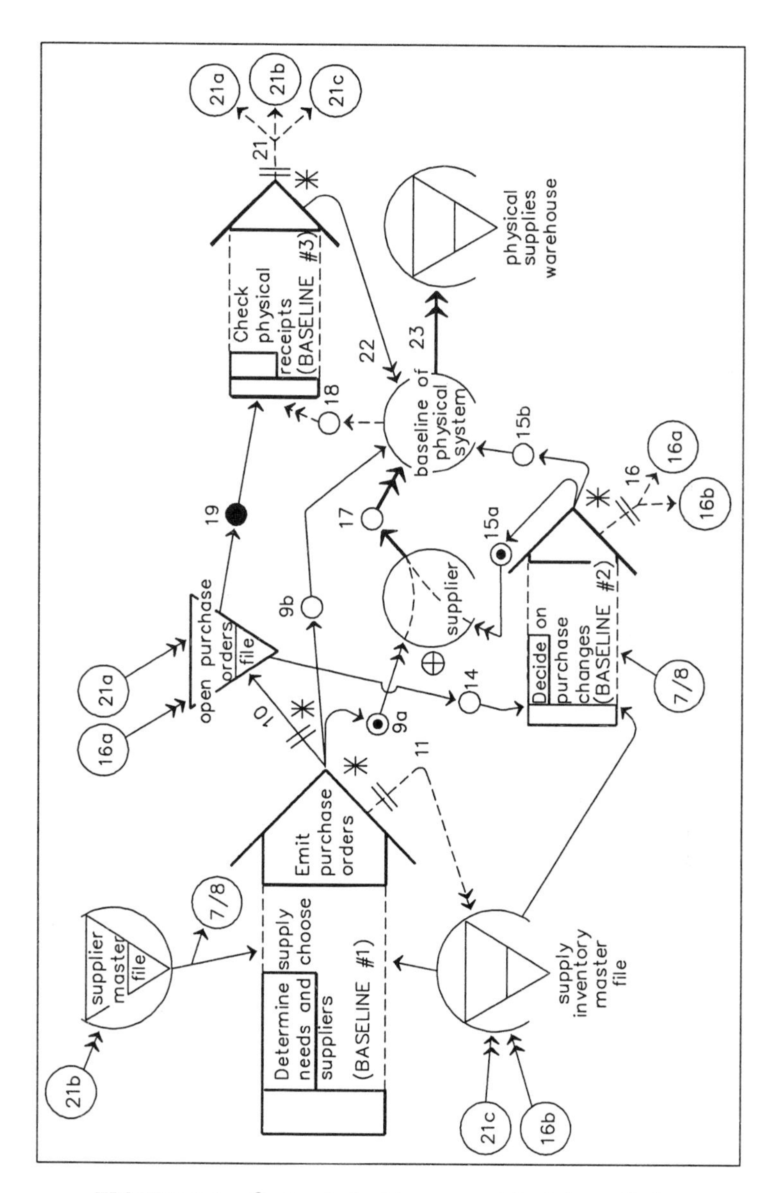

FIGURE 6-6: Computerizable areas of the information subsystem. *These are indicated by somewhat thicker lines enclosures.*

subsystem may be explored. Here, the information subsystem would again be divided into two internal subsystems of its own by a conceptual method, into computerized and non-computerized ones, each having a different instrument for carrying out its operations. This time the conceptual method, namely, procedural cohesion, incorporates the constraint of method by which functional tasks are executed.

Which System Operations Are Most Computerizable

Typically, the early and late flow functions of informational functions are those which lend themselves most easily to computerization, because these handle fairly routine communication tasks at both ends of the decision-making process. Of these routine tasks, the early flow functions detect potential problems or opportunities, while the late flow functions communicate (notify and record and report) decisions arrived at during the decision-making **process** flow functions. The **input** and **output** flow functions, respectively, are usually called "transaction-processing systems" (TPS) by business computer system specialists.*

The decision-making **process**—between the input and output flow functions of the informational function—has a number of internal steps of its own. Simply put, these steps are the following:

- —**intelligence**: collecting further data on the opportunity or problem;
- —**design**: study of alternative solutions;
- —**choice**: selecting the appropriate alternative to handle the opportunity or problem

Of the preceding internal steps, the easiest to computerize are generally the more structured (routine and repetitive) ones, especially the problem or opportunity detection steps.

In addition to steps in the decision-making flow, one should consider the **level** of decision-making function in which the flow occurs. For instance, supervisory (operational control/tactical) level decision making in an organization is more structured than managerial (strategic) decision making, and is therefore more computerizable.

How All This Applies to the Case System

Figure 6-6 indicates likely areas of the computerized operations in the supply acquisition information subsystem. These functional elements

---

* See discussion in related book, by Ed Baylin, *Conceptual Prototyping of Business Systems—A Templating Approach to Describing System Functions.*

are surrounded by heavy lines. Of course, more general levels of current adaptation functions will exist in the information subsystem, along with certain non-current functions. However, these are not yet indicated in the diagrams. If they were, one might find very little computerization of these functions, since they are less structured.

CONCURRENT STUDY OF INFORMATION SUBSYSTEM ENVIRONMENT

Introduction

The development of detail of any system requires attention to the constraints imposed by the system environment. In conjunction with developing detail via a conceptual prototyping approach, a concurrent study of the existing environment of the system is needed, since the specific, actual nature of this will determine how the conceptual prototype is to be tailored/expanded. In the case of the supply acquisition information subsystem, such a study might focus on the following:

- the physical part of the supply acquisition subsystem, in which material flow occurs, including that part in which material frameworks are established and maintained;
- flows between the buffer information subsystems of the business with which the supply acquisition system interfaces, e.g., the baseline of the warehouse operating system on the material flow side, or the master file operating system on the data flow side.

Following are a few examples of some of the relevant facts which might be uncovered by such a study.

Example of Flows Between Buffer Systems

Extra details might be added to show certain flows between buffer information systems in previous figures, such as the input/output items numbers 25 and 26 which appear in Figure 6-7. These items flow between the physical supplies warehouse and the supply inventory master file system, and are circled by non-continuous lines. Number 25 is a methodological framework item, giving the regular bin location for storage of inventory types in the warehouse. Number 26, a symbiotic output of the warehouse system, indicates changes to on-hand inventory due to stolen, spoiled, etc. inventory in the warehouse. While #25 goes from the supply inventory master file to the warehouse system, the flow of #26 is in the reverse direction between these two environmental entities. Both these flows indirectly affect the supply acquisition information subsystem.

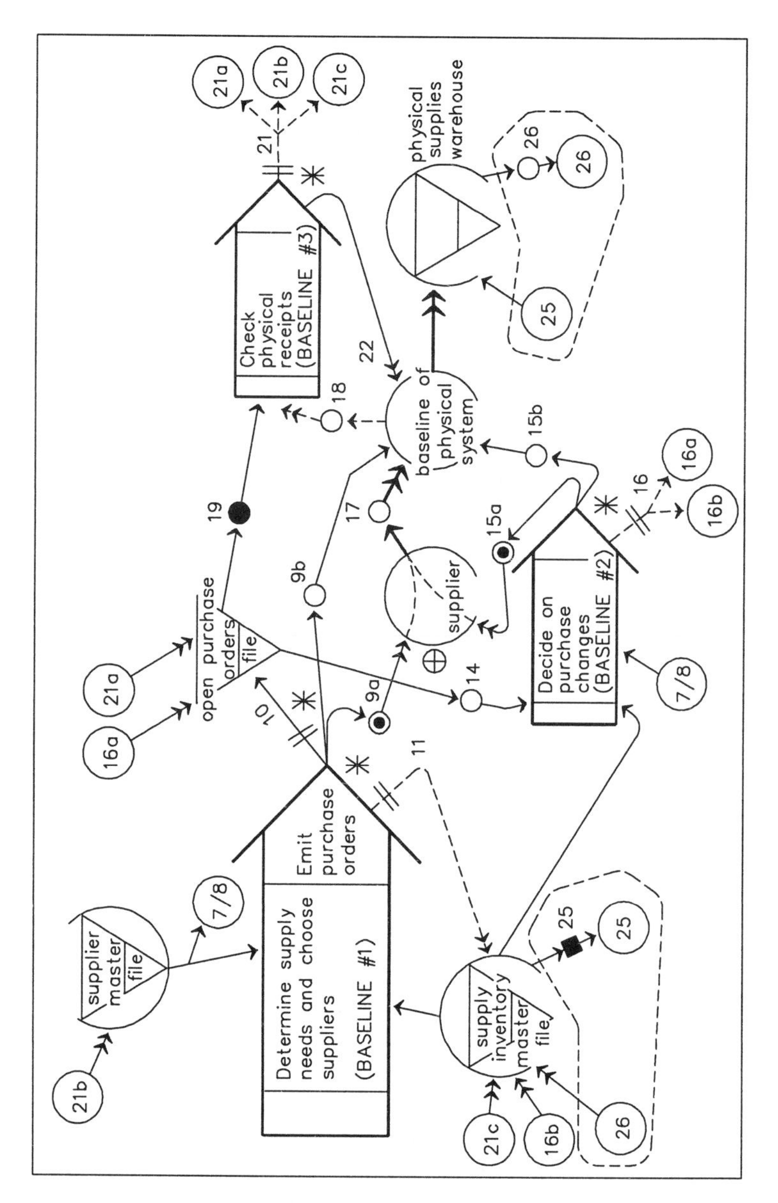

FIGURE 6-7: Some contextual aspects of the supply acquisition system.
Of note here are the areas surrounded by dashed lines, containing items 25 and 26.

Example of Long-term Physical-Base Functions

Various long-term operations needed within the supply acquisition system must eventually be added to the picture, as in Figure 6-8, which shows one planning and two framework functions. One of the framework functions is part of the physical-base system. It acquires machines needed in the supply acquisition system baseline, e.g., hand

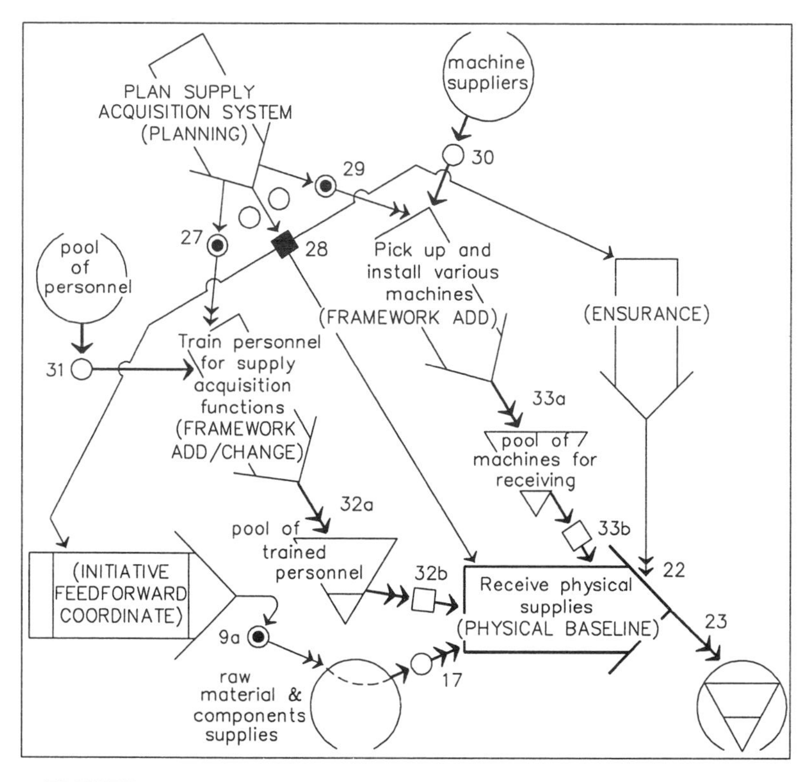

FIGURE 6-8:
Long-term operations added to the supply acquisition system.
This contains the following input/output items:

- *27: instructions on which personnel to train;*
- *28: ordering and receiving policies and norms;*
- *29: instructions on which machines to obtain and how to obtain them;*
- *30: machines to be made into frameworks of this system;*
- *31: personnel to be trained;*
- *32: personnel frameworks (32a) and their allocation for use (32b);*
- *33: machinery frameworks accumulated (33a) and their allocation (33b) for use.*

trucks, lift trucks, cranes, unloading dock equipment, trolleys, etc., all used to unload incoming supplies. The second of the two sample framework functions implements procedures, by training warehouse personnel on how to receive and check incoming supplies. These two framework functions are invoked by the planning function, which makes decisions concerning these frameworks, and which is also part of the supply acquisition **information** subsystem, in addition to the parts of the information subsystem which have been included in the umbrella of system elements to this point.

As indicated by item #28 in Figure 6-8, this planning function establishes policies for purchasing and receiving, in addition to the already mentioned functions of invoking decision-implementing functions to add/change/delete machinery and personnel. As seen in the figure, the outputs of the planning function are related to one another by a Boolean algebra symbol which looks like the transactional input circle, but is placed between the output arrows. This is the INCLUSIVE OR symbol, meaning "any combination, except zero, of those items involved." Thus, in any given functional run of the planning function, the planning operations may invoke any one, or any combination of, the two framework functions, and may also output the methodological framework item #28.

THE ANALYSIS/DESIGN LEVEL OF SYSTEM DEVELOPMENT

GENERALITIES

Scope of the Discussion in this Section

The following development of the supply acquisition system covers the interface area, between the pure analysis and pure design levels of system development. This is what some authors refer to as the "logical design" stage.

To keep developing details of the system, first, the information subsystem is divided into subsystems of its own, using the functional cohesion method. These modules are referred to below as sub-subsystems, since that is what they are in relation to the supply acquisition system as a whole. Each of the information subsystem baseline functions becomes, from the perspective of the previous level of system division, the base function of a separate sub-subsystem identified along functional lines. Specifically, four informational sub-subsystems are identified within the supply acquisition system, namely, the initiative, interruptive, ensurance and open purchase orders file reservoir ones. The latter forms an environmental buffer with respect to each of the three other sub-subsystems.

As they are external (to the supply acquisition system as a whole) buffer systems, the master file systems for maintaining the supplier and supply inventory data are not further expanded in the following discussion. In fact, certain suggestions for fields to be included in these files would undoubtedly arise at this level of system detail. However, these files are maintained by operations external to the supply acquisition information subsystem. Similar considerations apply to the elaboration of certain other environmental files to be identified below, namely, the product structure and sales forecast master files.

Descriptions of Certain Inputs/Outputs

The descriptions of certain framework input/output items briefly referred to above will be needed below in discussing one or more sub-subsystems. These are items which flow from the supply inventory and supplier master files, as follows:

- **First**, #5 represents guidelines (methodological framework) data for determining supply needs, this being the data to which #4 is compared. Number 5 is established by the planning function for determining purchasing needs. Examples of #5 data include lead time, safety lock level, maximum stock level, carrying cost parameters for economic order quantity calculations, lot sizes for reordering, prices, reorder units, etc. These serve to determine the needs for new purchase orders, and to decide changes to outstanding purchase orders, by the initiative and interruptive coordination functions, respectively. An on-page connector symbol in the below-mentioned figures disperses #5 to both the initiative and interruptive functions, as well as to the material requirements planning function associated with the initiative sub-subsystem.
- **Second**, #6 also shows a framework data item. However, unlike #5, this is not a *methodological* framework, but an "instrumental" one.

For example, the supply item description, stored in the master file record, is an instrumental framework, since it makes supply item codes from various files readable to personnel (another type of framework in the agents/instruments category) who receive reports about supply item status. The item code in the supply item records is itself an instrumental data framework, since it is a key for accessing other data about the item. Thus, instrumental frameworks may include data, as well as tools, machines, physical work locations, etc. needed to carry out operations. Number 6 goes to the middle flow functions of both the initiative and interruptive functions. Both numbers 5 and 6 travel "simultaneously" in these representations, i.e., flow via the same flow line.

- **Third**, #7 is an instrumental framework. It consists of data needed in both the initiative and interruptive functions to access supplier data through supplier codes used as record keys, and to make various reports and documents more readable to personnel, e.g., by providing supplier names instead of codes.
- **Fourth**, #8, also a framework data item, is, however, a methodological framework. It consists of the following:
  - —data used to select suppliers, e.g., items supplied, lead times, pricing parameters, supplier ratings, etc.;
  - —data used to send the purchase order, e.g., supplier address, supplier contact's name, etc.

Both numbers 7 and 8 are employed in both the initiative and interruptive functions. On-page connector 7/8 (see Figures 6-3 to 6-7) transmits this data to the interruptive function.

INITIATIVE SUB-SUBSYSTEM

Figure 6-9 represents the new purchase orders sub-subsystem of the supply acquisition system. The baseline of this sub-subsystem was formerly (see above figure) classified as the initiative coordinate of the supply acquisition system, and baseline #1 of the supply acquisition information subsystem. Certain details have now been added around this baseline function (from the perspective of the sub-subsystem), both to expand the umbrella of functional elements accounted for in the diagram, and to reflect choices of what would appear to be **physical** constraints. The details added include the following:

- new functions and associated input/output items;
- new files associated with both the new functions and the baseline function;

- specific input/output type designations for items already identified earlier in the flows between the information subsystem and the supplier and supply inventory master files.

Following is a discussion of each of the sub-subsystem details not seen in the higher level diagrams.

Planning Functions for Reorder Guidelines

Of the four additional functions shown surrounding the initiative sub-subsystem's base function, two are planning functions, while the other two represent an additional level of ensurance current functions. The

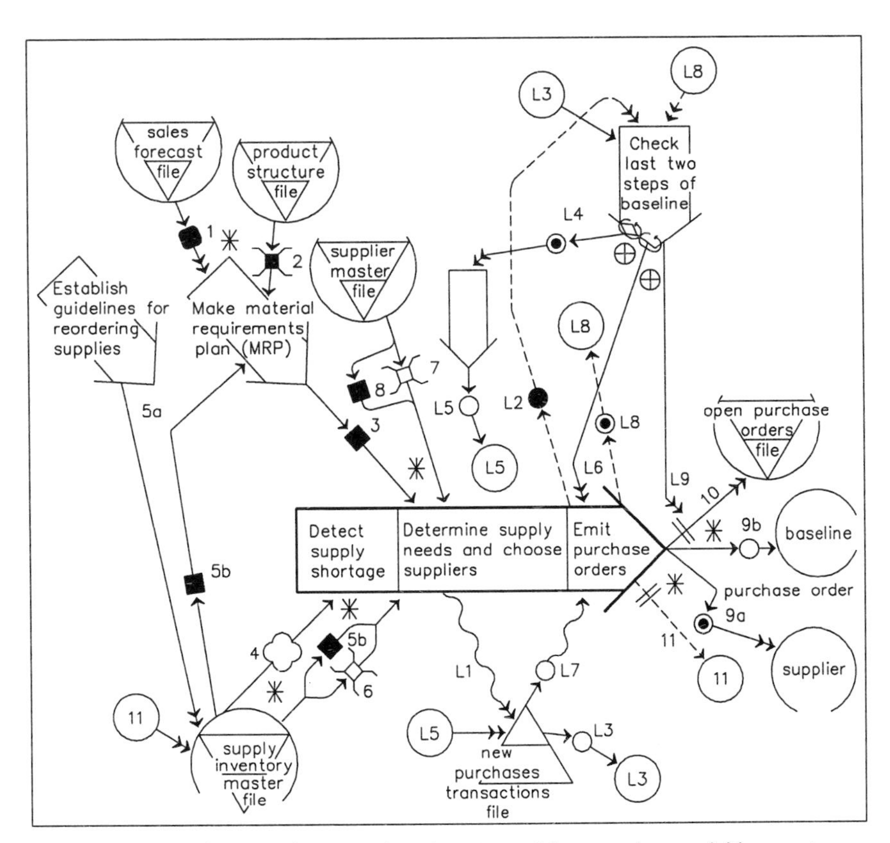

FIGURE 6-9: New purchases sub-subsystem of the supply acquisition system. *The items identified by "L" followed by a digit are local to this chart. Otherwise, the same numbering scheme is used here as on the earlier diagrams of the supply acquisition system.*

identification of planning functions implies the incorporation of scheduling constraints into the model, since planning operations have a longer time frame than do current operations.

Number 5a represents data established by the planning function for guiding the supplies reordering decision making. This has been described earlier in this section, as #5. Number 5b represents this data exiting the master file for use in the current supply orders decision-making process. Number 5b uses symbols for **medium-term** methodological frameworks. What it contains has been explained earlier in this sub-section.

Planning Function for MRP

The MRP (Material Requirements Plan), represented by #3 as a medium-term methodological framework, is a period-by-period plan for reorder of raw materials and component parts by a manufacturer. To calculate this, various methodological frameworks are needed, as follows:

- **First,** #1 shows the period-by-period sales forecast for finished products. This data is obtained by the general planning subsystem of the business via the sales forecast master file, as seen in an inverted triangle surrounded by an unclosed circle in the diagram. Number 1 is symbolized as a temporary framework (although not a one-shot one), since the sales forecast tends to change somewhat from month to month. It is a temporary framework with respect to its immediate destination, the planning function. However, it would be a medium-term framework from the viewpoint of the more frequently enacted sub-subsystem **baseline** function.
- **Second,** #2 is also a methodological framework for working out the MRP. The parameters represented by this item change very infrequently, as indicated by the shape of the framework symbol. These parameters are the formulae, or recipes of raw materials, component parts, and sub-assemblies making up the finished products sold by the business. These are referred to as "bills-of-materials," i.e., product structures in terms of their component make-up. The MRP is largely derived by factoring the finished goods sales forecast by the bills-of-materials.
- **Third,** #5c (see above explanations of item #5), a methodological framework also used to determine the MRP, helps to refine the crude MRP obtained by combining #1 and #2. By providing delivery lead times, reorder lot sizes, etc., more precise reorder timings and quantities may be obtained than would otherwise be the case.

Reservoirs Versus Stores

Number 3, the MRP, as it is now portrayed in the figure, is not retained in a reservoir. It is a fine point to note that "reservoir" in this book means an **accumulation/depletion point**, containing a **collection of like entities**, whereas a "store" refers to a physical location in which something may be left. A store may or may not be used for purposes of accumulation; also, reservoirs may or may not involve accumulation of things in stores, although stores are usually used to accumulate things in reservoirs.

Now, this example is admittedly a contrived one, since #3, the MRP, is undoubtedly in the form of sets of like records, because one record containing fields for each month of the sales forecast is needed for each supply item. However, #3 is not placed in a reservoir in this example, strictly for the purpose of discussing the difference between a store and a reservoir.

Accumulated like entities in a reservoir may or may not be operated upon by the functions which access them; i.e., either transactional supplies and/or frameworks may be retained in a reservoir. This applies to the supplier and supply inventory master files, where some of the record fields are used as framework, while others are used as transactional supplies by the functions which obtain data from these files.

Ensurance Functions

Two ensurance functions applicable to the initiative sub-subsystem are portrayed in Figure 6-9, one a validation decision-making function and the other a correction decision-implementing one. The latter implements corrections if any errors are found by the validation function. As will be explained toward the end of the next chapter, despite their apparent locality to the initiative system, the levelling dimension in which these are found is the external one, rather than that involving internal control. Internal control functions would only be of interest in cases where more physical levels of system development are reached, since they are of minor importance to the system as a whole.

Ensurance Control Loop Flow

The following paragraphs discuss the control loop in Figure 6-9, beginning with item #L1. A feedback loop as a whole occurs between L1 and L7. The actions or possible branches of this control loop may be accounted for by describing L2 to L6. L8 and L9 then apply to the feedback control loop affecting the final flow function of the baseline. L1 represents a tentative version of the purchase order, which may be revised before the final versions of purchase orders are emitted. It is stored in a temporary transaction file, where the current run builds a reservoir of more than one purchase order record before any one of these passes through the shown feedback loop. Once all data in this file has either been acknowledged as being correct, or has been modified, #L7 may be used by the final flow function of the baseline function to print a batch of purchase orders. The squiggly lines, connecting numbers L1 and L7 to the baseline function, symbolize an out-back-in relationship, in which an output at one point in the flow of an objective-defined function results in an input at a later point in the flow of the same run of this function.

L2 to L6 must be produced before L7 is reached. L2, a one-shot methodological framework, indicates to the ensurance function which of the flow functions of the baseline is to be checked in its current run. The transactional input data to be validated by this function is contained in the temporary purchase orders transactions file, and is represented by L3, which is connected to the function via an on-page connector. This data may be checked against various other inputs, which, although not documented in the diagram, may be made up of data similar to that used in the original determination of purchase needs.

If a purchase order represented by L3 is found by the ensurance function to have been wrongly determined, L4 will be produced. The use of a correction function to implement the decision change is likely unnecessary, since a decision-implementing function is only needed when communication of the decision does not suffice to cause its implementation. Nevertheless, the correction function in this example shows how such a function would look if it were to be needed. The correction function changes the purchase order data to reflect the revised purchasing decision. The record in question may then be passed via on-page connector L5 to the file subsystem for storage, or for cancellation from the file.

The base of the arrow leading to L4 from the validation function is circled by another arrow, in the form of a loop. This means "do until" all needed corrections have been placed in the hands of the correction function. This scheduling detail suddenly emerging in the diagram indicates that some flows in the diagram must be implicit. As has been seen, the purchase transactions are collected on the new purchases transactions file. The implied reason for this is that they are to be reviewed for changes and printed as a batch, rather than one purchase transaction at a time. Thus, on-page connector L3 involves the transmission of the **whole** of the file contents before any **one** of the records on the file is printed for review within the ensurance function. The records are then reviewed, and the review decision communicated, on a one-record-at-a-time basis, until all records have either been corrected or accepted in their original state.

After all purchase records have either been approved in their original state, changed or deleted, the purchase order emission step of the system base function is ready to be executed. This step is started by a continuation stimulus, labelled L6, which causes execution of the base function to continue. The loop symbol also surrounds the base of the arrow leading from the ensurance function to L6. This means that the purchase order emission step is executed again and again, until all purchase orders have been emitted. With each successive emission, the next record in the new purchases transactions file, represented by item #L7, is brought into the final baseline step.

Once all purchase orders have been emitted, feedback item L8 is transmitted to the validation function via on-page connector L8. L8 is a one-shot methodological framework, which tells the ensurance function **which** step has now been completed. This data is needed since the ensurance function checks two steps of the system base function. In the case of the purchase orders emission step, no corrective branch is portrayed in the figure.

Finally, the transmission of L9 to the end point of the system base function signals completion of the system base function's current run. L9, the completion authorization stimulus, has the same appearance as the continuation stimulus.

Conventions to Simplify Diagramming of Flow Scheduling

Incorporating all the scheduling and interaction of the function with files into the diagrams would make them considerably more complicated, and would involve showing the workings of **internal** levels of adaptation operations. Thus, many of the details are left out of the figure, on the assumption that they have been understood. These omitted details in the present example are as follows:

- The new purchases transaction file records are transmitted as a batch to the ensurance function.
- The sending of outputs outside the system base objective-defined function, or from step to step within this function, does not occur until successful completion of each step has been acknowledged by an appropriate ensurance function.
- The arrival of data from a file, or of any items from a reservoir of any sort, happens automatically, without a control signal first having to be sent to the reservoir operating system. Thus, numbers L7 and L3 go from the new purchases transactions file to their destina-

tions without a signal having been sent from the destination functional elements to indicate that the latter are now ready to receive input. As well, it is not shown that when data is to be stored in a reservoir, a communications "protocol" with the reservoir operations must first be established. (This communication will be seen below in the discussion of the open purchase order file operating sub-subsystem.)

- The symbols separating the various output arrows leading from the ensurance function in the figure have varying meanings, as follow:
  —**First**, the arrow leading to L4 is an optional output, as the small diamond (or square) symbol attached to its tail indicates. Also, the output is not produced in a definite sequence.
  —**Second**, the relationships shown by the EXCLUSIVE OR between the arrows leading to L4 and to L6, as well as the relationships between the arrows leading to L6 and L9, involve a sequencing logic in the order of output achievement. Thus, L4, if done, must be produced in a functional run which precedes the run which produces L6, and L6 must in turn be produced in a run which occurs before the run of the function which produces L9. In other words, a "round robin" logic is involved in the achievement of these different objectives of the ensurance function.

OTHER SUB-SUBSYSTEMS

Interruptive Sub-Subsystem

Baseline Inputs

Figure 6-10 represents the interruptive sub-subsystem. The meanings of the various input/output items have all been previously explained, except for numbers 12 and 13 and for those items (see below) associated with the two new functions used for ensurance control of this particular sub-subsystem.

Items 12 and 13 are connected with alternative sources for activating the sub-subsystem baseline operations. That these are alternatives is shown by the insertion of the Boolean EXCLUSIVE OR symbol between the arrows leading from these items to the start of the baseline.

Number 13 represents data from the supplier, e.g., about a delay, or some other problem, affecting an outstanding purchase order. This data is operated upon, and may result in change or cancellation of this order.

Number 12 represents any other source of data leading to cancellation or change of an outstanding purchase order. For instance, the customer orders, or forecasted market conditions, have changed since the time the supplies were originally ordered, counted upon previous supplies in the warehouse may have been lost, or suppliers are late in their deliveries. The latter condition could, for instance, lead to the

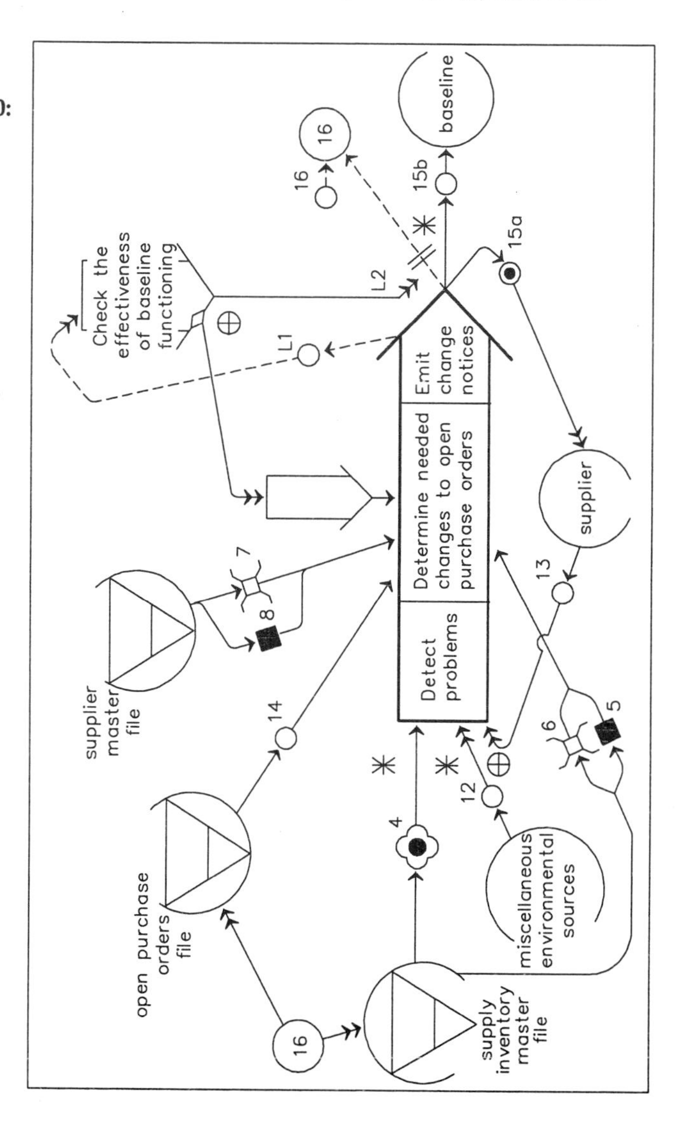

FIGURE 6-10: Interruptive sub-sub-system of the supply acquisition system. *The items identified by "L" followed by a digit are local to this chart.*

printing of a late deliveries report, based on the ETA's (expected times of arrival) in the open purchase orders file. Some of the late purchase orders may then be changed; e.g., some of the supplies could be requested from other suppliers. Also, the problem might be resolved by a decision-**implementing** function which might be associated with the interruptive system. In effect, this decision-implementing function would form a second baseline function, although it is not indicated in the present diagram.

Ensurance Control Functions

Items L1 and L2 are connected with a feedback loop for controlling current system operations. L1 represents data about what has happened in the operations just completed in the baseline. If the outcome is satisfactory, L2 will be the objective output of the ensurance function. Otherwise, the ensurance function will invoke the correction function shown in the figure. The exact nature of what is involved in correcting any errors is quite indefinable, or too complex to show at this level of system development. Thus, the correction function's output arrow is at right angles to the sub-subsystem base function, and the input/output types going into and out of the correction function are not indicated.

Ensurance Sub-Subsystem

As seen in Figure 6-11, two ensurance functions are shown **within** the ensurance sub-subsystem. These functions are "one degree removed" from the baseline of the **physical-base** supply acquisition system, while the baseline of the ensurance sub-subsystem is zero degrees removed from it. It is the zero degrees removed function which checks the effectiveness of receipt of physical supplies, while the one-degree removed function checks the effectiveness of the checking of the receipt of physical supplies. In other words, both the sub-subsystem baseline and its own ensurance functions are at some perceptual level ensurance functions, the difference being that the baseline can only be viewed as an ensurance function when seen from the perspective of the supply acquisition system as a whole.

Along with the ensurance function in Figure 6-11, certain other details are now added to the picture. These new features are explained in the following, working from L1 to L7.

L1, the verified form of the feedback from the physical baseline (which itself is #18), is collected in the form of records in a temporary transactions file, the receipts transactions accumulator file. Records are added

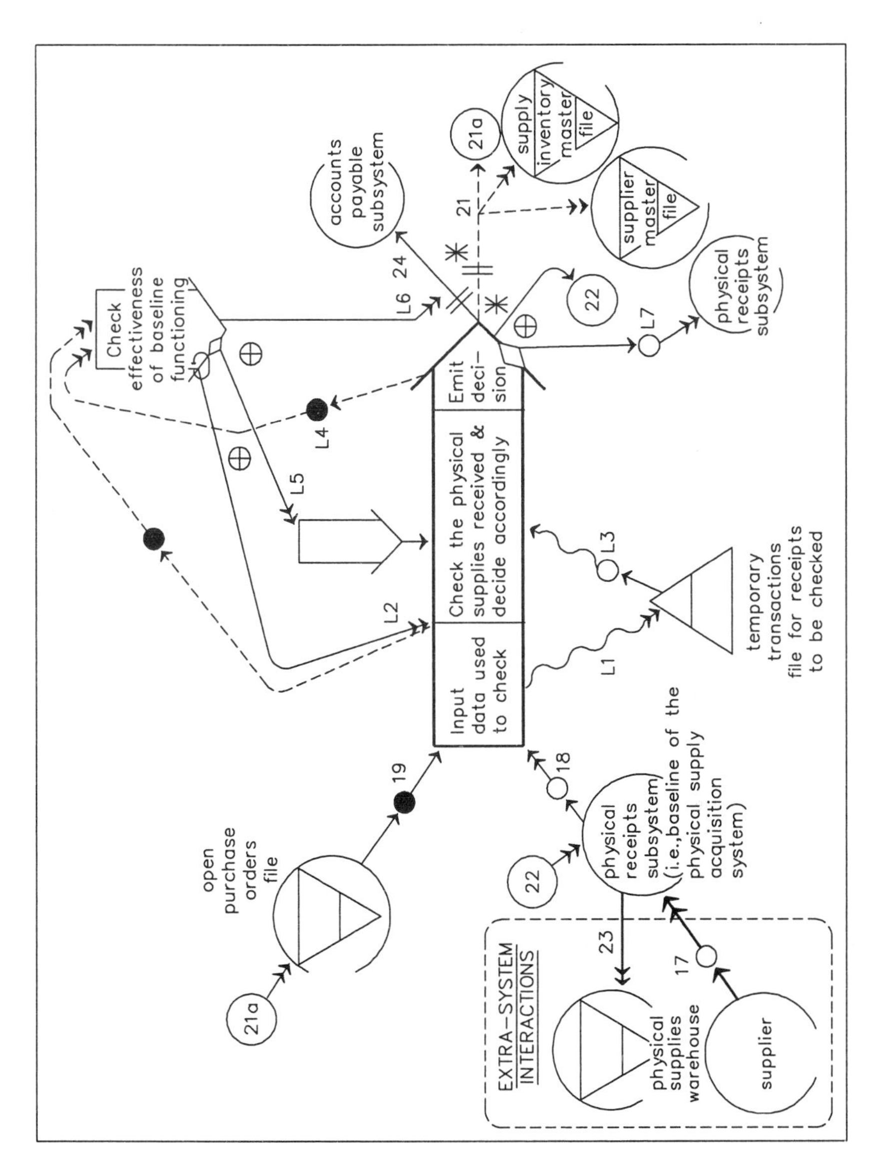

FIGURE 6-11: Ensurance sub-subsystem of the supply acquisition system. *Items identified by "L" followed by a digit are local to this chart. Extra-systems interactions (interactions occurring between entities outside the system) are enclosed by a dashed line.*

to this file after successful passage by certain *internal* ensurance functions not diagrammed in the figure.

As seen in Figure 6-11, the ensurance function performs validations both at the end of the first and the third of the sub-subsystem baseline flow functions. The last two steps (flow functions) are performed iteratively; i.e., L2 is produced again and again, in the form of a loop, until each transaction in the accumulator file has been processed by the next two steps of the baseline function.

At the end of each successive execution of the baseline function's last two steps, feedback item L4 is sent to the validation function. The latter validates the effectiveness of validation of the physical supplies done by the sub-subsystem baseline.

If the validation function decides that sub-subsystem baseline function has been completed successfully (effectively), then it outputs L6 to signal successful results; otherwise, L5 is the output. If L6 has been produced, the sub-subsystem baseline function outputs #22, indicating that the physical supplies have been successfully received by the baseline of the supply acquisition system as a whole. Otherwise, the output produced by the sub-subsystem baseline is L7, which invokes a correction function of the **physical-base** supply acquisition system as a whole. The correction function of the physical-base system operates upon physically tangible supplies of raw materials and component parts; e.g., it ships parts and raw materials back to suppliers when they are defective. This function of the physical-base system is not grouped with the informational functions when the supply acquisition system is broken down into subsystems using the logical cohesion. Rather, it forms part of the physical subsystem.

Open Purchase Orders File Operating Sub-Subsystem

Items Received by the Reservoir Sub-Subsystem

The inverted triangle representing the open purchase order file in previous diagrams implies the existence of a reservoir operating sub-subsystem. The essential structure-flow chart needed to describe this system is given in Figure 6-12. RS1 in this figure provides the record key, i.e., the purchase order number, and the update action to be followed, i.e., whether the record is to added, replaced, re-keyed (transferred), copied, etc. RS1, which is operated upon by the initiative coordinate, is the starting point in the diagram for following all sub-subsystem flow in Figure 6-12.

Initiative Function

In addition to data on the update action and the record key, the initiative function requires the current storage locations of records on the file. This data is stored on the storage locations file, which is internal to the reservoir operating sub-subsystem. DF2, the current data on storage locations, is combined with the other data in the initiative function to calculate either the desired storage location for new record addition to the file, or the current record location for records on the file to be replaced, removed, deleted, etc. The desired location data is represented by A1, in the case of new record addition. In the case of the other baseline streams, the current record locations are represented by C1, D1, T1, R1, and RC1, respectively. The T1 data must also contain the new location to which the record is to be moved. These items are all transactional inputs, since they are part of the data requesting action in the functions to which they are sent.

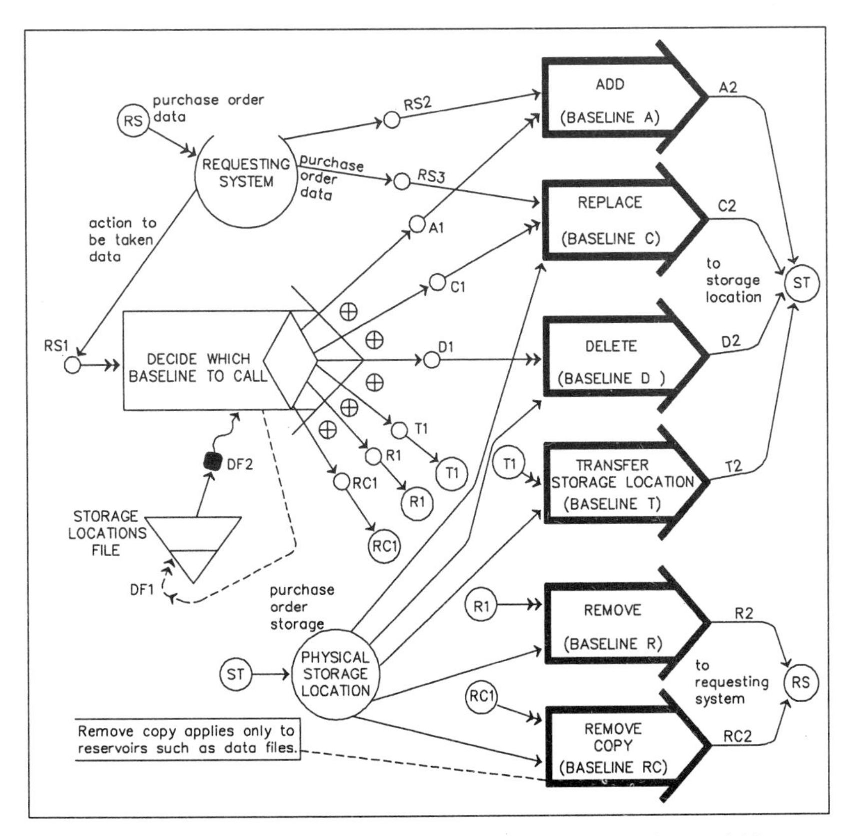

FIGURE 6-12: Purchase order file sub-subsystem of supply acquisition.

Baseline Functions

When new records are to be added, or new versions of records are to replace old versions, the purchase order record data is sent to the add and replace functions, respectively. This data, labelled RS2 and RS3 in Figure 6-12, goes straight to the baseline of the sub-subsystem. RS2 represents data sent from the initiative sub-subsystem of the supply acquisition system, while RS3 represents data from the interruptive or ensurance sub-subsystems.

The interruptive and ensurance sub-subsystems also request copies of existing purchase order records, via the RC baseline. This data is sent back to these sub-subsystems via on-page connector RS. As a result of actions by these sub-subsystems following the receipt of record copies, the record may later be replaced or deleted by baselines C and D, respectively.

When a purchase order record is added, replaced, deleted, or transferred, the data in the physical storage location used for the file is modified. These modifications flow via on-page connector ST, which links to the physical storage location, represented by a large **closed** circle.

When a record is replaced, deleted, transferred, removed, or copied, the current record in the physical storage location becomes an input to the baseline function involved. This is represented by the various arrows exiting the circle used for the physical storage location and going towards the baseline functions.

When a record is added, deleted, transferred, or removed, the physical location of the record changes or disappears, as the case may be. Thus, the storage locations file within the reservoir system must be updated. This is done by future-oriented feedback output DF1, sent from the initiative function to the storage locations file, from which the one-shot methodological framework item in question is retrieved in future executions of the initiative coordination function.

Summary

This chapter has partially developed the supply acquisition system. Many more details remain to be added, and certain parts of the system, such as the long-term functions and levels of control, have hardly been mentioned. However, enough has been presented to exemplify and allow further developments of the concepts associated with the structure-flow charting method. Subsequent chapters will continue to explore these concepts.

CHAPTER SEVEN

MORE ABOUT LEVELS OF CONTROL IN THE STRUCTURE-FLOW CHART

This chapter discusses the structure-flow chart's levels of control features. It compares their handling in the structure-flow chart to their handling in current levels-of-control charting methods (i.e., structure and function charts). In order to provide a background for the comparisons, to be made in the third section of this chapter, the first section of this chapter reviews levels of control concepts, following which the second section discusses current levels-of-control charting methods.*

---

* Various excerpts from the first two sections of this chapter are based on an article by Ed Baylin, "A Comparative Review of Popular Information System Charting Methods," Auerbach Information Management Series, Auerbach Publishers, Oct., 1985.

FUNDAMENTALS OF CURRENT LEVELS-OF-CONTROL CHARTING

Levels-of-control charts have hierarchical shapes. Nevertheless, they do not show **hierarchical functional** decomposition. Instead, levels-of-control charts illustrate single-level only functional decomposition, just as do single-phase flow charts. The hierarchical shape graphically represents the control relationships between different sets of operations within a system. Many of the underlying principles of these relationships are described next.

Allocation of Functions in the Hierarchy

Cohesion Methods for Identifying Modules

Although there is no required pattern in levels-of-control charting for the partitioning of tasks among different modules, optimal task partitioning in levels-of-control charts has been extensively discussed in the computer literature, in connection with the subject of cohesion methods. In fact, the study of cohesion methods up to this time has been almost exclusively made in conjunction with levels-of-control charts.*

Class of Function and Place ment in the Hierarchy

Commonly Established Relationships Between Modules at Different Levels — Although higher level modules may contain baseline operations (functions), single-phase levels-of-control charts are generally composed so that the higher level sets of operations in these charts "control" the sets of operations which are attached to them at lower levels of the same hierarchy chain. Since the word "control" is often used in different senses by different persons, it is necessary to explain that what is meant here by this term is that the controller modules perform functions such as planning, organizing, coordinating (directing), and ensuring (often called "controlling," when control is used in a more limited sense) with respect to the lower level modules attached to them. For instance, the "main routine" in a computer program may contain

* especially in the works of Yourdon and Constantine; e.g., Structured Design, Prentice-Hall, 1979.

organizing operations to open and close files, initialize variables, and print overall run totals, as well as ensurance control operations to detect and correct errors made by subroutines, in addition to coordinative control operations to invoke subroutines.

Coordination Relationships between Controller and Controlled Modules — Confusingly, controller modules are often referred to in levels-of-control charting as "superordinates," or even worse, "coordinates," in spite of the fact that, with respect to the modules within their control, they may perform operations other than coordination tasks. This is undoubtedly because the subordinate modules, i.e., the modules being controlled, are always at least coordinated by their superordinates, although the superordinates may or may not contain operations to plan, organize, and ensure operations within their subordinates. The terms superordinate and coordinate are used below instead of the term controller, on the understanding that the reader realizes that controllers may perform planning, organizing, ensuring, and even baseline tasks, along with coordinating ones.

To restate this, the only **necessary** relationships between modules at different levels in currently used levels-of-control charts are coordination ones, and the distribution of tasks between controller and controlled modules does not otherwise have to follow any set pattern. Thus, baseline operations may be assigned to the higher levels, in the same way as assembly line work may be assigned to a supervisor in a manufacturing company when an absent worker must be replaced.

Line and Staff Concurrent Hierarchies — Generally speaking, in a well organized single-phase levels-of-control chart, two types of hierarchies are present concurrently, namely, "line" and "staff" types. Staff subordinates are created when higher level modules delegate certain non-coordination control functions. That is, coordination of baseline processing is still done by the main routine, although the latter calls upon staff subroutines for assistance with **special** tasks such as planning and organizing. An example of such a staff function is the cash requirements planning subroutine in the structure chart of the accounts payable system shown below. As seen there, two essential branches exist in the hierarchy, one for staff, and the other for line. Within each kind of branch, any number of levels may be present.

Special Sub-Formats to Highlight Partitioning of Functions by Class — Levels-of-control charts generally have inherent basic formats. In addition, special sub-formats may be designed to show the partition of tasks in levels-of-control charts. One of these, designed by the present author, is based upon functional class/time-orientation. This sub-format places the baseline steps of the system in one alignment near the bottom of the chart, and distinguishes current from non-current operations, and baseline from non-coordination current control (decision-implementing) operations. Details of this special sub-format may be found in a related publication on procedural diagramming.

Other Symbols for Identifying Task Allocation

Finally, mention might be made of the fact that operations in levels-of-control charts can be categorized in terms of the symbols of the computer systems flowchart. Thus, the rectangular operations symbol normally used for operations in levels-of-control charts may be replaced by other symbols from media type flow charting. For instance, a parallelogram may be used for an input/output operation, or a rhombus may be used for a manual operation, in a computer system.

Module Design and Shape of the Hierarchy

Depth, Span, and Scope of Controls

Levels-of-control charts may theoretically have an infinite number of control levels. The analyst has to decide how much control depth (number of control levels) are to be shown. Also in connection with the choosing of the number of levels, the analyst may assign different spans of control to coordinates, e.g., one coordinate for steps 1 to 6, or one coordinate per three steps, or even one coordinate for every step.

To illustrate, Figure 7-1a (which, as is explained shortly, has the "transform-centered design") has two coordinate levels. The afferent (input) and transform (processing) coordinates each directly control two subordinates, while the efferent (output) coordinate directly controls only one. The executive coordinate directly controls four subordinates, including the corrective block for irregular errors. This executive coordinate not only has a wider **span** of control than the other coordinates, but also a wider **scope** of control. These are two separate ideas, since span of control refers to the number of subordinates directly connected to a controller, while scope of control refers to the number and importance of all subordinates of a controller, both directly and indirectly connected to it, down to the bottommost level of the hierarchy. Therefore, it is possible to have a wider **scope** of control, but a smaller **span** thereof. For example, the executive director of a hospital may have a small span of control compared to a head nurse, since the former directly controls just a few persons, whereas the head nurse directly supervises many nurses.

Transform-Centered Versus Transaction-Centered Design

In the transform-centered design, the subordinates of a controlling module are all activated by this controller in a sequential arrangement working from left to right. In contrast, in a transaction-centered design, subordinates are invoked in no particular order, nor is it required that all subordinates be activated by a controller in a given run of the control routine.

Examples of the transform-centered design are present in Figures 7-1a, 7-2b, and 7-3b, while examples of the transaction-centered design are contained in Figures 7-2a and 7-3a. These examples will be further discussed below, in relation to each of function and structure charts.

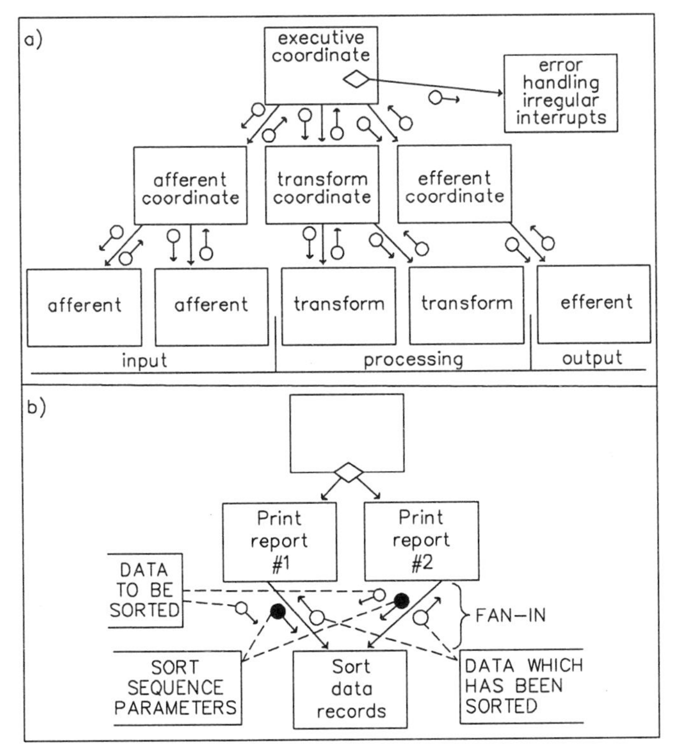

FIGURE 7-1: Aspects related to the ways subordinate modules are established in current levels-of-control charts.

a) multi-level levels-of-control chart with transform-centered design;

b) fan-in to eliminate duplication in a levels-of-control chart.

Fan-In

Fan-out refers to the spreading out of the hierarchy in the horizontal direction, based on different spans of control and design logics (transform versus transaction-centered). However, portions of the hierarchy may "fan-in." Thus, as in Figure 7-1b, two modules may have the same subordinate. Fan-in automatically creates a lower level of module in the hierarchy.

Different Reasons for Fan-In — One of the basic principles of the functional cohesion method is to avoid duplication of functions. On a levels-of-control chart, this avoidance is accomplished by fan-in. For instance, the same data sorting program may be used by two or more modules which produce different reports.

The design of the modular structure may be such that fan-in has nothing to do with eliminating duplication. This is the case when non-functional cohesion methods are used to group functions by criteria such as a classification scheme, time of occurrence, communication channel involved, etc. This is explained more fully in Chapter Ten, which discusses cohesion methods.

INDIVIDUAL METHODS

FUNCTION CHART
(Levels of Control Without Flow)

The only type of system element shown in a function chart are functions, which are shown along with their control levels. All other types of system elements are absent. For example, the hierarchical relationship between a number of different programs, or subprograms, is shown in Figure 7-2a. Here, the top module selects a second module which is to be executed next.

Representing the Two Design Logics

The transaction-centered design is shown in Figure 7-2a, where the various programs at the bottom level are called in random order. In contrast, Figure 7-2b demonstrates the transform-centered design, even though this diagram is identical in morphology (shape, and diagramming details, except for text) to Figure 7-2a. In order to recognize which design is involved, it is necessary to read the text in the diagrams. Thus, the text in Figure 7-2b tells us that the bottom level of three programs executes in a left to right order, i.e., input-processing-output.

Furthermore, as indicated in Figure 7-2c, each different level or branch of the hierarchy may have a different one of the two basic designs. In fact, it is possible to create a hybrid of the transform-centered and transaction-centered designs in the very same branch and level of the hierarchy.

Evaluation of the Function Chart

Function charts are limited to outlining the control levels of functions in a system. They cannot be used for very logical description of systems, since they not only lack all of the necessary symbols, but also because they force the showing of control aspects, i.e., control levels. For very physical depiction of systems, they lack all essential symbols **except** those for illustrating control levels. Their one use for perceiving systems is to provide an extremely simple way of depicting control levels at the logical design level of detail.

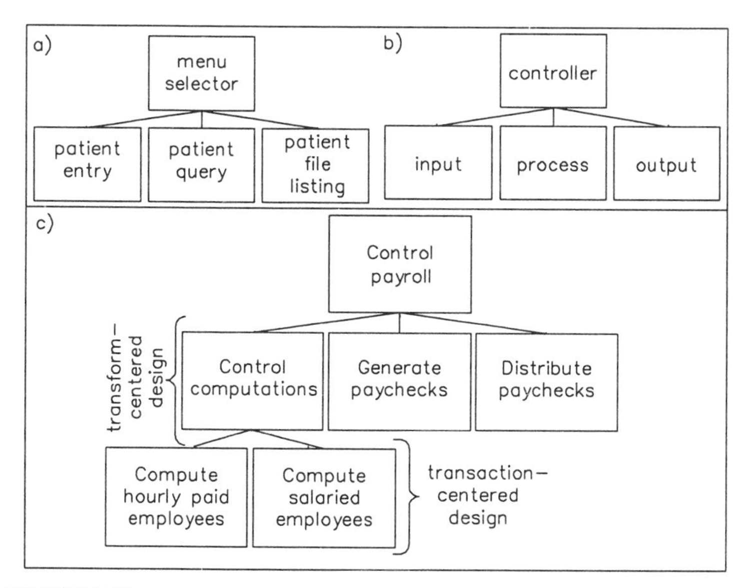

FIGURE 7-2:
Design logics for activating subordinates in function charts.
a) the transaction-centered design in a function chart; [Auerbach©]
b) the transform-centered design in a function chart; [Auerbach©]
c) different designs at different levels in a function chart.

STRUCTURE CHART
(Levels of Control With Flow)

The structure chart, unlike the function chart, shows inputs and outputs plus details of flow scheduling and routing, in addition to levels of control. No store symbols are provided, not to speak of specialized accumulation/depletion symbols. However, iteration is used to show accumulation in structure charts, just as in computer program flowcharts.. Although the system environment is not usually explicitly shown, an unclosed circle is sometimes used in structure charts to represent the environment, thereby allowing flow to and from the environment to be illustrated. Because it shows the above-mentioned system aspects, the structure chart is both a flow and a levels-of-control chart. However, it should be noted that the structure chart is poor at showing flow, while it ably demonstrates levels of control.

It should be noted that a type of chart in between function and structure charts can be drawn by showing the inputs and outputs in a function chart, but without providing the flow routing/scheduling details.

Representing the Two Design Logics

Figures 7-3a and 7-3b are the structure chart equivalents of Figures 7-2a and 7-2b, respectively. Since the structure chart symbols make the flow and branching logic quite evident, it is easier here than in the function chart to identify which of the two basic design logics is involved for invoking subordinate modules.

Transaction-Centered Design — In Figure 7-3a, the transaction-centered design is indicated by the fact that more than one control line emanates from the decision branching point symbol, i.e., from the diamond on the lip of the coordinate block. Therefore, any one of the lower level ("atomic level") blocks may be executed on a mutually exclusive basis, and in any sequence. Patient query may be executed first, then patient entry used twice, then patient query again, and so on, at random.

Transform-Centered Design — In Figure 7-3b, in contrast, every time the coordinate module is executed, control first passes to the input, then to the processing, and then to the output step. Here, the bottom row is executed on a round-robin basis. By convention, the sequence is from left to right in this transform-centered design. Nevertheless, whenever the left to right convention is to be overridden, numbers may be drawn near the tails of the hierarchical relationship arrows in order to indicate the sequence of subordinate module execution.

Demonstrating Flow

Illustrating Normal Flow

The normal flow in the structure chart is from superordinate to subordinate, and then back to the calling superordinate. The return of control from an invoked (or "called") module, i.e., the subordinate, back to the invoking (or calling) module, i.e., the superordinate, is implicit. However, it can be overridden by relationships explicitly shown between peer level subordinates, or between a subordinate and a superordinate which is not its direct superior. Thus, in the normal case, as in Figure 7-3b, execution begins in the coordinate module. Since Figure 7-3b is the transform-centered design, control first passes to the left-most subordinate, then back to the superordinate, then to the middle subordinate, then back to the superordinate, then to the right-most subordinate, and then, finally, back to the superordinate.

Illustrating Detailed Flow

Although "AND" (parallel) and "INCLUSIVE OR" (and/or) gating logic are not available, optional steps, simple and nested loops, and combinations of loops and optional steps may be demonstrated by structure chart symbols. Thus, in Figure 7-4b, steps 2 to 4 form an external loop, while step 3 is a loop inside of this. Steps 2, 3, and 4, in that order, are executed one or more times, in succession, for each execution of steps 1 and 5, while step 3 could be executed more than once during each execution of steps 2 to 4. Figure 7-4c shows an optional step within the transform-centered arrangement, indicated by a decision block on

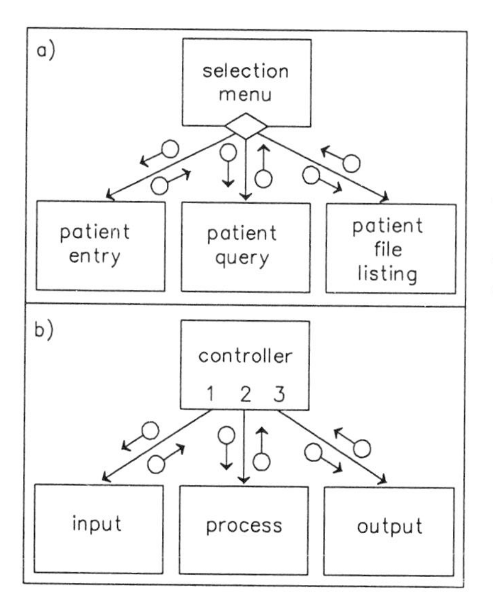

FIGURE 7-3:
Design logics for invoking subordinates in structure charts. [Auerbach©]
a) the transaction-centered design;
b) the transform-centered design.

the tail of a single hierarchy connecting line. In this situation, steps 1 and 3 are mandatory, while step 2 is optional.

Sometimes, the transaction-centered arrangement is found within the transform-centered one, and vice versa. Figure 7-4d contains a transform-centered arrangement in which steps 1 and 3, and either step 2a

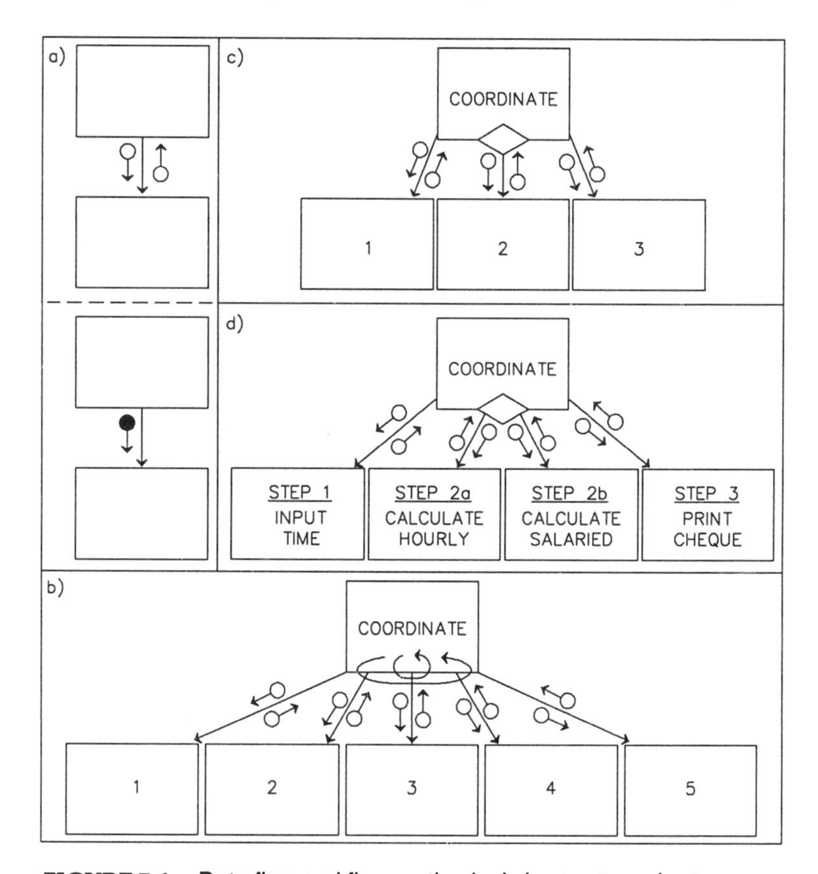

FIGURE 7-4: Data flow and flow routing logic in structure charts.

a) *control and other data input/output symbols in a structure chart. An arrow with a small, darkened circle on its tail represents so-called "control" data, and is usually passed in one direction only. The small arrow with the non-darkened circle on its tail represents other types of data; [Auerbach©]*

b) *showing of loops in a structure chart; [Auerbach©]*

c) *optional step within a transform-centered structure chart; [Auerbach©]*

d) *hybrid of transform- and transaction-centered designs in a structure chart.*

or 2b, are mandatory. Step 2 is thus not an optional step, but one having alternatives, as it involves transaction-centered logic.

Showing Inputs/Outputs and their Functional Significances

Symbols Used

Accompanying, and parallel to, the hierarchical relationship arrows are other small arrows. These symbolize the inputs and outputs.

The shading of the small circle on the tail of each arrow shaft categorizes the functional significance of the input/output item. Only two categories of these are differentiated, as indicated in Figure 7-4a. Here, the darkened circle indicates "control data," while the non-shaded circle represents other types of data, referred to below as "non-control data." Actually, these two types are groupings of a number of varieties of functional significance.

Specific Varieties of Functional Significance Included Within the Structure Chart's Two Types of Input/Output Symbols

Varieties of Control Inputs/Outputs — First, with respect to control input/output, the following different varieties may be involved:

- guidelines output from a superordinate and input by a subordinate, e.g.,
  - —instructions to the subordinate on which choice(s) to make from a number of possible alternatives;
- feedback output from a subordinate and input to a superordinate, affecting the flow of control, e.g.,
  - —indication of an error, referred back to the superordinate for handling;
  - —the payment scheduled date in the structure chart of the accounts payable system, which affects the timing of the invocation of the cheque production subordinate.

Varieties of Non-Control Inputs/Ouputs — Second, with respect to non-control input/output, the following varieties are possible:

- data input which is operated upon, and/or which serves as the immediate cause of execution of the function, i.e., transactional input, e.g.,
  - —the invoice received from the supplier in the accounts payable system;
- data output from one module to be used eventually as transactional input by another module. Such data output

often represents the objective (the reason for existence) of the sending module, although it may just be something passed along to another module for use there or elsewhere.

- data input by a module, other than control data, needed as an instrument in performing the operations in that module, e.g.,
  —a reference code used to access a storage location.

Limitations of Structure Charts in Specifying Functional Significances of Inputs/Outputs

Of course, the structure chart's simple distinction between control and non-control input/output types hardly arrives at the above finer distinctions, with the result that the meanings of control data and other data are seldom understood by those using the charting method. The accounts payable system write-up in the related Ed Baylin book, *Conceptual Prototyping of Business Systems—A Templating Approach to Describing System Functions,* classifies the inputs and outputs of each step in a way which does a rather better job of distinguishing their functional significances. A reading of this prototype example should immediately reveal to the reader the value of making these further distinctions of functional significance.

Structure Chart of an Accounts Payable System

Overall Description

Figure 7-5 gives one version of a structure chart of an accounts payable system. It may be noted that cash requirements planning is retained **within** the system boundary.

Step One: Establish Payment Scheduling Guidelines — Planning of cash requirements (establishing payment scheduling guidelines) is a non-current function. Its non-current time frame is symbolized by the loop symbol indicating iterative processing of the current operations (coordinating invoice processing, and coordinating payment processing). In this way, the non-current cash requirements planning operations are shown as being executed less frequently than the remaining subordinate modules.

Step Two: Receive Supplier Invoice — The current processing of the system begins with the receipt of the supplier invoice, requesting payment for supplies and services rendered.

Step Three: Schedule Payment of Invoice — The third step decides whether and when to pay the invoice. This block receives the invoice from the superordinate, and then, if it is to be paid, passes it upward, along with payment scheduling instructions. The payment instructions will be passed along via the highest current operations coordinate (which happens to be the highest level coordinate) to the payment coordinate, i.e., to the superordinate of cheque production. However, if the invoice is not accepted for payment, the payment refusal, if any, is indicated in the same control data item as is the payment scheduled date (the same control data item being used for both purposes). If payment of the invoice is refused, the payment scheduling step will return the invoice to the environment, i.e., to the supplier.

Step Four: (Optional) Reschedule Payment of Invoice — If the invoice is to be paid, the next mandatory step is to produce the cheque and to send it to the supplier. However, preceding this mandatory step is an optional payment rescheduling step, which may change the invoice payment scheduling instructions.

Step Five: Produce and Send Cheque — The cheque is produced on the scheduled payment date, and then delivered to the supplier.

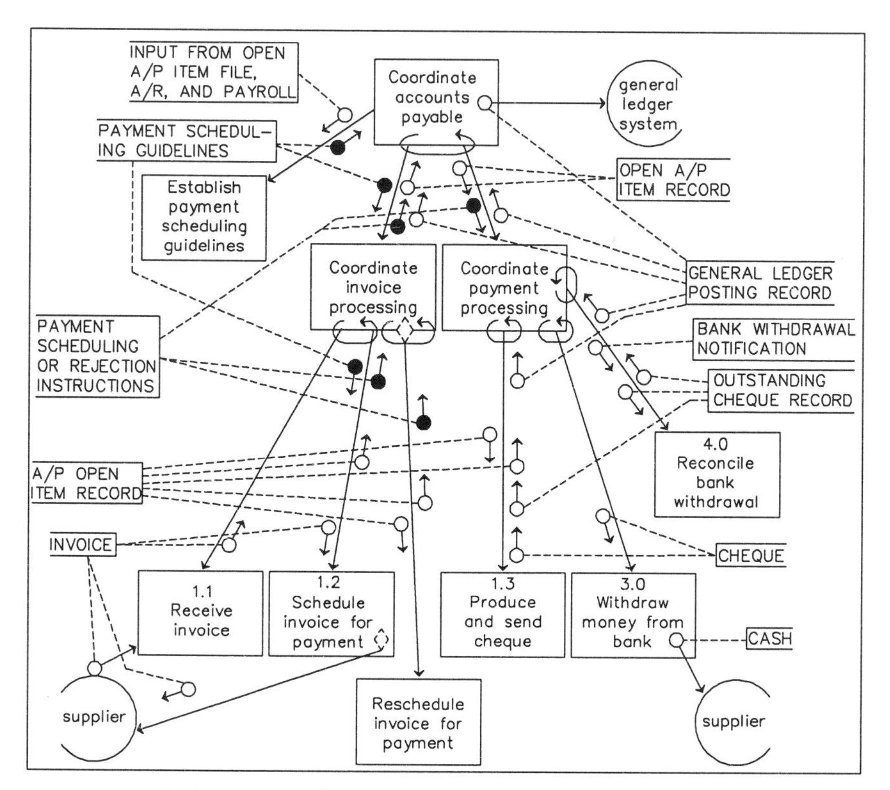

FIGURE 7-5: Structure chart of th

Step Six: Withdraw Money from Bank — After the cheque is sent, the supplier is supposed to cash the cheque in a bank, which eventually results in the money being withdrawn from the business' bank account.

Step Seven: Reconcile Bank Withdrawal — Finally, reconciliation is made of the intended money withdrawal transaction at the bank. This step inputs:

—the file from the bank showing cheques cashed;
—the latest version of the outstanding cheques file;
—any information from the supplier or elsewhere indicating a problem with the bank withdrawal, for instance, NSF funds, improper signature on the cheque, etc.

Note on General Ledger Postings Made at Various Steps — Data to be posted in the general ledger are indicated in the structure chart being passed back from the invoice acceptance and cheque production steps, as well as from the bank reconciliation. The G/L posting resulting from the bank reconciliation is optional, because it happens only in the event of problems occurring after presentation of the cheque at the supplier's bank. Such an irregular posting is needed in order to reverse the posting made at the time the cheque was sent to the supplier, since the latter posting assumed that bank withdrawal would go through without any problem. The actual general ledger postings are carried out by the general ledger system, which is in the environment of the accounts payable system.

Use of Loop Symbols

Iterations shown in procedural diagrams can represent three different concepts, as explained earlier. One is to distinguish operations occurring in different time frames. The other two concepts occur in the current time frame. One illustrates accumulation/depletion points (reservoirs) resulting from a repetitive step in the flow, while the other shows return for purposes of correcting errors. In the structure chart of the A/P system, iterations of all these types are represented by the loop symbol. This symbol merely compensates for the lack of better symbols which are needed to differentiate the three ideas involved.

Indicating a More Long-Term Time Frame — **First**, the loop symbol surrounding the hierarchical relationship lines leading to the two coordinate modules at the second level of the chart symbolizes the more long-term, less current, time frame of the cash requirements planning step, which is, in contrast, not surrounded by a loop. In this case, the long-term operations play a kind of "staff" role within the system.

Indicating Accumulations/Depletions — Second, the loops shown at lower levels, in this example, are used due to the lack of store symbols to show the relative scheduling of the different steps because of build-ups along the line of flow. Thus, invoice receipt and scheduling occur each week until all invoices have been received and scheduled (rescheduled in some circumstances) for payment. This accumulates records in the open items file. At the end of the week, cheque production is performed for each open item due for payment. Each iteration produces one cheque. Similarly, bank withdrawals occur in an iterative way; i.e., each iteration results in one cheque being cashed. Bank reconciliation, the next step, is not only iterative, but also involves an internal decision (not shown in the figure).

Indicating Redo Loops (Flow Routes) in Case of Errors — Third, any of the loops associated with the current processing may be interpreted as redo loops. This is impossible to tell in this diagram, as the same symbol is used for both accumulation/depletion and redo.

Evaluation of the Structure Chart

Factors Mitigating Against Use

Factors Applicable to All Levels of Perception — These factors are as follows:

- Structure charts have difficulty showing the flow of inputs and outputs, as these do not pass directly from subordinate to sibling (having the same superordinate) subordinate. Rather, the flow goes from the first subordinate, up to the common superordinate, then down to the second subordinate. In fact, this passage may involve more than one level of intervening superordinate.
- The lack of ability to illustrate parallel branches forces flow scheduling constraints to be incorporated.
- The lack of ability to illustrate stores is also undesirable.

Factor Causing Unsuitability for Analysis — The one detail forced in structure charts is the showing of controls, since the latter is the very purpose of the different levels of hierarchy. However, this inherent feature of the chart negates it for use at the logical level of perception.

Factors Mitigating Against Suitability for Logical Design — Structure charts are widely used in the logical design of computer programs. However, the following factors mitigate against their usefulness for this:

- Return loops for correction of errors, so important in logical design of computer systems, are inconvenient in structure charts, since the same symbols are used for "normal" iterations (for accumulation/depletion) in the flow as for return loops in the event of errors.
- Slotting of special correction-implementing modules into the hierarchical arrangement, where needed, is usually awkward.

STRUCTURE CHARTS VERSUS STRUCTURE-FLOW CHARTS

The only charting form other than the structure-flow chart which shows levels of control while still demonstrating flow is the structure chart—hence the reason for part of the name of the structure-flow chart. The structure-flow chart is unlike the structure chart in that it shows flow **clearly**. Also, the structure-flow chart handles many further aspects of levels of control than does the regular (or "conventional") structure chart. The following discussion of levels of control in the structure-flow chart makes comparisons to the regular structure chart.

Dimensions of Comparison of Structure and Structure-Flow Charts

Following are some of the dimensions of understanding captured by the structure-flow chart, which are **not** captured, or **less well** captured, by the regular structure chart:

—functional significances of inputs and outputs;
—direct flow towards objectives;
—flow routing branches;
—accumulation/depletion points along the line of flow;
—functional classes, time frames, and time-orientations;
—internal (as opposed to external) levels of control.

As will be seen in the following discussion, conventional structure charting techniques can be adapted by using new packaging conventions for page spatial arrangements and a few new symbols to achieve some of the above features. However, the structure-flow chart is definitely superior for depicting the above items.

INPUTS/OUTPUTS, FLOW, AND STORAGE

Channelling of Input/Output Flow

Differences between conventional structure charts and the structure-flow chart occur in the following areas:

—diagramming of transfer of control between levels;
—flow of inputs and outputs versus control transfer flow.

Showing Activation and Return (Transfer of Control)

The diagramming of activation and return (transfer of control) relationships in conventional structure charts is handled by a single arrow connecting a superordinate and a subordinate (Figure 7-6a), or even connecting two peer blocks (Figure 7-6b). Uni-directional transfers of control **may** be represented by placing a dot on the arrow tail as in Figure 7-6c. Uni-directional transfer of control means that one block activates another, but the activated block does not return control to its activator after execution.

When wideshafted arrows are used instead of rectangles in the hierarchical arrangement, two arrows, rather than one, are needed to demonstrate both activation and return, because each wideshafted arrow is itself a flowchart of a single objective-defined function. Thus, one

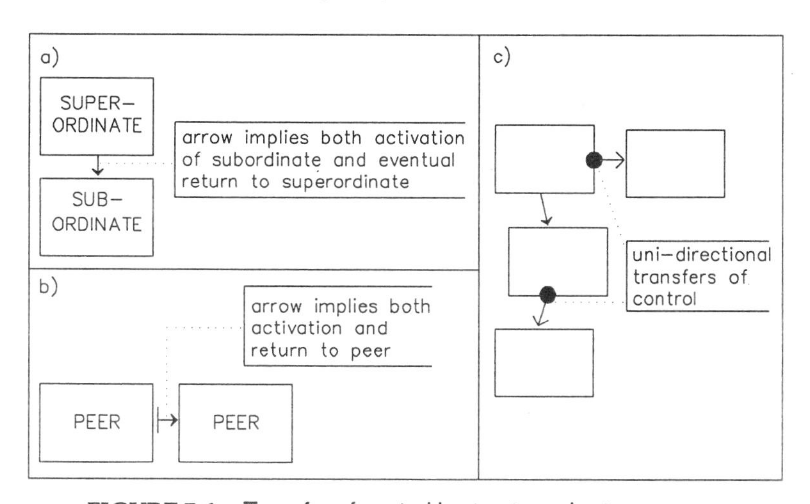

FIGURE 7-6: Transfer of control in structure charts.
a) normal control relationship;
b) co-routines (calling peers), an unusual situation in which a peer activates a peer, following which control returns to the activator.
c) uni-directional transfers of control.

arrow shows activation, while the other shows return of control, as in Figures 7-7a, 7-7b and 7-7c. Uni-directional transfers of control may be handled by a single arrow, as in Figure 7-7d.

In all of these diagrams, the transfer of control is represented by a **double-headed** arrow tip, as opposed to the single-headed arrow tip in the conventional structure chart. This fact is important in showing long-term operations in the structure-flow chart. This is because, as will be seen later in this section, the transfers from these operations to the system base operations involve no control transfer (activation).

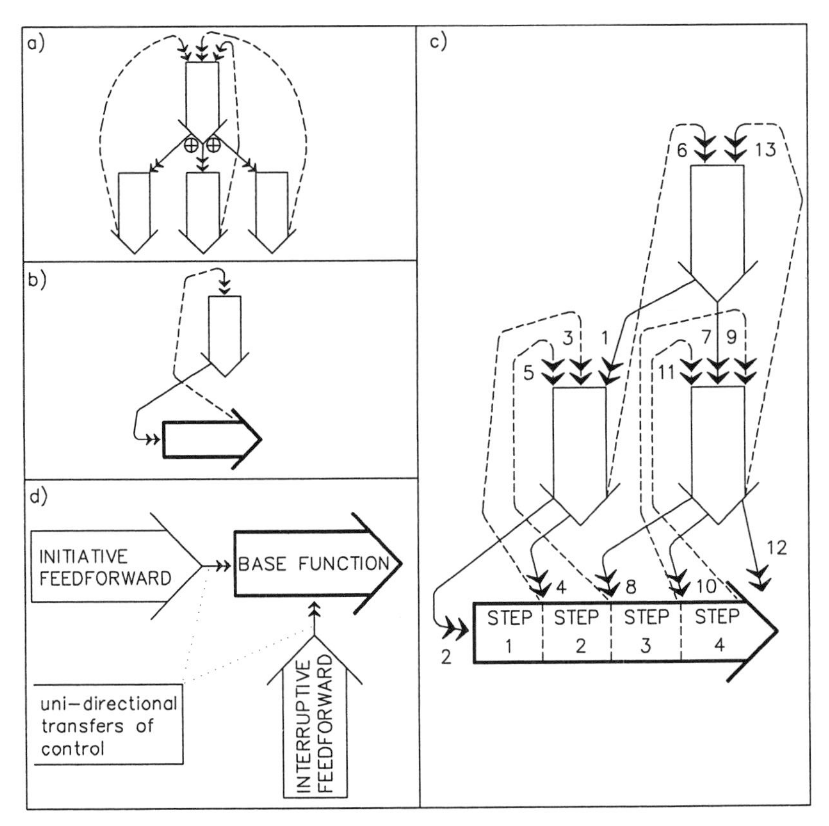

FIGURE 7-7: Transfer of control in the structure-flow chart.
a) activation and return in transaction-centered arrangement;
b) activation of base function, followed by return of control;
c) activation and return relationships at more than two levels in a transform-centered part of the structure-flow chart, with the sequence of activation given by the reference numbers;
d) uni-directional transfer of control.

(Note the use of "long-term" instead of the more precise "non-current.")

Showing Flow of Transactional Inputs and Outputs

Inputs and outputs accompanying the basic control relationship arrows in conventional structure charts are shown by small arrows, each having either a darkened or a non-darkened circle on its tail. These represent control and other types of data, respectively. Two differences from the structure-flow chart exist here, as follow:

- In the structure-flow chart, the circle, or other data symbol, is embedded **within** the basic level relationship arrow, as opposed to being attached to the tails of small **accompanying** arrows.
- The flow of operand transactional data from step to step at a peer level in the structure-flow chart is **usually** shown as following a path separate from that traced by the invocation and return relationships. However, the latter can also be obtained through conventional structure charts, as may be seen by comparing Figures 7-8a and 7-8b.

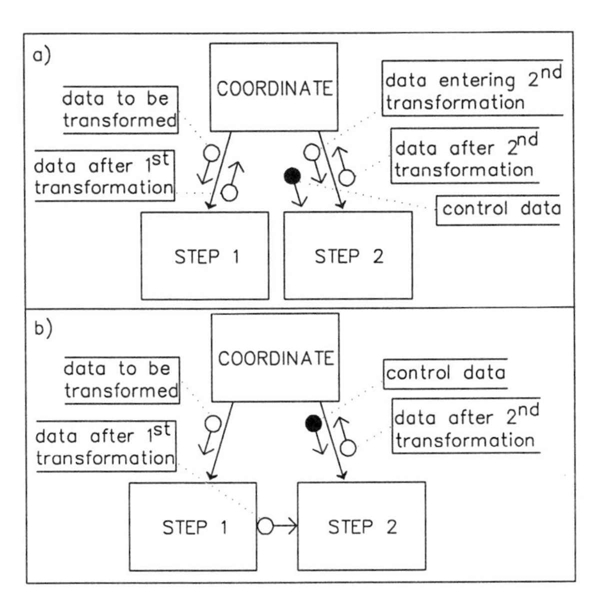

FIGURE 7-8:
Operand transactional input passed from peer in structure chart.
a) passed from step to step via the superordinate;
b) passed directly from step to step, as opposed to the more normal arrangement via the superordinate(s).

In contrast to the normal arrangement in the regular structure chart, Figure 7-9 demonstrates the separation of the flow of operand transactional input from the transfer of control flows within the structure-flow chart. For clarity, it is assumed that this is a physical-base system, rather than an informational one. Thus, we may speak of raw materials, rather than data, to be input and operated upon. The raw materials, once brought into the system base operations, stay within that wideshafted arrow until the end of the system base operations, at which time they have been transformed into delivered finished products.

In addition (Figure 7-9), the little non-darkened circles passed from the system base operations to the validating function after each flow function is feedback of **data** about what has happened to the raw materials in the baseline flow function. The transactional input symbol is used for this feedback data because this is the data which initiates the ensurance function. The feedback data is, in fact, a hybrid of transactional input and control data, with the control data being symbolized by the darkened circle within the transactional input circle. There appears to be no exact equivalent in the conventional structure chart, which merely assumes the existence of the feedback data, except that the structure chart uses control data passed upward when an error has been detected **by the subordinate**. In contrast, with the structure-flow chart, it is not the role of the base (subordinate) function to find the error, but the role of the validating (superordinate) one. Strictly speaking, however, even in the conventional structure chart, control data should be passed upward to a superordinate having more than one

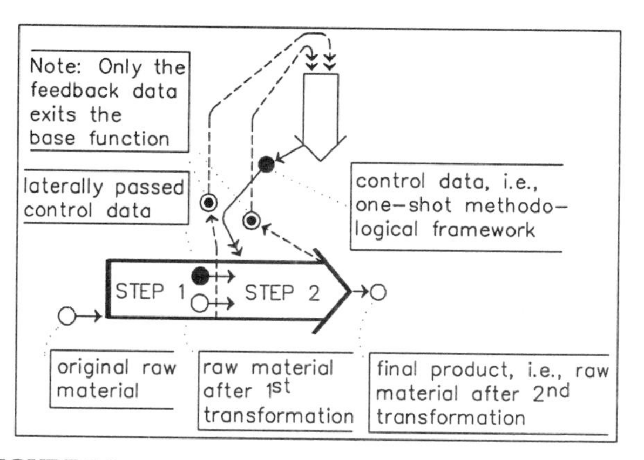

FIGURE 7-9:
Transfer of operand transactional input in structure-flow chart.
Here, in contrast to the regular structure chart, operand transactional input is received directly from the preceding peer level step within the structure-flow chart.

subordinate in order to signal which step has just completed execution. This is the only reason for the element of control data (the small dark circle in the middle of the transactional input circle) accompanying the otherwise transactional input feedback data in Figure 7-9.

Accumulation/Depletion Points

The staying of operand transactional inputs within the baseline wide-shafted arrow is convenient as long as these are not stocked in reservoirs during their flow. For instance, the raw materials obtained by step 1 in Figure 7-10 are output to an inter-flow function reservoir, to be drawn upon by step 2. Figure 7-10 also shows a reservoir established by one run of the function for use by a later run of this same function (see on-page connector "A"). The use of squiggly lines in demonstrating outputs to, and inputs from, these reservoirs symbolizes either out-back-in (in the same functional run) or recursive interactions (between different functional runs) of an objective-defined function with itself. Reservoirs may also be involved in relationships **between** objective-defined functions, as in Figure 7-11. Reservoirs are accumulation/depletion points for which regular structure charts have no symbols.

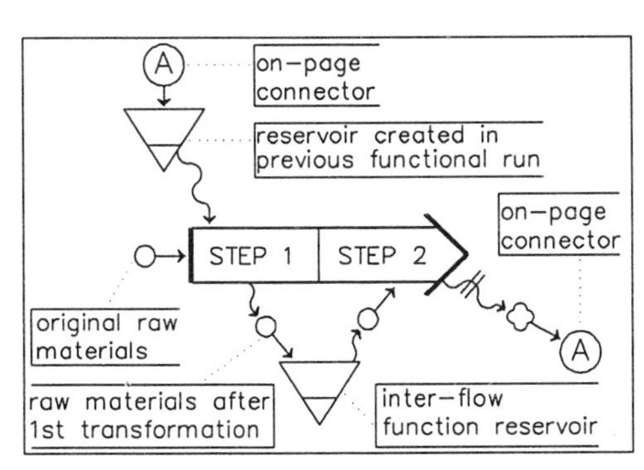

FIGURE 7-10:
Flow via reservoirs in structure-flow chart.
The squiggly lines indicate either recursive interaction involving different runs the objective-defined function interacting with itself, or out-back-in interaction between different steps of the same run of the objective-defined function.

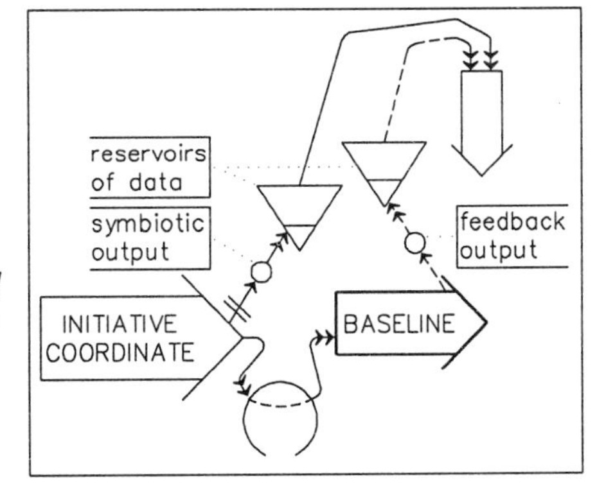

FIGURE 7-11:
Reservoirs in relationships between objective-defined functions.
Interactions between different objective-defined functions via reservoirs in the structure-flow chart.

Branching in the Direct Flow Towards Objectives

Flow Branching and the Wideshafted Arrow

The patterns of data flow between the functional elements within the wideshafted arrow can be interrelated by the various Boolean flow routing logic symbols, as in Figure 7-12. This, of course, is not found in regular structure charts, which in any case cannot show parallel flow routing branches. However, regular structure charts can show **decision** branches, although not always in a way which is clear.

Bringing Direct Operations Into Alignment

Certain specially formatted flowcharts provide a way of aligning those base system operations flowing directly toward the same objective. In many cases, this can also be achieved with structure charts, although the page formatting (packaging) conventions do little to require such packaging of blocks on the page. To illustrate, the input, processing, and output steps of the same function are not aligned in Figure 7-13a, but may be brought into alignment, as shown in Figure 7-13b. While it may appear that this is nothing more than a matter of visual esthetics, improper alignment fails to make visible the all-important common relationships to an objective. In contrast, use of the wideshafted arrow emphasizes objectives-oriented thinking. To make this point, the wideshafted arrow is superimposed over the three steps in Figure 7-13b.

Alignment Involving Flow Branches

Alignment using conventional structure charting techniques is not possible when flow routing branches occur.

What if a simple transaction-centered relationship is found in a basically transform-centered design, as in Figure 7-14a? In this example, step 1 is followed by either step 2a, or step 2b. Both steps 2a and 2b are shown side-by-side, which means that no longer can a linear flow from

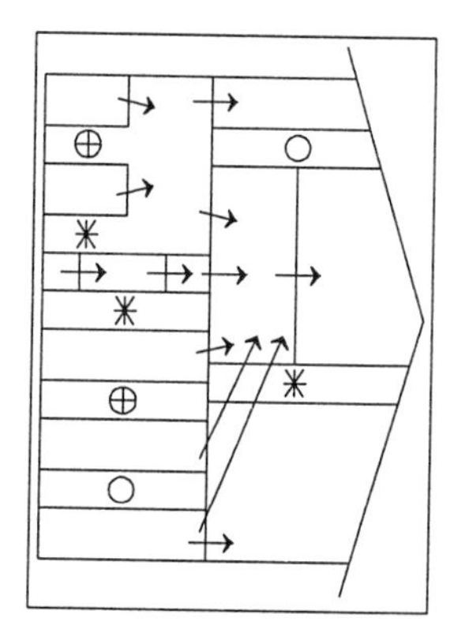

FIGURE 7-12: Flow branching within the wideshafted arrow.
The various relationships between inputs and outputs and/or functional elements are represented by Boolean symbols. As an alternative to the type of representation given here, a specialized hierarchical functional decomposition flowchart (Warnier-Orr, Action, HOS, etc.) may be drawn cross-cutting the left-to-right flow of the wideshafted arrow.

left to right at the bottom of the chart be assumed. In contrast, Figure 7-14b demonstrates the same idea using the structure-flow chart. Here, the decision branches are shown going with the flow, separated by an EXCLUSIVE OR sign, to indicate either one or the other but not both. In this figure, it is possible to clearly see a continuous flow from left to right, despite the alternative paths. Would the reading of conventional structure charts such as Figure 7-14a be easier if all the steps were brought into alignment, as in Figure 7-14c? The non-continuous lines in the upper part of this latter figure show how things would look if all these "atomic level" (i.e., baseline) modules were aligned in a straight way. However, the two optional steps which appear from a distance to be **sequentially** related to one another is still a problem.

Indeed, the alignment problem of the regular structure chart becomes more and more severe in the case of complex decision branches in the flow. Alternative operational paths within the wideshafted arrow using Boolean flow routing symbols is a way of handling what would require multiple levels of the transaction-centered design in conventional structure charts, as in Figure 7-15a. Two ways of handling this problem are used in the structure-flow chart. **First**, where objective-defined functions at different control levels are involved, the problem is handled in a way which is similar to that used in the regular structure chart, i.e., by levels of transaction-centers involving controller functions. However, the flow here is from left to right, as in Figure 7-15b. **Second**, Boolean flow routing symbols may be used for branching arrangements within the same objective-defined function, as is also demonstrated in Figure 7-15b. This second alternative involves hierarchical functional decomposition of the wideshafted arrow, as opposed to levels of control, and is applicable when different parts of the flow are not at different levels of control.

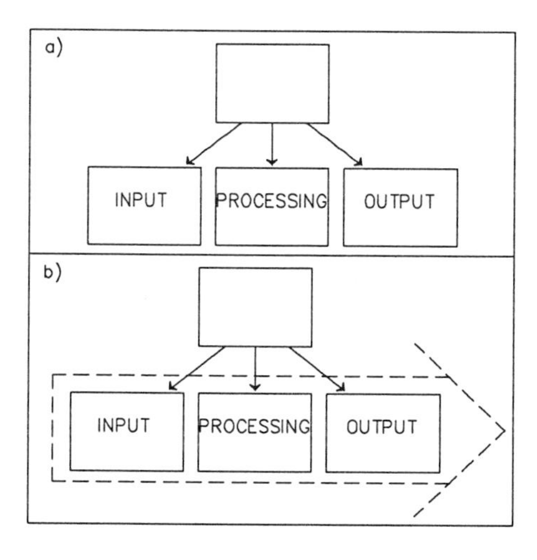

FIGURE 7-13: Baseline ("atomic" level) steps aligned in the structure chart.
a) example of non-alignment of steps;
b) steps in "a" brought into alignment.

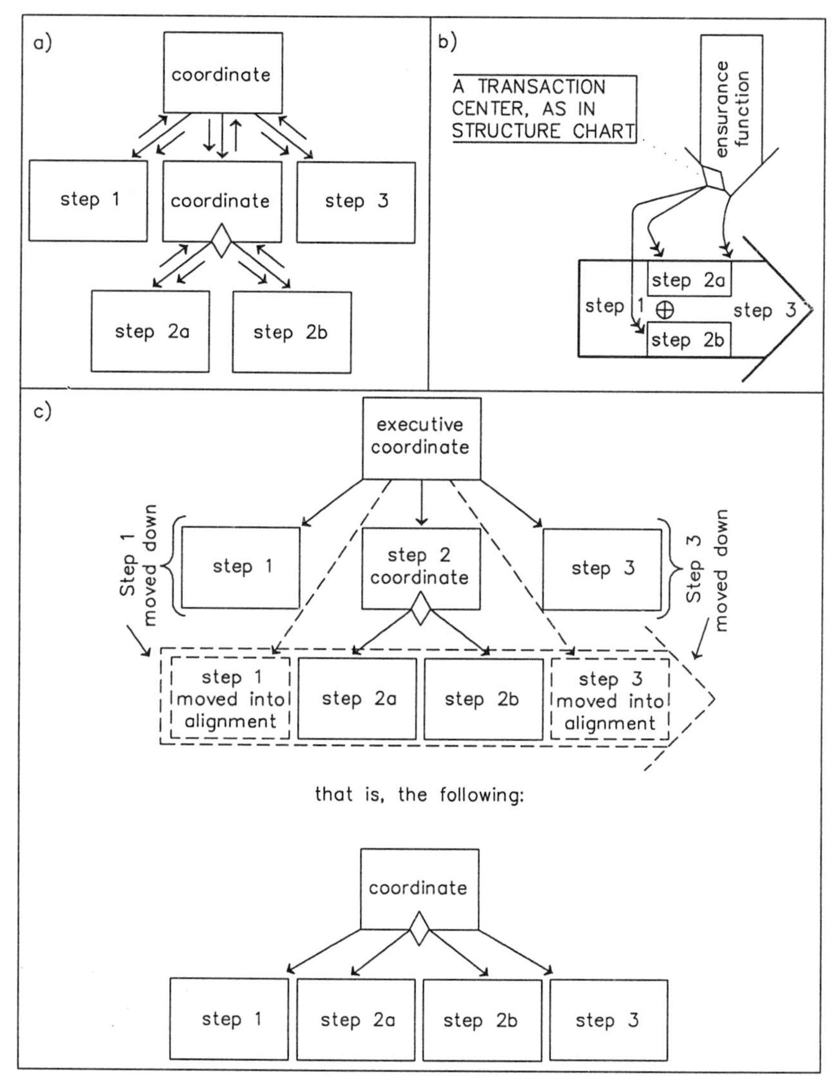

FIGURE 7-14: Alternative steps and alignment in structure charts. *This shows the effect of alternative paths in the flow on the alignment of steps in structure charts, and comparison to structure-flow chart:*

a) transaction-center disrupting simple left to right flow pattern of transform-centered arrangement in structure chart;
b) structure-flow chart equivalent of previous figure, in which left to right flow is not disrupted by alternative branches;
c) another arrangement of the regular structure chart to handle the same problem.

Iterative Branches to Redo Operations in Case of Error

Backward Looping Using Flowcharts

For purposes of having a basis of comparison of both structure and structure-flow charts, Figure 7-16a shows a form of specially formatted flowchart rather than a structure or a structure-flow chart. First step 1 is executed, then the results are checked. If there is no error, execution continues with step 2. If there is an error, however, a correction is performed on step 1, which is repeated and then rechecked, and so on. What is shown here is an ensurance relationship, since the validation and correction functions pertaining to a step are executed **after** the step has completed (successfully or unsuccessfully). A loop is formed if an error occurs by going back to the erroneously executed step.

Backward Looping Using Structure Charts

Figures 7-16b to f all deal with the same problem as Figure 7-16a, except that they use a structure chart. Differences exist in each case. Figures 7-16b to d use hierarchies, which, although neater in appearance than the flowchart, are harder to follow in terms of their flow routing logic than is Figure 7-16a.

At least Figures 7-16b to d do not distort the levels-of-control relationships. However, corrections in structure charts are usually shown in the fashions shown in either Figure 7-16e or Figure 7-16f. Both of the latter distort the levels-of-control relationships, for reasons which will be discussed towards the end of this chapter in connection with the

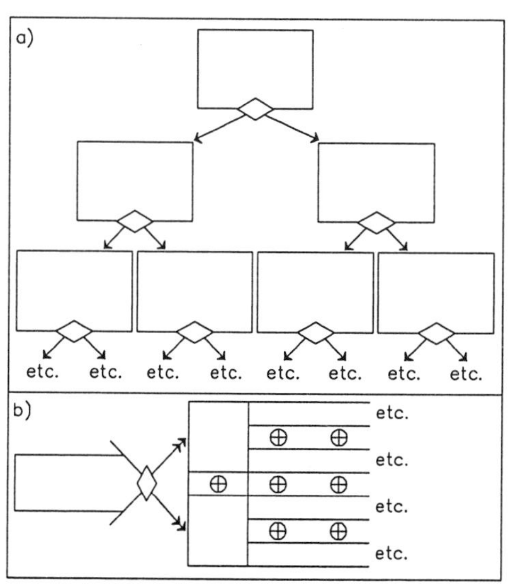

FIGURE 7-15: Decision-trees, structure vs. structure-flow charts.
- a) *decision-tree in structure chart, obtained by successive levels of transaction-centers;*
- b) *decision-tree in structure-flow chart, equivalent to structure chart except that it does not show the various* levels *of control after the first.*

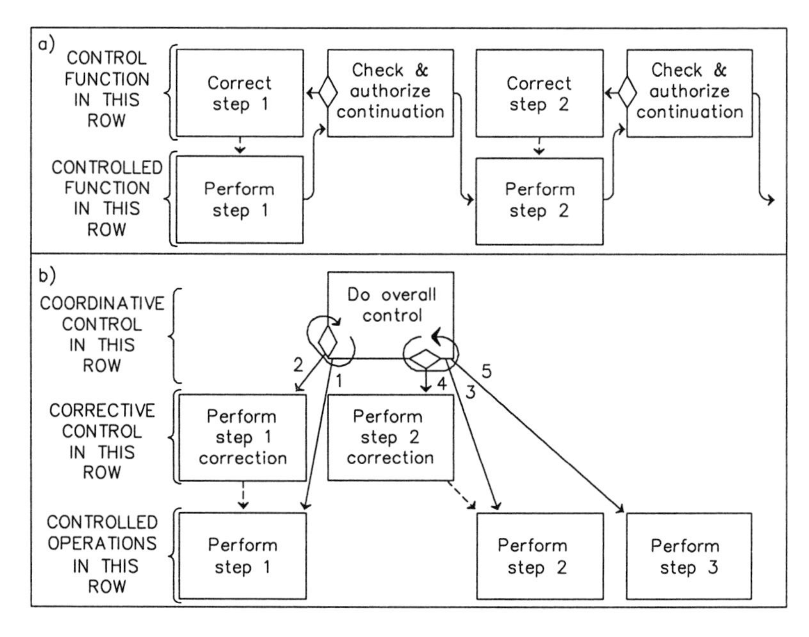
a)
CONTROL FUNCTION IN THIS ROW
CONTROLLED FUNCTION IN THIS ROW
Correct step 1
Check & authorize continuation
Correct step 2
Check & authorize continuation
Perform step 1
Perform step 2
b)
COORDINATIVE CONTROL IN THIS ROW
CORRECTIVE CONTROL IN THIS ROW
CONTROLLED OPERATIONS IN THIS ROW
Do overall control
2
1
4
3
5
Perform step 1 correction
Perform step 2 correction
Perform step 1
Perform step 2
Perform step 3

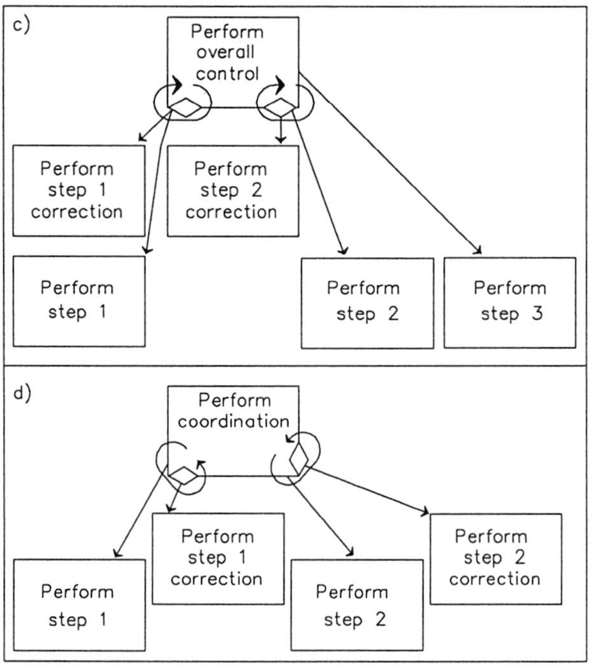
c)
Perform overall control
Perform step 1 correction
Perform step 2 correction
Perform step 1
Perform step 2
Perform step 3
d)
Perform coordination
Perform step 1 correction
Perform step 2 correction
Perform step 1
Perform step 2

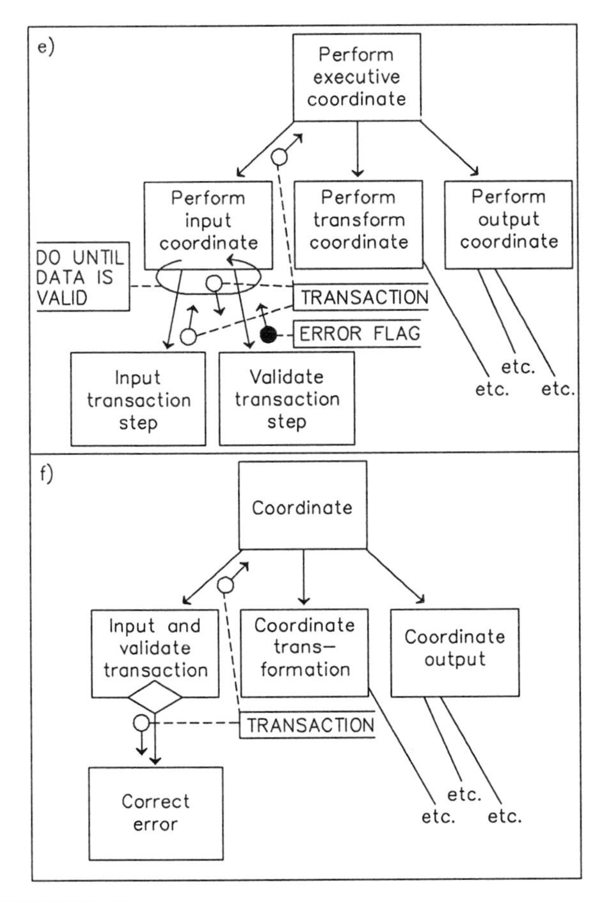

FIGURE 7-16:
Backward loops for error correction in structure chart.

a) ***(facing top of page)*** *flowchart formatted to highlight direct line of flow, with ensurance relationships in a different alignment;*
b) ***(facing top of page)*** *ensurance control versus the direct line of flow in the structure chart;*
c) ***(facing bottom of page)*** *version of the same structure chart, but with transaction-centers for corrective loops;*
d) ***(facing bottom of page)*** *another version of previous figure, with corrections handled in regular left to right flow;*
e) ***(above)*** *error handling done in structure chart by repeating the input step until transaction is free of error;*
f) ***(above)*** *error handling done in the structure chart by a conditional call to a subordinate to handle the error.*

subject of internal levels of control. In Figure 7-16e, one step inputs data and the next step verifies (edits) it. The two steps are repeated in that sequence again and again until the data passes the edit checks; i.e., the loop surrounding the two modules in the chart is a DO UNTIL type of a loop. In Figure 7-16f, a subordinate is called to handle an error, if it occurs.

Backward Looping With the Structure-Flow Chart

Figures 7-17a and 7-17b demonstrate the use of wideshafted arrows, instead of rectangles. In Figure 7-17a, each validation function is diagrammed as having two possible mutually exclusive objectives; while, in Figure 7-17b, the single validation function has a mutually contingent set of six specific primary objectives. Thus, the link to step 2 of the base function is made in the first validation function's run, once the base function's step 1 has executed correctly; the link to step 3 of the base function is made by the run of the validation function, which deems that step 2 of the base function has completed without error, and

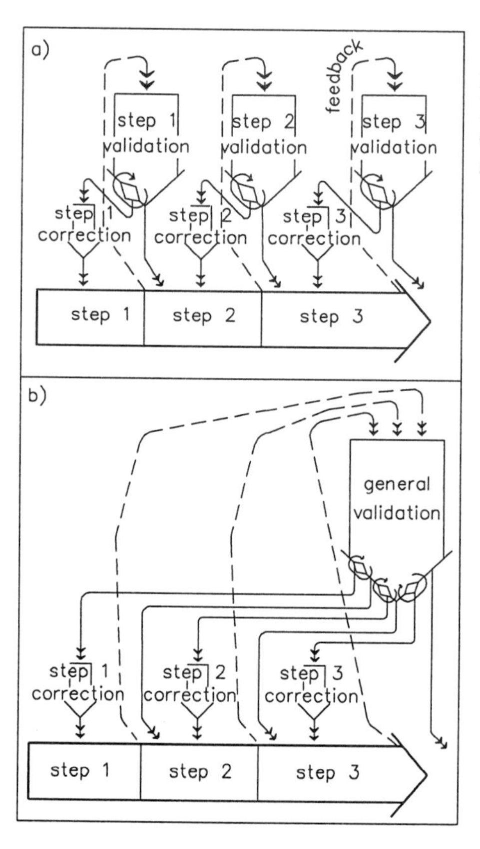

FIGURE 7-17: Backward looping to correct errors with structure-flow chart.
a) *structure-flow chart having no apex, i.e., no overall controlling validation function for its ensurance operations;*
b) *structure-flow chart with apex validation function in place.*

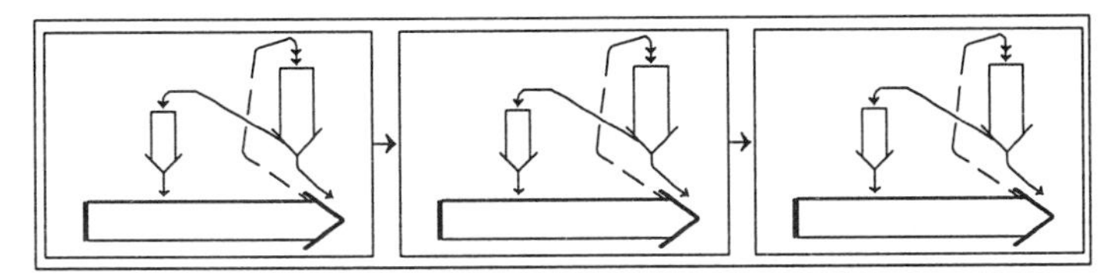

FIGURE 7-18: Homologous (flat) structure chart. *This is intended to capture relationships which are similar to those captured by the preceding series of flow, structure, and structure-flow charts.*

so on. The handling of errors, if they occur, is portrayed by including the corrective function for each step in the figures. Figures 7-17 provide more page space for backward looping to correct errors than do Figures 7-16 b, c, and d.

Finally, Figure 7-18 may be compared to Figure sets 7-16 and 7-17. This is an homologous, or flat, structure chart, similar to a flowchart. In effect, the validation operations are contained **within** the same modules as the validated ones. In fact, because of its homologous nature, another way of looking at Figure 7-18, as well as Figure 7-17a, is to think of it as a form of specially formatted flowchart, like that in Figure 7-16a.

Regardless of form, the same problem may be represented, in different arrangements, in all of Figures 7-16, 7-17, and 7-18. That decision-making operations may be grouped in the same module as the operations about which they make decisions is explained by the concept of "internal adaptation operations," which is discussed toward the end of this chapter.

FUNCTIONAL CLASSES, LEVELS, AND TIME-ORIENTATIONS

Introduction

Although alignment of the bottommost level of modules across different legs of the hierarchy is often desirable in the structure chart, the modules at the bottom of different hierarchy branches should sometimes be kept at **different** levels. That is, when using the functional approach, only the modules belonging in the same objective-defined function should be brought into alignment. For example, as in Figure 7-19, when different levels of summarization of data are produced by a "control break" reporting program, the operations pertinent to each level should be left at that level in the structure chart. Otherwise, the

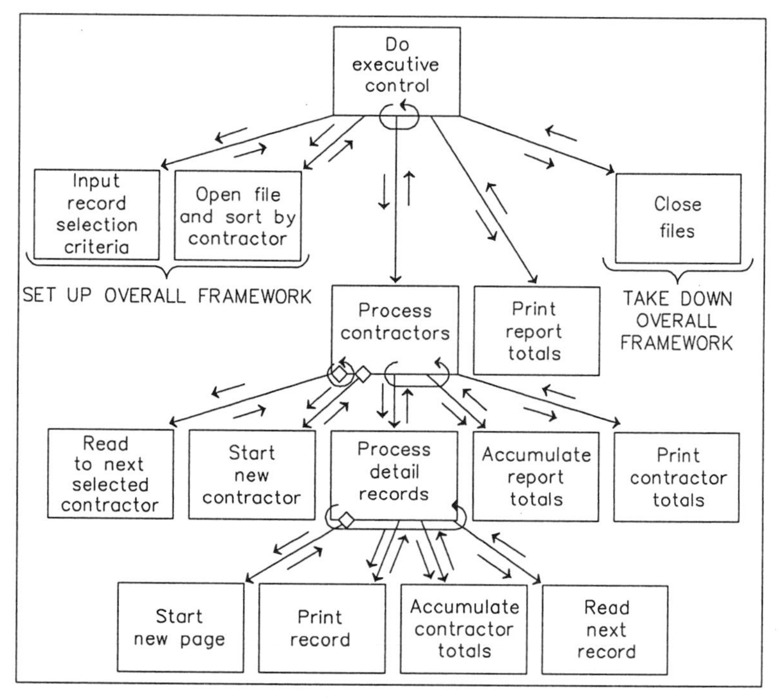

FIGURE 7-19: Structure chart in which alignment is not desirable. *This chart of a control break reporting program demonstrates that alignment of bottom levels of all branches is not desirable in all cases.*

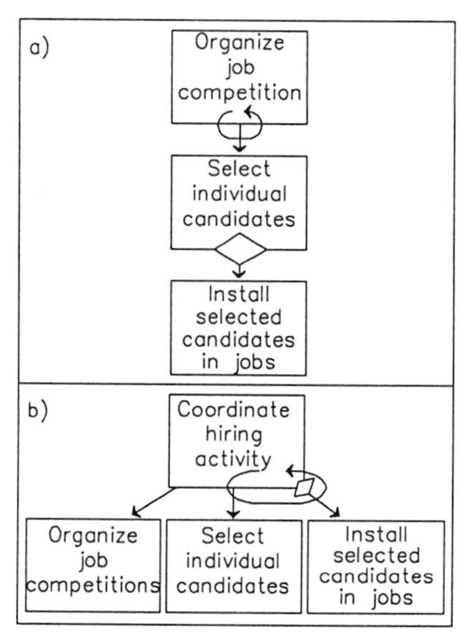

FIGURE 7-20: Another structure chart in which alignment is not desirable. *This shows two ways of capturing sequential relationships versus different functional levels by alignment or non-alignment of steps in structure charts:*

a) personnel hiring activity highlighting different levels of control;

b) personnel hiring activity highlighting sequential relationships.

result would be a flattened structure chart emphasizing sequence over different functional levels. Another example of this idea is presented in Figure 7-20.

The idea being conveyed in the above is that functions in different classes, and/or at different control levels, and/or with different time-orientations, should **not** be aligned in the structure chart when attempting to specially format the chart for conceptual prototyping. This should be remembered in connection with the following discussion's dealing with possible use of the structure chart for conceptual prototyping.

Another issue to recall in connection with the following discussion of using the regular structure chart for conceptual prototyping, is that interruptive feedforward and environmental interactions are extremely

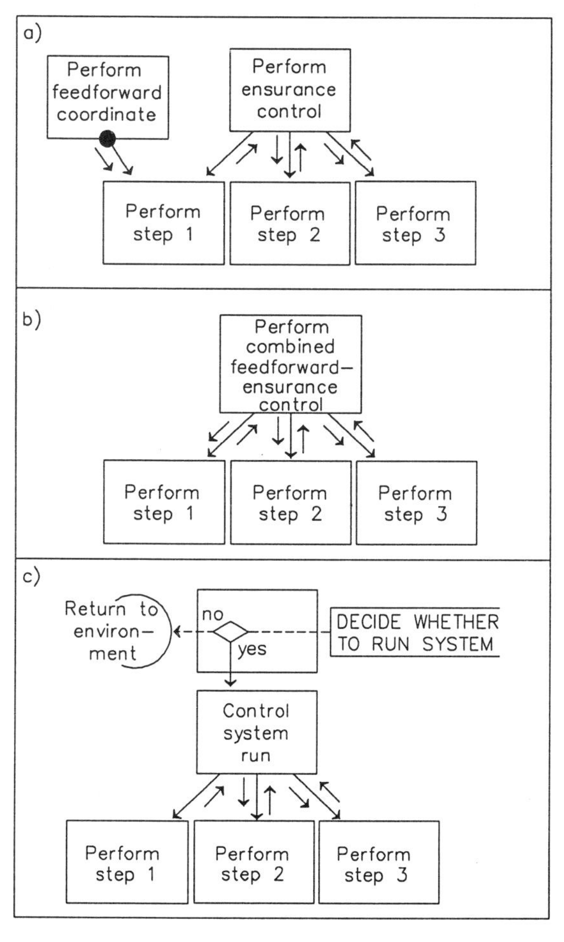

FIGURE 7-21: Ways of showing initiative relationships in structure charts.

a) ***(left)*** *one way, albeit a very unorthodox one, of showing process initiation;*

b) ***(left)*** *a more accepted way of showing process initiation;*

c) ***(left)*** *initiative function in structure chart contained in executive coordinating/validating function;*

d) ***(next page)***

e) ***(next page)***

f) ***(next page)***

important in complex, probabilistic, open systems, and that, unlike the conventional structure chart, the structure-flow chart is able to clearly indicate relationships when it conceptually prototypes a system.

Now, on with the discussion about use of the regular structure chart for purposes of conceptual prototyping.

Current Operations

Figure sets 7-21 to 7-22 show various models which may represent both feedforward and ensurance current relationships in the same chart.

The model used in Figure 7-21a is not found in the computer literature. In contrast, in Figure 7-21b the task initiator sees the task through

FIGURE 7-21 d, e, and f:
d) initiative function done by subordinate;
e) another unorthodox way of showing initiation, in which increasingly clear distinction of initiative and ensurance time-orientations is made;
f) further expansion of the idea in "e," by placing operations on underside of basic set of steps in structure chart.

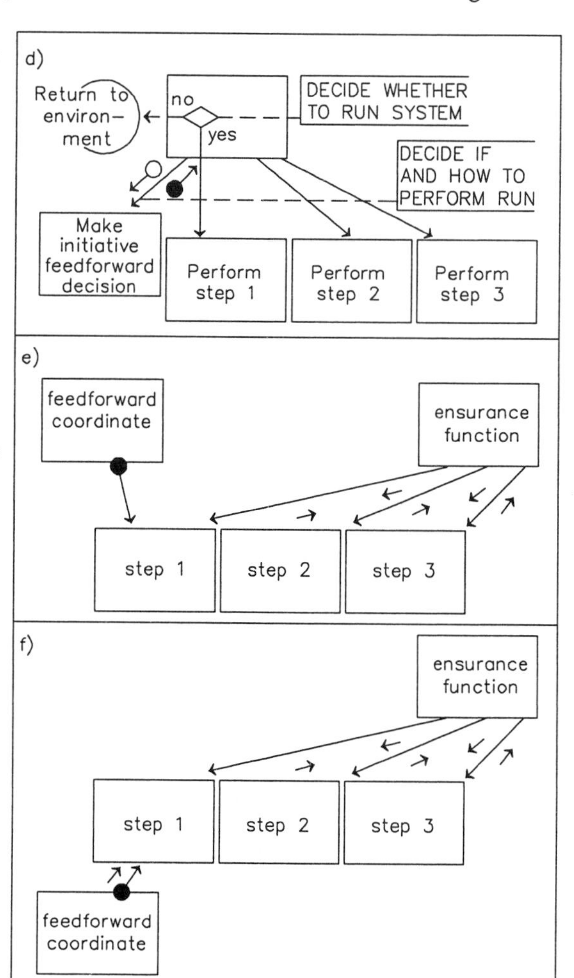

from start to finish, using conditional transfers of control in all cases of activation. The latter model simplifies charting by reducing diagram space requirements. Figures 7-21c and 7-21d are two other examples of how initiative feedforward may be diagrammed in conventional structure charts.

Figure 7-21e, another variation of Figure 7-21a, has the ensurance function placed just over the final step to which it applies. The spatial arrangement clearly designates the ensurance relationships, in that it shows the validating function after all steps. Although those control relationships forming part of the ensurance cycle are not shown in their correct spatial proportions, the placing of the controller functions in this fashion is advantageous in other ways, one of which is that the corrective blocks (see above) associated with ensurance may easily be "hung over" the steps to which they apply.

Figure 7-21f is similar to Figure 7-21e, except that an interruptive coordinate is placed on a side of the step to which it applies opposite from that of ensurance function. This makes use of another dimension of the diagram, and avoids overlapping of interruptive and ensurance current decision-making functions. Thus, it is easier to "hang" the corrective blocks associated with the ensurance function over the steps to which they apply. Another advantage is that any current decision-implementing blocks specifically associated with the coordinate may be shown without cluttering the upper part of the chart.

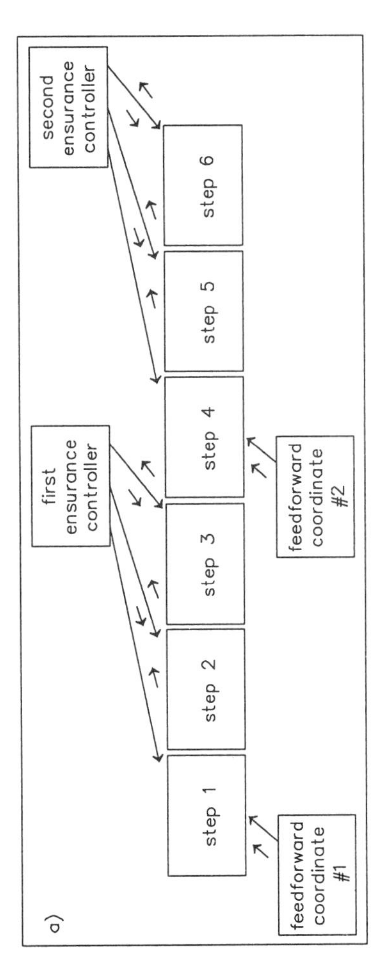

FIGURE 7-22a:
Avoiding time-orientation interferences.
By placing feedforward operations on the underside of basic steps, interferences of different time-orientations are avoided:
a) as captured in the structure chart.
b) ***(next page)***

Also, where there are many ensurance functions, the interferences between interruptive feedforward and ensurance functions shown on the same side of the steps to which they apply are avoided. For instance, in Figure 7-22a, coordinate #2 would interfere with ensurance function #1 if they were not on opposite sides of the base sequence steps.

In the model of Figure 7-22b, ensurance and interruptive feedforward controller functions are distinguished and placed on opposite sides, while initiative coordinates are at the front, with validation functions being located at the ends of the series of steps to which they pertain. Why such an arrangement is preferred over those which visibly demonstrate an apex to the structure chart hierarchy is explained beginning with Figure 7-22b, in which various senses of feedforward are shown. This diagram clarifies the preference for using both sides of the controlled sequence when both ensurance and feedforward relationships are involved.

However, before proceeding with the type of page arrangements adopted in this book, it is a good idea to regress a little, to first attempt further ways of capturing the needed relationships in the regular structure chart used in a conventional fashion. To begin with, Figure 7-23a, a rectangular structure chart of a simpler example, involves relationships in which something is inserted prior to the next step. Figure 7-

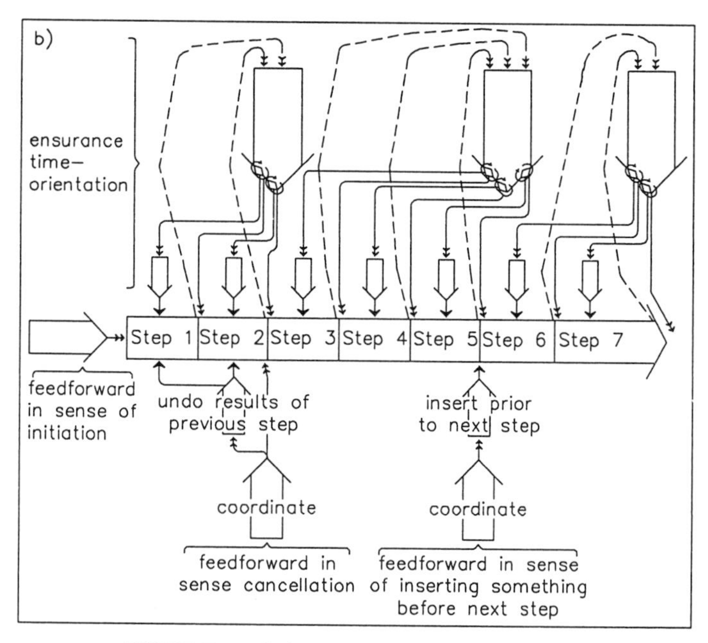

FIGURE 7-22b (See "a" on previous page):
b) as captured in structure-flow chart.

23b results from combining the ensurance and feedforward functions of Figure 7-23a into one rectangular block, and inserting the operations invoked by the coordinate as a step in the sequence. This gives a series of steps in which one is not a controlled step, but a control step aligned below the controlled ones; i.e., the figure is again packaged so that the control step is not in the same line as the controlled ones. In this case, the control step is inserted in the base sequence on a conditional basis, as symbolized by the decision block attached to the tail of the arrow pointing to it from the controller. Figure 7-23b regresses even further than Figure 7-23a from the page arrangements portrayed in Figure 7-22b, as the structure chart is once again brought into its conventional hierarchical format with a single apex.

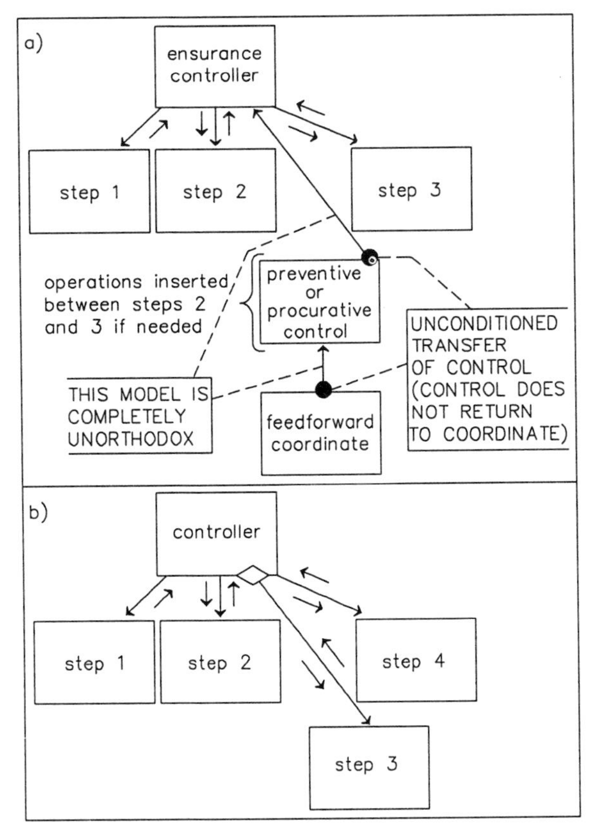

FIGURE 7-23: Interruptive current operations in the structure chart. *This shows two alternatives, as follow:*

a) an unorthodox way of showing step insertion interruption;

b) structure chart having ensurance and interruptive roles in the same controller.

Notwithstanding the simpler appearance of the arrangement in Figure 7-23b, the arrangement given in Figure 7-22b is preferred. The reasons for this are essentially the same as the reasons for having a pyramidal (three-dimensional) representation of system levels of control, as opposed to the simple triangular (two-dimensional) one. These reasons should be somewhat clearer by the time this chapter has been completed. Meanwhile, the following continues to add to the discussion of how different functional classes, time frames, levels, and time-orientations may be captured by special sub-formatting of the regular structure chart.

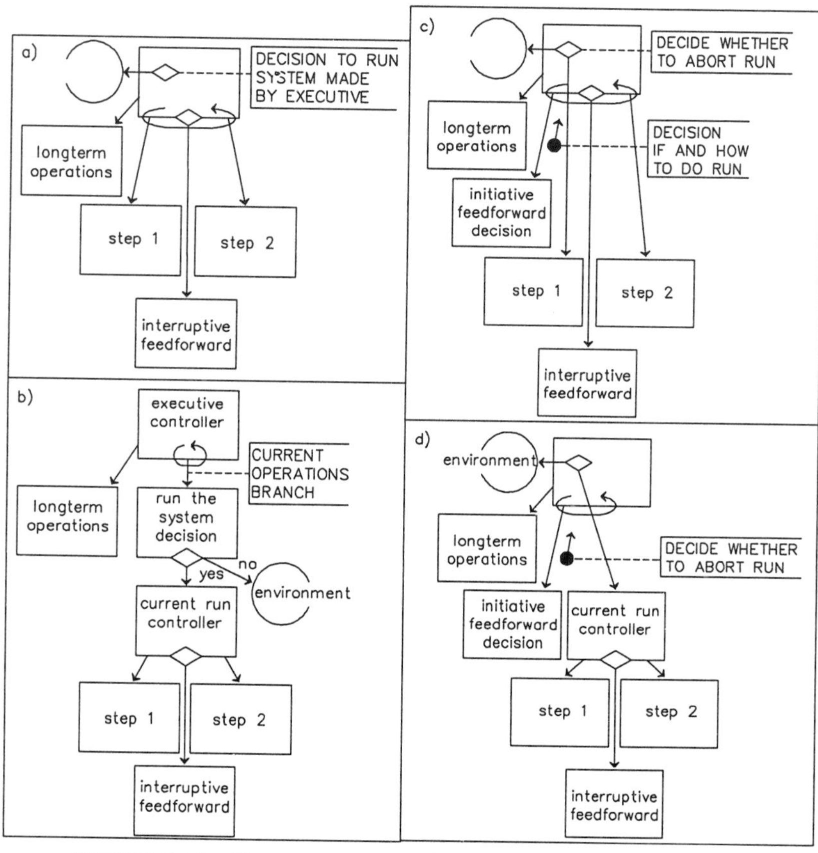

FIGURE 7-24: Non-current operations included in structure chart. *This provides different ways of showing long-term (non-current) operations in structure charts along with short-term (current) ones. In all of these figures, the short-term operations are within a loop, to symbolize their greater relative frequency.*
a) model #1; b) model #2 ; c) model #3; d) model #4.

Non-Current Operations

Long-term (i.e., non-current) operations make and implement decisions concerning framework inputs affecting the long-term of system existence. As such, they differ from the system base and current operations, which affect only the current system run, or which affect only transactional inputs. Long-term operations, by definition, are feedforward in nature, in that they affect only the future; i.e., they set up the conditions for the current operations. Since the long-term operations affect only the future, they may precede the current operations in time. Moreover, they are executed less frequently than the current operations. These concepts are reflected in the structure charts in Figures 7-24.

Long-term operations are pictured in another fashion in structure-flow charts, as in Figure 7-25. As seen, the long-term operations relate to the base operations through a single-headed arrow, to indicate the lack of stimulus transmitted. In other words, the system base operations are not activated, or otherwise currently controlled, by these operations. In contrast, all other relationship arrows in the diagram are double-headed since they activate the system base operations. The long-term operations simply provide frameworks in whose context the activation relationship and current system flow take place. The distinctions between activation and non-activation relationships and between uni-directional and bi-directional control flows can also be captured in conventional structure chart, as in Figure 7-26. However, what is seen in Figure 7-26 is a completely unorthodox structure chart.

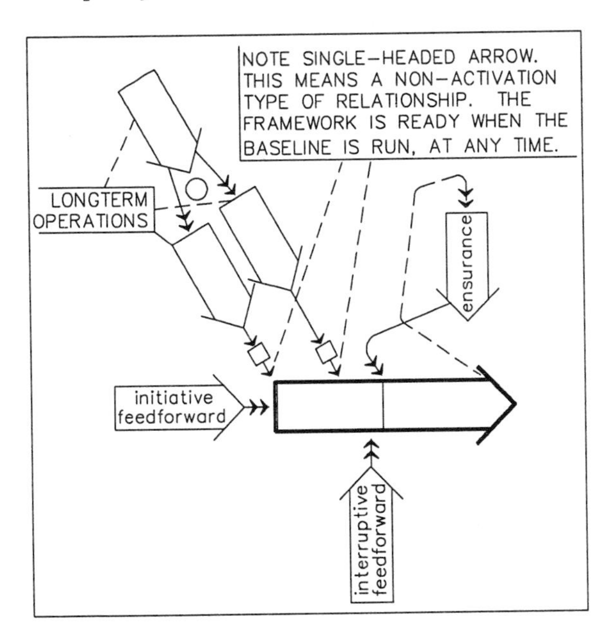

FIGURE 7-25: Long-term operations in the structure-flow chart.

FIGURE 7-26: Model including all time-orientations in a structure chart. *This shows a comprehensive, but completely unorthodox model.*

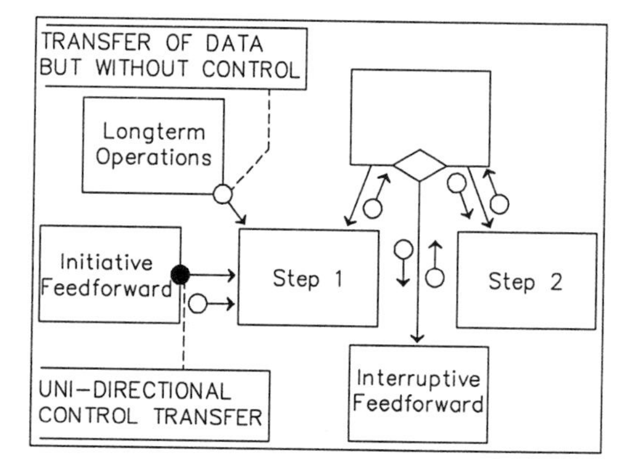

LEVELS OF CONTROL IN THREE DIMENSIONS

THE STRUCTURE-FLOW CHART AND HIERARCHICAL FORM

Structure-flow charts appear to be more like flow charts than like levels-of-control charts, since hierarchical arrangements are not immediately evident. However, different hierarchies may be envisioned for non-current, and for each of initiative, interruptive, and ensurance, current adaptation operations. This was illustrated in Figure 5-2 (Chapter Five).

For those who feel more comfortable with having an apex, the arrangement into different hierarchies may be visualized in terms of a pyramid. In the pyramid, each hierarchy may be represented by a triangular side, and in the case of a four-sided pyramid, the system base operations by the square pyramid base. Such a pyramid is laid out in "flattened" form in Figure 7-27a.

Within each hierarchy of wideshafted arrows (represented by one triangular face of the pyramid), the control relationships of different levels are similar to those in conventional levels-of-control charts, i.e., the higher level objective-defined functions perform **all** the different types of adaptation operations with respect to their subordinates. This includes the roles of all the different functional classes, time frames and time-orientations. Of course, as a result of using structure-flow charting conventions, flow representation between controllers and their subordinates becomes very complex in this type of a relationship.

Finding the Center of Authority in the Pyramid

Since the apex of the pyramid is nothing but a point in three-dimensional space, where is the center of power when a pyramid is used? To answer this, there is no "boss" in the system, conceptually speaking, or, at least, the structure-flow charting method leaves this flexible. Notwithstanding, a single validation module, one having the transaction-centered design, might be conceived. This would be possible as long as parallel branches are allowed within this transaction-centered design, which would require allowing and/or logic in the transaction-centered design, as in Figure 7-27b.

Alternatively, if parallel branches are permissible, one could "elect" one of the hierarchy executives as overall coordinator/validator. Thus, for example, cross-communication could occur between the executive ensurance function and the coordinates on the other sides of the baseline wideshafted arrow, such as to obtain a system in which there is but one ultimate source of power. This could be done by showing inputs and outputs connected by on-page connectors relating the executive ensurance function to the coordinates on the various other sides of the baseline. Return to the executive in-

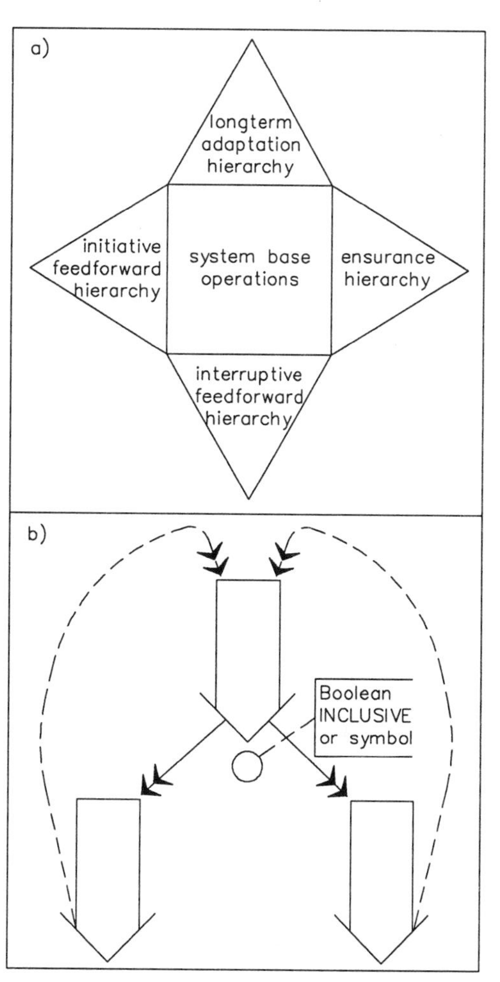

FIGURE 7-27: Pyramidal representation of system levels of control.

a) *four-side pyramid representation of hierarchies in different time-orientations. The above triangles may be folded up to form a 4-sided pyramid.*

b) *AND/OR gating logic in transaction-centered arrangement in structure-flow chart.*

voking source would then occur through normal feedback channels, as each subordinate in the chain reports back to its direct superordinate after completion of its tasks. The relationship can, to simplify things, be arranged through executives on each side of the base system operations, as demonstrated in Figure 7-28.

Alternatively, a **feedforward** executive could take the role of the chief executive. This latter arrangement is possibly preferable to that in Figure 7-28, since it removes all feedforward duties from the ensurance executive controller, and emphasizes that in open systems in dynamic environments, the feedforward executives are perhaps of greater importance in system functioning than ensurance ones. In fact, the best idea might be to place the center of power in the hands of the chief planner (i.e., the non-current operations). To know which hierarchy is the most important in the pyramid is really a value judgement (although, as discussed in Chapter Nine, placing the long-term operations on an equal footing with the short-term ones goes intuitively opposite to the normal conception of system levels of control, such as in the Anthony triangle).

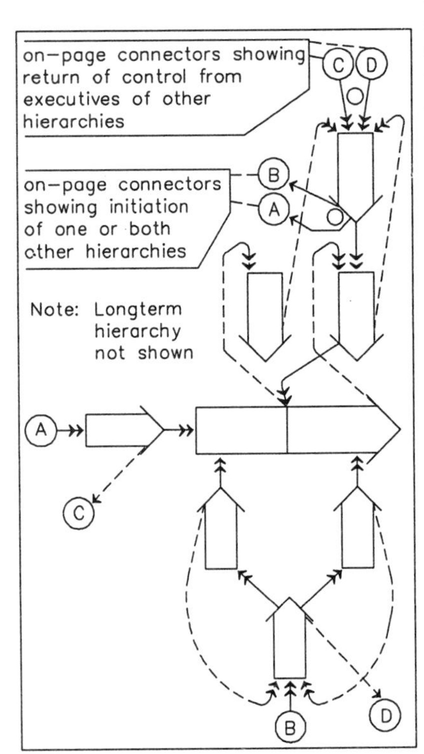

FIGURE 7-28: Allocation of overall control to a particular hierarchy. *Executive controller of ensurance hierarchy given charge of system adaptation operations.*

Three-Sided Pyramids

As a compromise between the four-sided pyramid and a simple triangle, a **three**-sided pyramid might be attempted. For example, because of the straightforward flow routing dependencies which exist between initiative and the other two current operations hierarchies, the initiative feedforward hierarchy might be combined with one of the other two current hierarchies, as in Figure 7-29a, which merges the initiative and the ensurance current hierarchies. This idea could be used to generate a three-sided pyramid, having a triangular base, as may be obtained by folding up the flaps in Figure 7-29b. Alternatively, as in Figure 7-29c, the

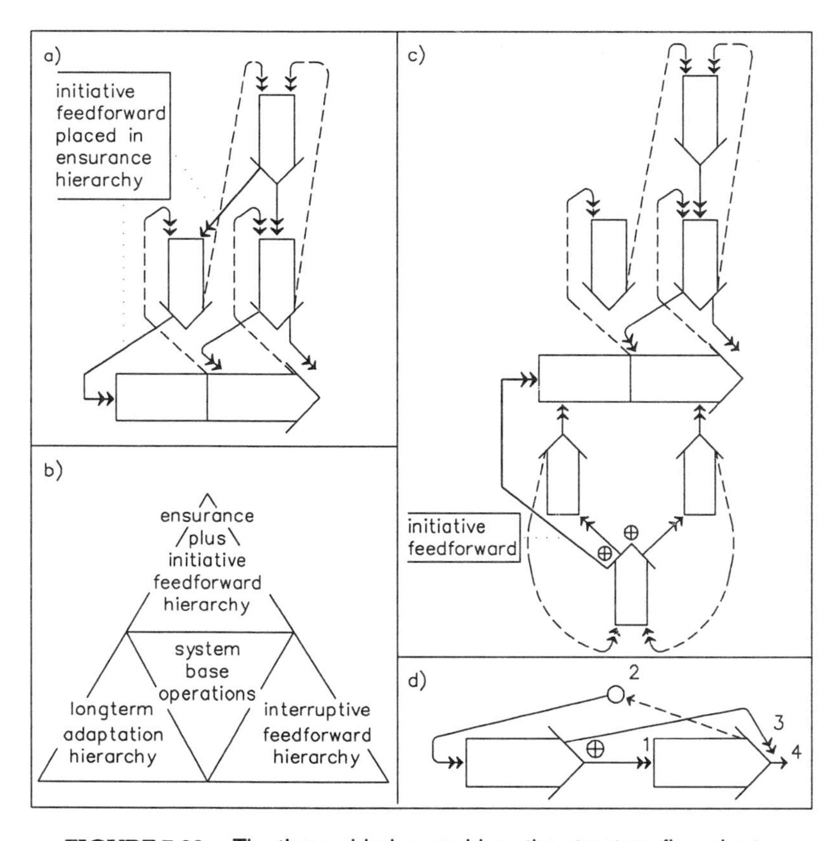

FIGURE 7-29: The three-sided pyramid vs. the structure-flow chart.

a) *With the three-sided pyramid approach, the initiative functions may, for example, be assigned to the ensurance hierarchy, as they are in this figure.*

b) *Three-sided pyramid representation of system adaptation hierarchies. The triangles surrounding the system base operations triangle may be folded up to form a three-sided pyramid. In this particular example of a three-sided pyramid, the initiative hierarchy of the four-sided pyramid is integrated with the ensurance hierarchy.*

c) *Integration of initiative and interruptive roles into the interruptive position in the structure-flow chart. Here, the interruptive feedforward hierarchy also assumes responsibility for initiative operations.*

d) *Integration of initiative and ensurance roles into the initiative position in the structure-flow chart.*

initiative feedforward operations could be placed in the interruptive feedforward hierarchy.

The Basic Need for a Three-Dimensional Form

So far, only two current adaptation hierarchies have been integrated. Further integrating of all the initiative, ensurance, and interruptive roles obtains a **single** triangle for **all** current adaptation operations. This still leaves a separate triangle for the non-current adaptation operations, with the result that flow can still be more clearly demonstrated than with a single triangle. Of course, given only two triangles to represent all adaptation operations, one could no longer use the pyramidal form. However, a three-dimensional representation of system levels of control is still present here, since there is more than one triangular face.

Notwithstanding the three-sided pyramid idea, or another three-dimensional form, the preference in the structure-flow charting method is for a **four**-sided pyramid. The reasons for this are the same as those which justify the shape of the structure-flow chart, as explained earlier in this chapter by comparing it to the conventional structure chart. Mainly, these reasons have to do with the need to clearly demonstrate both flow and levels of control in the same chart.

But why have any three-dimensional form if the only concern is to show levels of control? The answer is that a three-dimensional representation is needed to capture **internal** levels of control.

The Counter-Intuitive Nature of the Three-Dimensional Model

Much of science is counter-intuitive, and so, perhaps, is the three-dimensional form, since it places non-current operations at the same control level as current ones. The structure-flow chart's implied three-dimensional pyramidal layout of levels may be compared to the simple triangular models of system functional levels. In such triangles, planning operations are often understood to be on a higher **level** than current ones. In the structure-flow chart, in contrast, the planning operations slope in from the side of the control operations, and are in a triangle of their own.

That the structure-flow chart places the planning operations in a path different from that of the current adaptation hierarchies is not just a matter of showing flow dynamics in addition to levels of control in the diagram. To assume that scope always increases along with long-term orientation is to assume that a valid basis exists for comparing the scope of short-term and long-term operations. It seems that this is more like comparing apples to oranges. That the Anthony triangle appears to make such an assumption is based more upon subjective values reflected in the typical formal organization chart than upon pure conceptual logic. In the typical organization chart, long-term decision making is assumed to be more **important**, and therefore long-term decision makers are given more authority than short-term decision makers.

It is indeed likely that long-term decisions affect more parts of the system. However, this is not a hard and fast rule from which a general systems principle can be obtained. For instance, it may be that the general coordinator of daily nursing operations in a large hospital has more scope of effect in the hospital than a long-range hospital planning consultant hired to do ten weeks work as an adviser. Is this not reasonable? Thus, to create a truly universal model, the long-range adaptation functions must be placed in a control levels hierarchy different from that used for the short-term adaptation functions. In other words, the scope subject should be handled apart from the discussion of the functional classification scheme.

INTERNAL CONTROL LEVELS

Consequences of the Idea of Direct Relationship to An Objective

Is the conventional structure chart in Figure 7-30a the same as the one with the wideshafted arrow in Figure 7-30b? The answer is "approximately, but not quite." To make them more similar, the "internal" adaptation functions (shown by de-emphasized lines in Figure 7-30c) must be included. At the micro-level these build a structure-flow chart having a morphology similar to that of the overall structure-flow chart seen in this figure, using as a base each wideshafted arrow drawn with normally emphasized lines. That the morphology at the micro-level resembles that at the macro-level is an essential facet in the development of a paradigm of system conceptual structure. This paradigm is explainable, as seen in Figure 5-4 (Chapter Five), in terms of pyramids adhering to other more basic pyramids, with the less basic pyramids providing levels of control of an "internal" nature. This paradigm is explainable in heuristic terms, even though the fact is that the distinctions between the different types of levelling dimensions (internal versus external) are often extremely blurred.

Internal Control Levels in the Conventional Structure Chart

Since each rectangle may contain both direct and indirect operations closely related to the same objective, the internal adaptation operations are usually hidden in conventional structure charts within the umbrella of the charted externally related modules. Sometimes, however, the internal adaptation operations are charted with separate rectangles. Unfortunately, it is virtually impossible in the conventional structure chart to tell whether the internal or external sense of control levelling is being represented by a rectangle (assuming that it is possible in a case in question to clearly distinguish operations as belonging to different levelling dimensions).

For example, the earlier Figures 7-16e and f demonstrated the difficulty of distinguishing the different levelling dimensions in the conventional structure chart. If the "Validate transaction" operation in these diagrams involves nothing more than editing of input data, this apparently constitutes internal validation operations, as opposed to validation operations imposed by an external validator. External control operations would be constituted, for instance, by checking that data input is being done in the first place, using spot checks requiring intuitive human judgement. This type of external control activity is very different from the validation of individual fields of data within an input transaction by operations of the internal control levelling dimension.

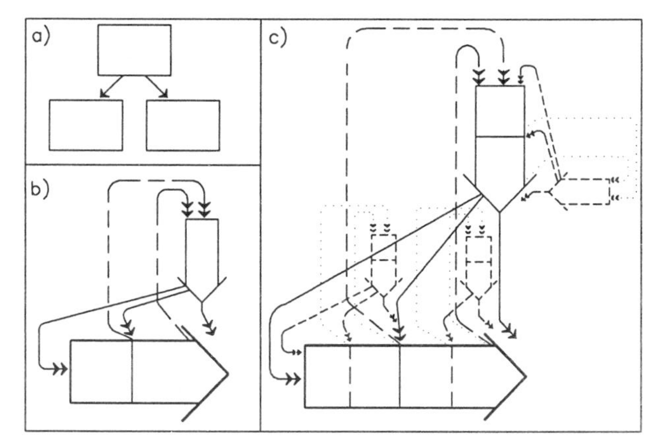

FIGURE 7-30: Internal functions, structure vs. structure-flow chart.

a) structure chart for comparison to structure-flow charts in the next two items;

b) rough structure-flow chart equivalent of structure chart in "a";

c) structure-flow chart bearing a somewhat closer approximation to structure chart in "a." Theoretically, by increasing to infinity the number of internal levels of adaptation operations (represented by de-emphasized lines), the operations represented in "a" would be totally included within the umbrella of functional elements represented in "c."

Internal Control Levels in the Structure-Flow Chart

Although demonstrating the internal levelling dimension in the structure-flow chart using a two-dimensional display medium may be almost as difficult as in the regular structure chart, the ideas at least have the potential of being clearly shown in three dimensions, and can be demonstrated to an extent even in two. The two-dimensional representation can be achieved in two ways. **First**, as in Figure 7-30c (or, preferably, Figure 5-3 in Chapter Five), the internal functions may be drawn in a de-emphasized fashion.

Second, the structure-flow chart of the broader parent system in which a subsystem exists may be reassembled to have a view of the levelling dimension (external or internal) in which given adaptation functions in the structure-flow chart of the subsystem exist. For instance, the supply acquisition system's structure-flow chart may be reassembled from the structure-flow charts of its information sub-subsystems (see Figures 6-9, 6-10, and 6-11 in Chapter Six). This is described in the next paragraphs. To follow this reassembly, the reader should begin by identifying the same items in the sub-subsystems and the reassembled parent system (Figure 7-31), based on the identical reference numbers having been used for the same input/output items.

When the supply acquisition system is reassembled, as in Figure 7-31 the externality of the roles of the sub-subsystem ensurance functions can be demonstrated. Thus, although the reference numbers for the corresponding items are the same in the sub-subsystem and the parent system figures, the angle and position of what was the ensurance function in relation to what was the baseline function of each of the sub-subsystems now shows that ensurance function as occupying the angle and position of an initiative function in relation to that baseline. In fact, what is being done in Figure 7-31 is to increase the role of what would have been an initiative coordinate in each of the sub-subsystems (no initiative coordinates were charted in Figures 6-9, 6-10, and 6-11) to include the role of the ensurance function. Had the ensurance function in each sub-subsystem been in an internal levelling dimension in relation to the sub-subsystem baseline, no change in position would have occurred in going from the sub-subsystem structure-flow chart to that of the parent system. However, this change has occurred, as the reader should be able to see by comparing corresponding items in the sub-subsystem and parent system figures. This change indicates that the ensurance functions in each sub-subsystem are in the external levelling dimension.

To add internal controls to Figure 7-31, wideshafted arrows may be drawn using de-emphasized lines. For example, an internal initiative coordination function is drawn with a non-continuous line directly in front of the baseline. This plays the role of deciding whether or not to begin receiving the goods from the supplier. This would, for example, be decided based upon the existence of a receiving copy of the purchase order applying to the goods in question, as well as upon a preliminary inspection of the goods for damage.

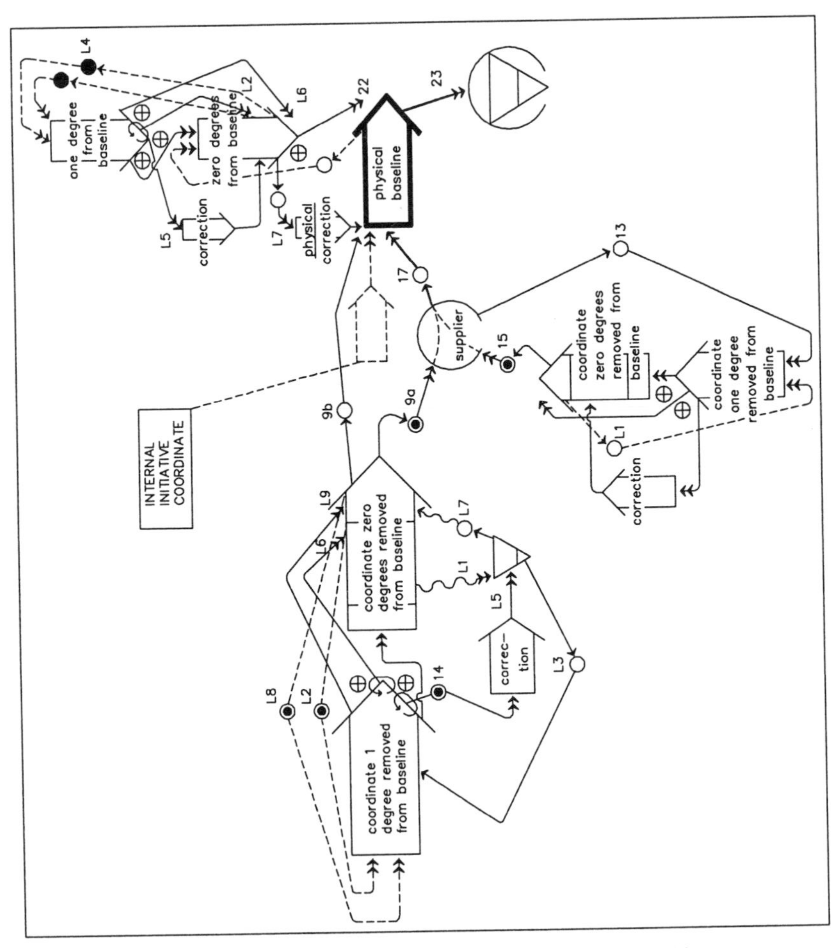

FIGURE 7-31: Reconstruction of levels in the supply acquisition system. *The reference numbers in this Figure are the same as those used for the corresponding items in the sub-subsystem figures (Figures 6-9, 6-10, and 6-11 in Chapter Six).*

PART THREE— FURTHER PROBING OF CONCEPTS

CHAPTER EIGHT

MORE ABOUT THE CONCEPT OF FUNCTION*

This chapter undertakes the complicated task of further exploring the assumptions, implications, and related concepts of the heuristics for identifying functions begun in Chapter Three. This discussion, broken down into two sections, includes the following elements:

- *in the first section*—identifying objective-defined functions:
  - —assumptions underlying the identification of what is eligible to be included in any direct line of flow which forms an objective-defined function;
  - —directness of relationship between functional elements and an objective;
- *in the second section*—functional objectives:
  - —the mutual contingency of functional objectives;
  - —functional objectives versus system objectives;
  - —distinguishing specific objectives within general ones.

---

* The materials in this chapter are based on an article by Ed Baylin, "Identifying System Functions," International Journal of General Systems. New York, Gordon and Breach Publishers©, vol. 12, no. 1, 1986.

MORE ABOUT IDENTIFYING OBJECTIVE-DEFINED FUNCTIONS

STARTING POINTS FOR IDENTIFYING FUNCTIONS

That a function contains all and only those operations directly related to the same objective is qualified by the fact that operations in the same direct line of flow cannot be considered to form part of the same objective-defined function when they belong in different basic functional classes, or are at dissimilar levels of control. As well, differences in time-orientation and distinct types of system components usually imply different objective-defined functions. Thus, the functional classification scheme, the level of the function, time-orientation, and the type of system component or area of the system affected all provide starting points for the identification of distinct objective-defined functions. Following are some notes about these starting points.

The Functional Classification Scheme as a Starting Point

When operations belong in different classes within this scheme, different objective-defined functions will have to be identified, except perhaps when dealing with those parts of the scheme identified in Chapter Three as being optional. Embodied in this functional classification scheme are certain assumptions in addition to directness of relationship to the system objectives (which is used only in the primary distinction between baseline and adaptation operations). Within the adaptation operations, certain secondary but still mandatory distinctions are made which are founded upon relative time frame (current versus non-current) and decision-making involvement (decision-making versus decision-implementing). Certain tertiary, optional parts of the functional classification scheme may, at the discretion of the user of the structure-flow chart, be utilized to separate otherwise direct lines of flow into distinct objective-defined functions. These tertiary factors are:

—focus in relation to the system boundaries (internal or external);

—time-orientation (ensurance or feedforward).

The focus factor has been applied only to decision-implementing operations.

Level of the Adaptation Function as a Starting Point

Difference in level is another criterion for distinguishing separate objective-defined functions. Level refers to two dimensions, namely, degree of generality in the control hierarchy, and the degree to which the controls in question are internal or external. In other words, the two essential factors are "degree of distance from the operating core" and "degree of internality/externality." With respect to the degree of internality/externality, whether a function is internal or external in level can be quite ambiguous. For example, using the pyramid, does the function belong at the bottom of the *primary* hierarchy, or, alternatively, near the top of one of the *internal* levels-of-control hierarchies represented by a smaller pyramid glued to the wall of the more primary pyramid? Deciding this is by no means easy, since the bottom-most level of functions in the primary pyramid may have a fairly local scope. Such ambiguity is represented in Figure 8-1.

Time-Orientation as a Starting Point

Certain time-orientation distinctions have been made in this book, other than those included in the functional classification scheme. The structure-flow chart, based on the four-sided pyramid model, is designed so that current decision-making adaptation operations, as well as current decision-implementing ones, may be placed in three different time-orientations, namely, initiative, interruptive, and ensurance. Distinct objective-defined functions are defined in the structure-flow charting method for each of these time-orientations.

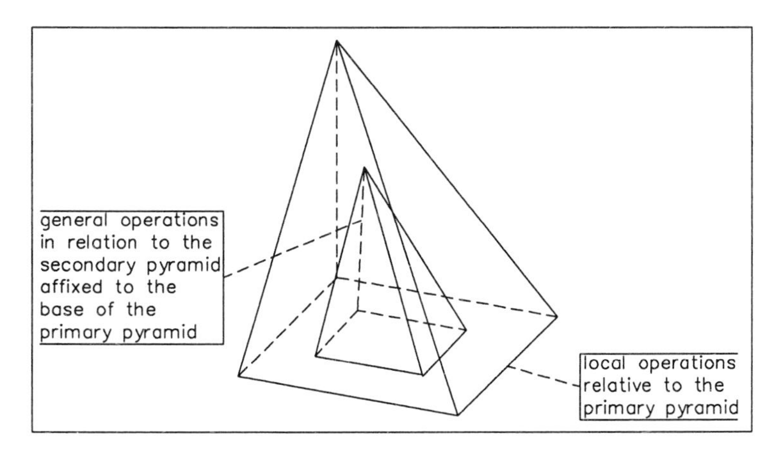

FIGURE 8-1: Ambiguity as to type of appropriate levelling dimension.

Area of System/Type of System Component Affected as a Starting Point

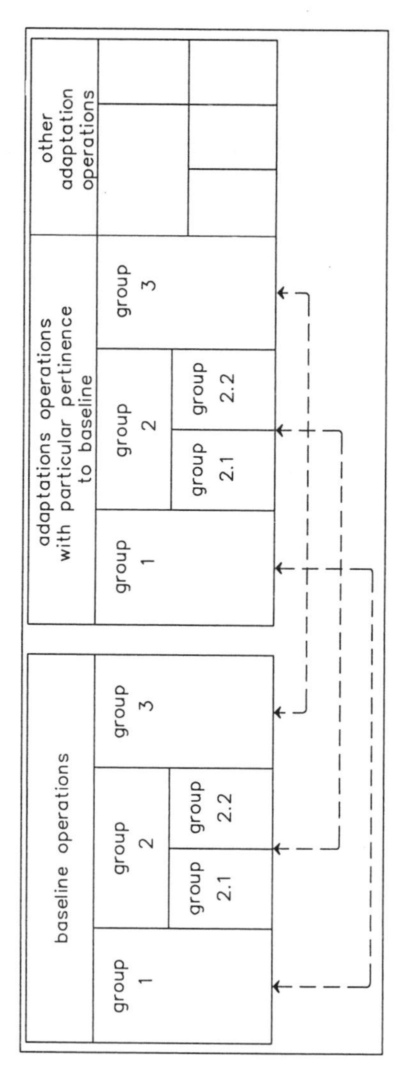

FIGURE 8-2: Different ways of configuring adaptation functions. *Configuring of the adaptation functions according to local areas of baseline.*

Other criteria for identifying distinct objective-defined functions concern the area of the system and the type of system component affected. These criteria are actually based upon the patterns of locality or generality with which adaptation operations in the system apply. Figure 8-2 demonstrates the general idea of basing the configurations of adaptation functions on the functional elements of the baseline. In this diagram, the objective-defined functions in groups 1, 2, and 3 correspond to groups 1, 2, and 3 in the baseline functions, while the remaining adaptation functions are not configured according to the baseline functional elements.

Also, the type of system component affected may be used to identify different adaptation functions, whereby each component type may be related to a group of adaptation operations local to it. Thus, certain planning operations may be related to procedures, while others are tied to machinery used in the business, these being different classes of frameworks. Alternatively, in a computer system, some planning operations may apply to software, while others pertain to hardware. However, if changes due to hardware and software are intertwined, it would seem that both should be decided upon by the same planning function.

DIRECTNESS OF RELATIONSHIP TO AN OBJECTIVE

This sub-section further defines what it meant by "direct relationship to an objective." Also, it shows that directness of relationship may be erroneously attributed, when operations are merely sequentially related. On the other hand, directness of relationship may be overlooked when the direct line of flow passes outside the objective-defined function; i.e., it is possible that functional elements which are really flow functions of the **same** objective-defined function may erroneously be viewed as belonging to **different** objective-defined functions. Deciding what operations should be placed inside an objective-defined function versus those that should be located in another such function can be a difficult matter.

Direct Versus Necessary Operations

The Meaning of "Directly Related"

The operations in an objective-defined function are all **directly** necessary to the achievement of the functional objective. Nevertheless, certain operations **necessary** to the fulfillment of this objective are not **directly** necessary for that purpose. These indirectly necessary operations are excluded from the objective-defined function even if they are executed every time the function is executed. Instead, they are placed in objective-defined functions of their own, which directly achieve their own objectives. The idea of directly and indirectly necessary operations is represented in Figure 8-3 by sub-sets of operations within other, "parent" sets of operations.

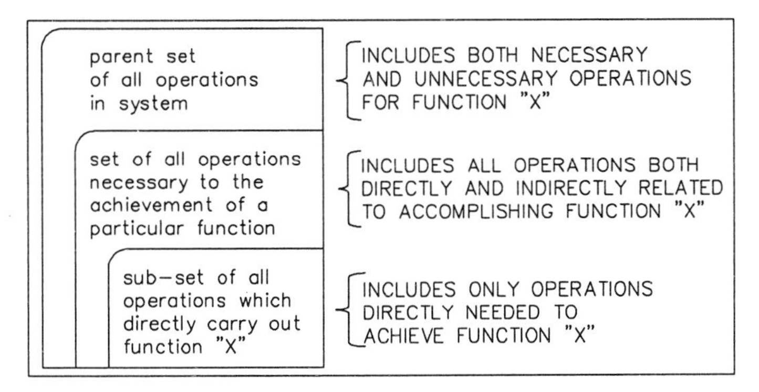

FIGURE 8-3: Necessary versus directly necessary operations. *Only the smallest sub-set in the above properly belongs in function "X." Most of the other operations are necessary, but not directly necessary.*

Operations are directly necessary in two different senses. **First**, for example, two workers and a supervisor are assigned to a task. Both workers and supervisor have as their common objective the carrying out of the task. In spite of this, only the workers **directly** achieve the task's objective. The supervisor's role is to attain the specified system objective indirectly, by directly achieving another type of objective, namely, that of coordinating and otherwise controlling the execution of the worker's task. Despite the highly cohesive nature of the tasks of both supervisor and workers, the supervisor's work belongs in a different objective-defined function. That is, for operations to belong in the same function, it is not simply enough to form a cohesive grouping related to a similar objective. They must all be **directly** related to the attainment of this objective.

Second, the notion of indirectness may be linked to a relationship between different sets of operations in which one set establishes a long-lasting framework for the other set, which directly performs the work on a frequent basis. For instance, in a business, the installation of a new piece of machinery, or of a new group of operating norms, require operations which mount the long-term frameworks for daily operations. The latter operations are only indirectly involved in achieving the frequently executed task.

Not All Direct Operations are Needed in Every Run

The operations in an objective-defined function are all directly necessary to the achievement of the same objective, but this does not mean that all these operations are executed **every time** the objective is to be attained, i.e., in every "run" of the function. To be more precise, it is only in the long-term of system operations that all operations in a functional grouping will necessarily be used in the accomplishment of the objective. Three possible reasons for this are as follows:

—a particular run may use energies and materials stored in reservoirs created during previous runs of the same functions, or may create reservoirs to be used in future runs;

—different runs may use different strategies of operation for achieving the same objective (the "equifinality" idea);

—different runs may achieve different options from the mutually contingent set of specific functional objectives (see below).

As a result, not all the operations in a function need be executed in every run.

Functional Elements of the Same Sequence Only

Despite the existence of a sequential relationship, various conditions determine that different objective-defined functions be identified. For example, problems often occur in recognizing the existence of different objective-defined functions when

—indirectly related functional elements occur regularly along the line of flow of an objective-defined function;
—two different objective-defined functions are related such that the first one must complete execution before the next one begins;
—functions exist at different levels.

Indirect Functions in Regularized Interaction

Indirectly involved operations may enable, or support, or direct, etc. another function's operations, either on a random (variable) or on a regularized basis. Where interaction is very regularized, it may erroneously look to the casual observer as if the two objective-defined functions are indeed one and the same, since they may be cohesed into a common sequence.

Regularized interaction along the line of flow may take a number of different forms, such as an indirectly involved function being called upon for help from a directly involved one, or an indirectly involved function interfering in the operations of a directly involved one. For example, a sequence of events in which a purchase order is 1) entered on a computer terminal, 2) printed and 3) sent, is a functional sequence. However, if validation steps are inserted, there is no longer just one function, although there still is a sequence. Thus, steps from one type of function, called a "control" function, are mixed in with steps from another type of function in relation to which the control function plays an "adaptive" role. What is seen here is a "sequential cohesion" involving more than a single function flowing towards an objective, although not all operations in the sequence may be **directly** necessary to the achievement of the technical objective of obtaining purchased materials for the business organization.

Definite Execution Sequence of Objective-Defined Functions

A sequential relationship being confused with a direct one also occurs when two different functions interact with a common base function (to which they both have an indirect relationship) at different points along the line of flow of the base function. When the two functions are separated into their own shared subsystem, apart from the base function, they may both become baseline functions in the subsystem thus constituted. For example, the supply acquisition *information* system seen in Chapter Six has three baseline functions in sequential relationship to one another, forming the base of the information subsystem. These baseline functions are in sequential relationship to one another, forming the base of the information subsystem. These baseline functions are sequentially related to one another, since they occur at different points along the line of flow of the baseline function of the **parent** system from which they are commonly derived. However, they remain separate objective-defined functions.

Different Levels In Sequential Relationship

Finally, when functions exist at different levels, one function invokes the other, creating a sequential relationship. The relationship of these levels is not, however, simply a sequential one, since reciprocal coupling between the levels is usually involved when the lower level feeds back to the higher level after completion of execution.

Functional Elements Really of the Same Function

Functional elements in the same direct line of flow being erroneously placed into different objective-defined functions may occur in two different situations. The *first* of these is dealt with towards the end of the next section, in connection with "mutual contingency." This is the case (see below) in which the objective of an intermediate **flow** function within an objective-defined function may also be viewed as an objective of the objective-defined function **as a whole**.

The *second* situation is one in which it is not apparent that two operations belong only to different **flow** functions, instead of to different **objective-defined** functions. For instance, in Figure 8-4, the direct line of flow towards an objective involves three steps. However, when the parent system is divided into subsystems, the second step is removed to a different subsystem, while the first and third steps are kept together in the same one. From the latter perspective, the objective of the first step should be seen as that of a flow function, since it still has the same ultimate relationship to the objective of the third step. That is, both steps one and three are still in the same direct line of flow. The first and third steps do not form different **objective-defined** functions just because the second step has been removed in the direct line of flow seen from the perspective of one of the subsystems.

To illustrate, a firm performs both extraction of iron ore and manufacture of iron ore finished products, but does not handle the shipping of the iron ore from the mine and the refinement of the ore. In this case, the organization in question should not be considered to have two baselines, one whose objective is to extract the ore from the ground, and the other whose objective is to manufacture iron ore finished products. That is, each of these are flow functions belonging to the **same** objective-defined function, the objective of which is to manufacture and dis-

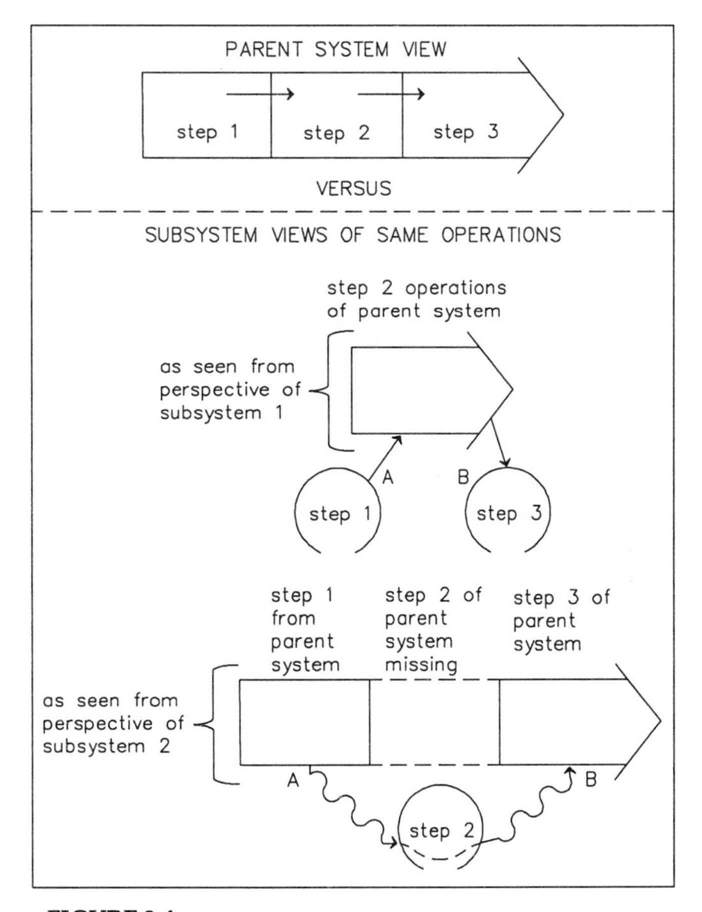

FIGURE 8-4:
Environmentally linked single objective-defined function.
This shows functional elements which should be placed in the same objective-defined function despite environmental linkage. The relationship between step 1 and step 3 in subsystem 2 is such that they belong in the same objective-defined function.

tribute finished iron ore products. The flow functions of this objective-defined function are, however, environment-coupled.

MORE ABOUT FUNCTIONAL OBJECTIVES

FUNCTIONAL OBJECTIVES VERSUS SYSTEM ONES

Mission rather than Objectives

The term "mission," rather than "objectives," is used here to describe what the system is all about. It is only relative to the system's mission component of its essence-identity that the system operations can be "functional." As an illustration, if the mission of a system is the technical objective of producing paycheques, then all functions of the system must be understood relative to this objective. All operations are either directly or indirectly— in some cases **very** indirectly— related to the carrying out of the objective of producing paycheques. Specifically, the input of time spent information, the calculation of pay, and the production of paycheques for employees, are directly related to the system's mission, while the control of system operations and the maintenance of the basic system methodologies and fixed assets needed to execute the payroll activity are indirectly related to the fulfillment of the system mission.

System Objectives As Technical Objectives

The relationship of the functional objectives to what the system is all about raises the question of whether functional objectives are technical ones, such as to produce paycheques, as opposed to "agent-based" objectives, such as to produce profit, power, or pleasure for the agents responsible for the production of paycheques. The logical analysis of a system requires that the objectives of functions **indirectly** related to the achievement of the system's mission be technical ones. Also, even when the **system** objectives are **not** technical ones, technical objective achievement is always assumed for directly related intermediate objectives, i.e., those of flow functions occurring at an intermediate point in the baseline flow. Baseline (system) objectives, however, need not be technical ones. That is, they may be agent-based (produce profit or other gain for those who carry out system operations).

System Mission Versus Functional Objectives

The system mission is fulfilled directly through the achievement of the objectives of the baseline objective-defined functions. This, however, does not mean that the system mission itself is given in terms of specific objectives, since the system mission may be stated at a more logical level. If this is so, the baseline objectives are **not** equivalent to the system mission. Rather, they constitute a strategy, or a methodological framework, for achieving the system mission, which framework may be altered by planning operations. Planning operations may make plans for both intermediate and final objectives, including those final system objectives which directly achieve the system mission. For example, a business may change models within a product line whose mission it is for the business to produce. Again, direct achievement of mission does not mean system objectives are the same thing as the system mission. It is only where the mission is defined in a specific enough sense that this would be true.

Possible Clash of Various System Objectives

As long as the system's mission is that of a single, clearly stated technical objective, such as to produce paycheques, it is easy to break the system's operations into different functional groupings. However, functional analysis of socio-technical systems in particular is much more complicated, since it usually requires that certain simplifying assumptions be made in order to carry out the analysis. For instance, a complex, but still manageable, case for functional analysis is that of a business organization. The mission of such a system may, in general terms, be understood as a combination of product objectives and profit objectives of the business owners. Since the two types of objectives complement each other in this case (see Figure 8-6c)—or since they can at least be fulfilled in such a way that they can peacefully co-exist—the functional breakdown relative to the product objectives can be understood in relation to the breakdown in terms of the profit objectives element of the system's mission. However, one element of the system mission may be **dys**functional relative to another, such that the system has a degree of internal conflict (without this necessarily meaning that the system is bound to self-destruct). In such cases, the performance of functional analysis may prove to be impossible unless each element of the mission is considered in isolation.

Terminology Related to Objectives

To summarize,

—system objectives are specific in nature;
—baseline objectives are system objectives; in contrast, all adaptation function objectives are intermediate ones in relation to the system as a whole;
—system objectives may or may not be equivalent to system mission, depending upon how specifically the essence-identity has been defined;
—depending upon whether system objectives are equivalent to the mission component of the system essence-identity, planning operations may or may not establish final system objectives, but they may always establish intermediate objectives;
—system objectives may or may not be technical ones; however, adaptation function objectives, as well as objectives of all flow functions which do not themselves achieve system objectives, are necessarily viewed as technical objectives, since they are seen as means, rather than as ends.

DISTINGUISHING SPECIFIC OBJECTIVES

The distinction between logical and physical levels applies to functional objectives, which should be at a level of specificity appropriate to the level of detail from which the problem is viewed. Thus, they may be seen as logical, although from a more general perspective these same objectives might be seen as physical.

Functional objectives are specific ones. What constitutes a specific objective is not always a straightforward matter, because even a so-called "specific" functional objective actually refers to a range of more specific objectives. Another problem is that some apparent differences in objectives are actually nothing more than differences in the way the same objective may be achieved, i.e., are not really differences in objectives. The following four examples highlight these matters:

Case #1 — An Objective Within a Range of Objectives

A system has as its mission to produce nails and hammers. Assuming the existence of different basic sizes/shapes of hammers and nails, does this imply that each different size/shape corresponds to a different objective? Moreover, since, in a purely scientific sense, each **individual** hammer or nail has no precisely identical counterpart, e.g. to the micrometer, does this imply that a different objective is attained **each time** an individual hammer or nail is produced? Obviously, the minute differences should make no difference; that is, the analyst should not be overly sensitive to small differences in the functional product, where this sensitivity does little to enhance the analysis. However, different basic sizes/shapes may be useful to identify in the analysis. Of course, what is a basic and what is a minute difference is a matter of judgement. Nevertheless, the fact remains that by objectives, the reference is to a range over which objective achievement is considered to be valid.

Case #2 — Methodological Frameworks, Not Different Objectives

A customer places an order for a particular shape/size of nail, and orders a quantity of 100 boxes. Is the objective of the function which produces 100 boxes identical to that of the function of which producing only a single nail of this size and shape? It would seem that it is, unless the greatness of the difference in quantity actually alters the perception of the object produced. Otherwise, the quantity itself has nothing to do with the objective. In fact, the quantity here is a one-shot methodological framework input to each run of the function.

Similarly, in another type of case, a bank makes loans to consumers and to businesses. It classifies all of its loans into either "consumer" or "commercial." Within each of these categories, it identifies each type of loan by the terms it uses for payments and interest. Do either the terms of the loan or the customer type variations correspond to different functional objectives? Are these items not just long-term methodological frameworks telling the function what strategy of operation to use, that is, the terms and the general type of party to which the functional product is to be delivered?

Case #3 — An Objective as a Slice of One or More Other Objectives

The subject of defining objectives relates to identifying the elements going out of functions. In an information system, this subject is connected with data definition. When data is referred to, is this just one field, or some grouping of fields at various levels? With respect to grouping fields, are these physical records, as stored in files, or are they simply logical groupings of data elements into data structures, whereby the same data field may belong to more than a single logical group? How to define such logical groupings is connected with the subjects of data base and file design, and setting up data dictionaries. In particular, in setting up files and defining their physical record structures for access efficiency (of the same record by different functional elements

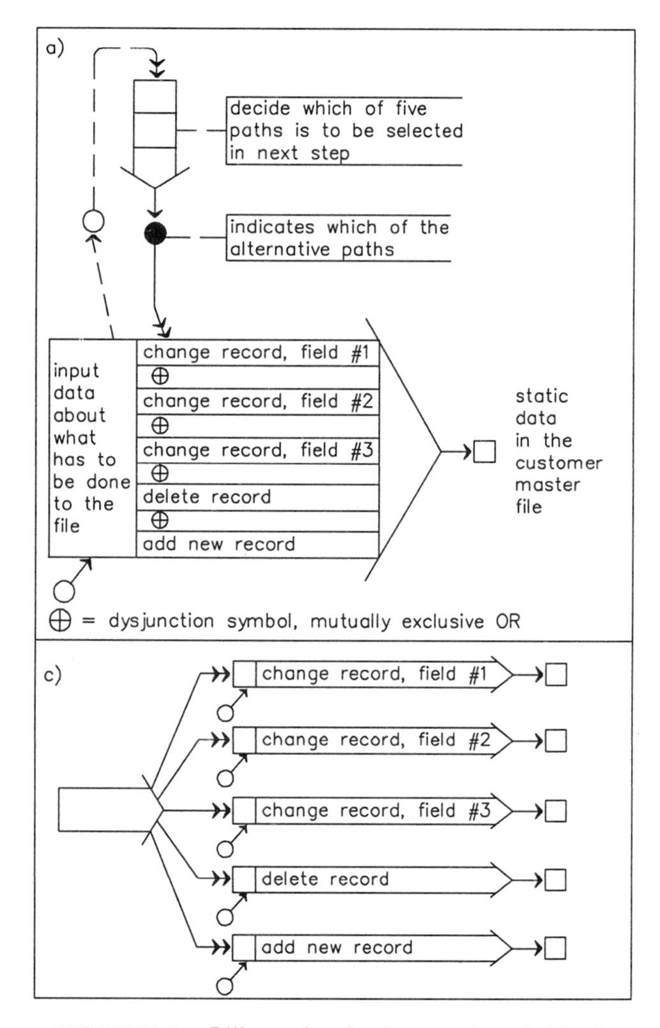

FIGURE 8-5: Different levels of perception of objectives

a) structure-flow chart #1 of customer master file maintenance system, in which all the different mutually exclusive objectives have been combined into a single objective, and a single objective-defined function;

b) ***(facing page)****;*

c) a still more detailed breakdown compared to "a."

in the information systems of an organization), much attention must be paid to set theory. Certain concepts known as "decomposition" and "normalization" have been developed to optimize the design of file record layouts.

For example, in a data flow diagram of an accounts payable system, **two** arrows, instead of one, are used to show the placement of data into the open item file. One arrow represents all the invoice data in the open items record, while the other represents the payment scheduled date (added to the invoice record at time of receipt and approval of the invoice). In other instances in this same data flow diagram, the data in the open items record (of which the payment scheduled date is a part)

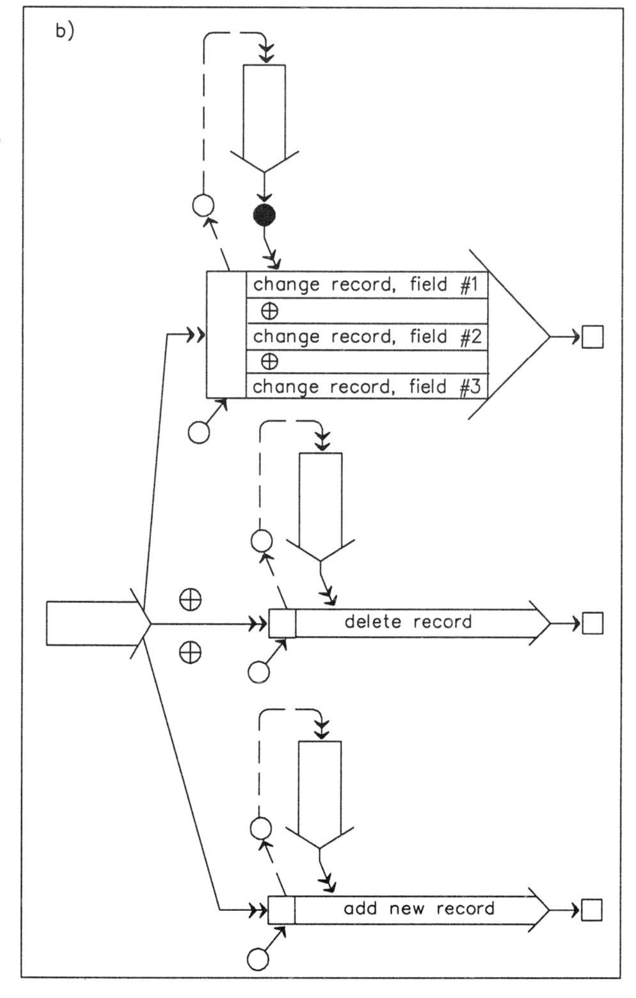

FIGURE 8-5b: *b) a more detailed breakdown compared to "a";*

is indicated only by a **single** arrow flowing between the open items file and the process which creates/updates/uses the data in the record.

Moreover, an arrow can represent a slice of more than a single data structure, based on logical views of the data which cross-cut different physical data structures involved in an information system.

Case #4 — An Objective as A More General Statement of Different Mutually Exclusive Objectives

The mission component of the essence-identity of a computer information system is defined as the technical objective of maintaining static data in the customer master file. This may involve adding new records, or changing or deleting existing records. The following questions relate to identifying the objective(s) involved:

- Should a different functional objective be identified depending upon whether the file records are added, changed, or deleted?
- When existing records are changed, should a different objective be identified depending upon which of the fields is changed?
- Each time a different objective is identified, should a different objective-defined function be identified as well, or should there simply be different, mutually exclusive objectives for different runs of the same objective-defined function?

Figures 8-5a, b, and c show some of the results of these different alternatives chosen by the analyst. All figures represent the same thing, but the functional breakdown and/or the degree of specificity of the functional objective is different in each case.

The level of specificity represented should be appropriate for purposes of analysis of the system in question. Thus, the very general representation of a single function with the single objective of maintaining the static data is useful when the file maintenance system is viewed as nothing more than a subsystem within a much larger context. On the other hand, when the file maintenance system is seen as a system at its own level of hierarchy, a full explosion of detail necessitates using one of the more detailed representations.

THE MUTUAL CONTINGENCY OF FUNCTIONAL OBJECTIVES

The general objective of an objective-defined function can be fulfilled by achieving a number of more specific objectives. The specific objectives will be achieved by the same objective-defined function if they are "mutually contingent," or by different objective-defined functions if they are not.

The concept of mutual contingency means that specific objectives are mutually contingent if they satisfy any one, or any combination of the following criteria:

- They are achieved in parallel (perhaps simultaneously), and are necessarily both achieved if either one is achieved.
- They are achieved in a mutually exclusive way, such that either one or the other is achieved in a particular functional run. One variation of this is that they are achieved in a certain "round-robin" order in different runs of the objective-defined function; e.g., run #1 achieves specific objective "A," run #2 achieves specific objective "B," run #3 achieves "C," run #4 fulfills "A," run #5 attains "B" and so on, in that pattern. In the round-robin idea, the actions to be taken in the next run are determined by a "state variable," giving the state as of the previous run. Also, a step may be optional within the round robin.
- One objective's achievement is a direct step towards the achievement of another objective.

The phrase "mutually contingent," used in relation to anything except objectives achieved in parallel, is perhaps an unusual choice of phrase, since mutual contingency usually means an AND type of relationship. However, in the present use of this phrase, the objectives are mutually contingent in that they combine to achieve the same function, even if this is done in a mutually **exclusive** sense.

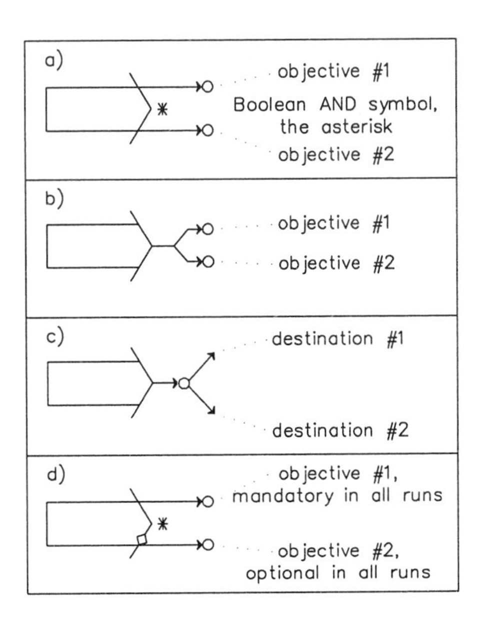

FIGURE 8-6:
Mutual contingency of parallel objectives.
a) objectives achieved in parallel, but not necessarily simultaneously in the same run;
b) objectives achieved simultaneously in the same functional run;
c) a single objective, but with the output going in two different directions;
d) objectives produced in parallel in the same run, although one of the two is optional.

Parallel and Mutually Exclusive Objectives

The various ideas on functional objectives being **sets** of specific objectives are represented diagrammatically in Figures 8-6 to 8-10. Figure 8-6 covers parallel or simultaneous outputs, each of which is a different variation of the same essential notion. Figure 8-7 shows various senses of mutually exclusive outputs, including the basic round robin and non-round robin ideas, and variations in which optional outputs exist within both of these basic arrangements. The various symbols used here are from Boolean algebra (as also used in data flow diagrams) and/or from structure charting. By combining both sets of symbols, new combinations not possible if either is used alone can be achieved.

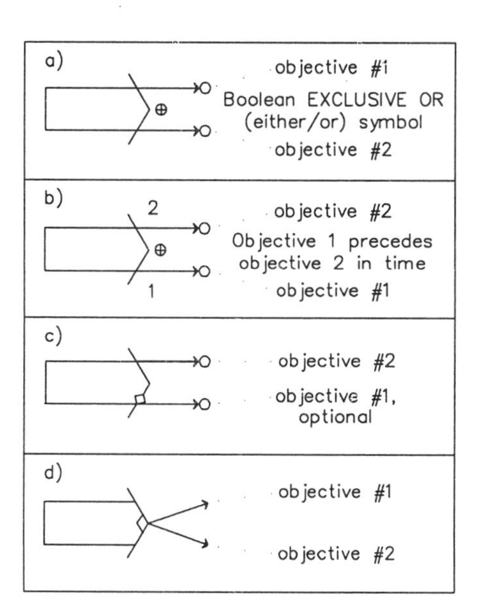

FIGURE 8-7: Mutual contingency of mutually exclusive objectives.
a) either objective, but not both, produced in the same functional run, and the objective achieved in any next run is not determined by which objective is achieved in the current run;
b) round-robin sequencing arrangement, i.e., objectives take turns in a definite order in being fulfilled by successive functional runs;
c) round-robin arrangement, but in which one of the objectives is optional;
d) same logic as "a," except different symbols used.

Step as well as System Objective

Figure 8-8 shows the mutual contingency of objectives due to one being a step in the direct line of sequential flow towards the achievement of the other. The step (flow function) objective is also seen as an objective of the objective-defined function **as a whole,** because the analyst perceives the system in a certain way relative to the system environment, and has equated a final system objective with the objective of the step in question. Otherwise, such a **flow** objective would not be referred to when speaking of the objectives of the

objective-defined function. For example, in the business organization, the products delivered to customers are delivered before the collection/disbursement of funds. Should this product delivery be shown as an objective output of the business baseline when profit is seen as the system mission, the particular type of product which the business creates is in fact part of the system mission as well, and the system is logically constrained to achieve profit by this type of product.

And/Or Related Objectives

Another issue related to the mutual contingency of objectives concerns those which may be related in an "and/or" fashion, e.g., two objectives which in certain functional runs are achieved in parallel, whereas in other runs they are achieved in mutually exclusive fashion. For example, optimizing personnel resources may involve hiring and/or firing of staff, depending upon circumstances. This "and/or" logic may, in fact, relate **more** than two objectives, as when optimizing personnel resources involves hiring and/or firing and/or promotion and/or retraining of people, any combination of which may be fulfilled by a

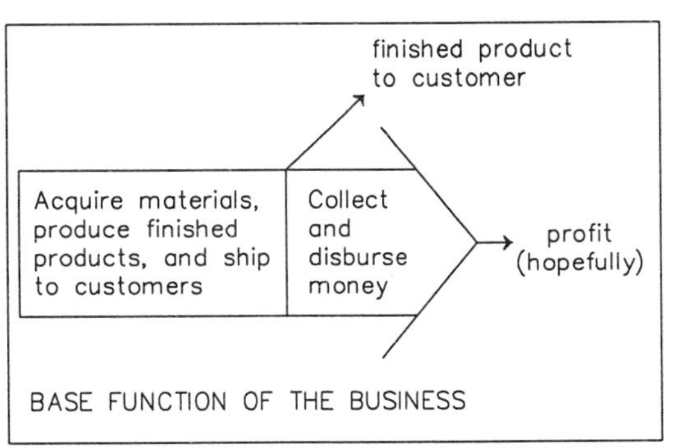

FIGURE 8-8: Mutual contingency between sequentially related objectives. *The two objectives are mutually contingent in the sense that one is a step on the way to achieving the other. This occurs only in the system base functions, and only when the system is defined in such a way that a manner of achieving an objective also becomes a system objective. Note that no assumption is made about the relative priorities of these objectives, although the profit objective is shown as occurring after the product one.*

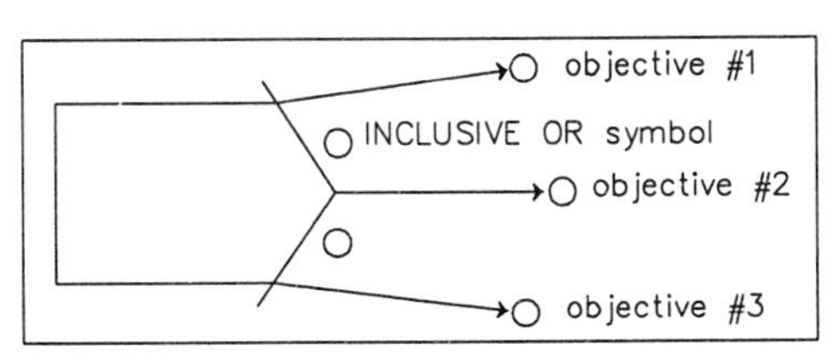

FIGURE 8-9: Mutual contingency involving an AND/OR relationship. *Any one, or any combination except zero, of the objectives may be achieved in a functional run.*

single run of the personnel optimization function. This and/or type of situation may be diagrammed using the Boolean INCLUSIVE OR symbol, as demonstrated in Figure 8-9.

The INCLUSIVE OR idea is, in itself, too loose to support an axiomatic definition of the concept of function. Actually, it should only be used to provide a higher level statement of a number of more specific objectives which are related in one or more of the fashions already described. Otherwise, "and/or" related objectives are not what we would call "mutually contingent."

Hybrids of Different Types

Finally, various combinations of mutually contingent objectives may be achieved by any one function. Figure 8-10 gives various examples of this.

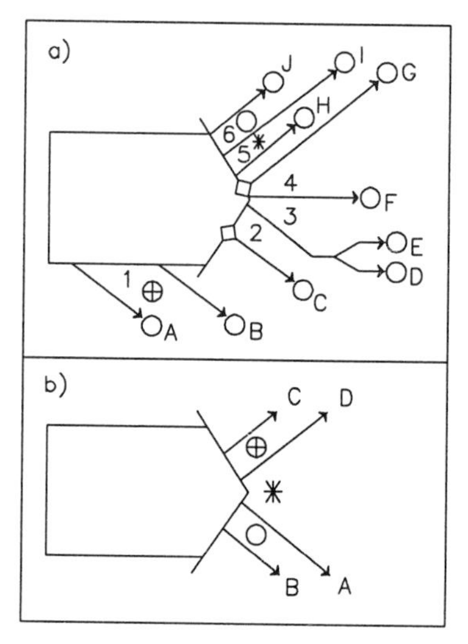

FIGURE 8-10:
Mutual contingency—
combinations of various senses.

a) - First, either A or B, but not both, is produced.
- In the next run, C may optionally be produced.
- In run #3, D and E are simultaneously generated.
- Next run, one of F or G is mandatorily output, thereby completing run #4.
- In run #5, H and I are produced in parallel.
- Finally, I and/or J are output in the round robin.
Now, the second round robin may be begun, beginning with another production of either A or B.

b) A and/or B is/are produced in the same functional run, along with C or D.

CHAPTER NINE

MORE ABOUT CLASSES AND LEVELS OF OBJECTIVE-DEFINED FUNCTIONS

The discussion of the new functional classification scheme and functional levels begun in Chapter Three will now be completed by exploring certain underlying facets and implications of the ideas presented. Specifically, the following is dealt with after a brief review of some preliminary issues of terminology:

- functional classifications (dealt with in the second section):
  - —grey areas between the different classifications;
  - —sub-classes in the scheme not yet described in any detail;
  - —further aspects of the incorporation of the functional classification scheme into the structure-flow chart;
- subtleties behind the different types of functional levels (discussed in the third section);
- the relativity of the functional classification and levelling schemes (handled in the fourth section).

FURTHER EXAMPLES AND SPECIFICS

SOME PRELIMINARY ISSUES OF TERMINOLOGY

Degree of Routinization of Adaptation Operations

Adaptation operations may or may not be of a routine, programmable nature, depending upon the type of system and sub-class of adaptation operation. Planning operations, in particular, are seldom highly programmable in organizations, except to the extent that procedures for strategy-setting (e.g., the corporate planning process) may be standardized and formalized. In contrast, current and framework adaptation operations in organizations are often highly routine in nature. As an example, local supervisory operations, involving current decision making, are usually highly standardized. These characteristics of adaptation operations may not be present in all systems, although they usually do exist in socio-technical systems, such as organizations.

Use of the Term "Managerial"

"Managerial," as opposed to "supervisory," generally refers to non-routine, wide-scope (general, or strategic level), long-term decision-making operations. These attributes of decision-making operations cannot all be assumed to occur together, even in organizations. For example, long-term decision-making operations may be routine in some systems, or may even be very local (tactical level) in scope. Thus, it is preferable here to avoid the "managerial" designation as much as possible in using a generalized approach to functional classification, except perhaps when discussing organizations.

Use of the Terms Informational, Data, and Decision Making

The subject of informational, or decision-making operations may be connected to the subject of levels in the Anthony triangle, and to the matter of data versus information. Any level of decision-making operations, be it called "managerial" or not, is considered here to be "informational" in nature. "Data," as opposed to "information," consists of signals and representations of isolated facts, and may be transmitted by any class of function, whether or not it is an "informational" (decision-making) function. However, only an informational function handles this data as information, by combining data items into a configuration called "information," useful in decision making. Information is used in all levels of decision-making functions, whether or not the decision-making process is structured, regardless of the triviality and routinization of the decision in question.

Classes of Computer Information Systems

The discussions of data versus information and level in the Anthony triangle tie into the subject of classifications of business computer information systems. These classes may be explained as follows:*

- **transaction-processing systems**: These are computerized aspects of routine input and output (but not processing) flow functions of **all** levels of informational objective-defined functions. The input flow function detects opportunities or problems for the decision-making process, while the output flow functions transmit data to communicate decisions rendered by the decision-making process.
- **structured (or programmed) decision systems**: These are computerized aspects of more structured and routine processing (but not input and output) flow functions of informational functions. These flow functions constitute the decision-making **process**, during which data is transformed into information, and ultimately into other data representing decisions.
- **DSS, or decision support systems**: These are computerized aspects of processing (but not input and output) flow functions of **non-**routine, wide-scope informational functions, including those at the management control level, but especially including those at the strategic planning level.

Of the above three categories, based on one usage of the term "management information system," only decision support systems are "management" information systems (MIS, according to some authors usages of this acronym), while structured decision systems are often part of the MIS. Although transaction-processing systems may play roles in higher level management control functions, or may even be involved in strategic planning functions, transaction-processing systems are not classified as part of the "MIS."**

* The classification of computer systems in business is discussed more fully in the related Ed Baylin book, *Conceptual Prototyping of Business Systems—A Templating Approach to Describing System Functions.*

** A fuller discussion of application file classes is contained in the related Ed Baylin book, *Conceptual Prototyping of Business Systems—A Templating Approach to Describing System Functions.*

MORE ABOUT FUNCTIONAL CLASSES

Framework Adaptation Operations

Frameworks

Illustrations of frameworks are given in Table 9-1. Framework operations establish (add, set-up), modify (change, replace), take down (destroy, delete), and upkeep (repair, restore) internal or partially internal frameworks, and influence largely external long-term frameworks. The framework notion is fairly wide-ranging.

A framework may be defined as something either inside or outside the system with which the system operations interact, which is not a stimulus, and which is not itself operated upon by the operations which use it or interact with it, except in the case of those system functions whose very purpose is to operate upon frameworks (which are thereby maintained for use in other system functions). A frequent accompanying, but not defining characteristic of a framework is that it is usually relatively permanent, although it may in certain cases be depreciated or otherwise transformed by use.

The notion of a framework being a method of doing things extends to time and technical objectives, as follows:

- **First**, with respect to time, a schedule provides a framework. For instance, the opening and closing of a place of business each day is according to a schedule, which constrains when the business may or may not be open.
- **Second**, technical objectives in a system may provide a frame of reference for achieving the system mission. While the system mission is unalterable, the specific technical objectives which achieve it may from time to time vary, provided that the mission is stated in a flexible enough way. For example, the mission of a system is "to manufacture consumer household metal products." From time to time, the specific product lines may be varied. These objectives are usually stable frameworks, since the product lines change relatively infrequently.

As seen in Table 9-1, there are three types of frameworks.

Stability of Many Frameworks

Frameworks vary in their degree of stability. For example, in an information system, framework data is usually static, compared to transactional data like sales orders, purchase orders, etc., and maintenance programs for static data usually constitute framework operations. For example, in a banking deposits computer information system, infrequently changing framework data include branch names and addresses, and parameters describing the options available to depositors in the different types of deposit accounts. Less static framework data would include the names and addresses of the bank's key customers. Framework data which are neither highly static nor highly dynamic would, perhaps, be currency exchange rates. These illustrations show that frameworks may be classified on a continuum, from very stable to temporary.

Despite the possibility of one-shot frameworks, a framework is, as mentioned above, usually stable relative to the frequency of a run of a given objective-defined function which uses it as a framework. Thus, the same method of achieving a functional objective may be used for many different runs of a function. This method of operation is, therefore, a stable framework in relation to this function. Also, the same "agent," machine, person, etc., may carry out this function from one run to another, and so the agent is a relatively permanent framework. Moreover, the function may receive input from the same external system source, or may deliver its product to the same external system destination. These sources, e.g., suppliers, customers, are relatively stable frameworks in relation to the objective-defined function in question.

CATEGORY OF FRAMEWORK	COMPUTER SYSTEM	ORGANIZATION
INSTRUMENTAL agents/instruments who/which carry out operations	computer operators and dataentry personnel, computing hardware, computer programs	personnel, machinery, real estate, catalysts
METHODOLOGICAL methods involved in achieving objectives	program parameter data, operating procedures	policies, standards, schedules, mandates
ENVIRONMENTAL environmental entities with which system interacts	sources/destinations of data flows, i.e., other systems	customers, suppliers, governments, economy

TABLE 9-1: Framework examples from organizations and computer systems. *NOTE: The above examples from computer systems assume that the boundaries of the computer system encompass equipment, personnel, and software change and maintenance.*

Whenever a framework is relatively stable, a specific class of system operations may be identified to maintain or influence it. Such framework operations are indirectly related to the achievement of the system objectives, since they occur less frequently and deal in that which is not operated upon by the system base operations. Despite the indirect nature of this relationship, they will be grouped in certain cases below with the system base operations, into what might be described as the system "operating core."

Framework Inputs Versus Reservoirs of Transactional Inputs

Because of their frequent characteristic of relative longevity, framework inputs are easily confused with reservoirs of operand transactional inputs, such as stores of raw materials and component parts in a business, since both are long-lasting in comparison to the run frequencies of the functions which use them or set them up. The tendency to confuse the two is also present because, in some cases, frameworks, like reservoirs of operand transactional inputs, may sometimes be consumed as a result of usage. This usually applies to frameworks in the sense of "depreciation" (as opposed to "depletion"). For instance, the depletion of a stockpile (a transactional reservoir), and the depreciation of machinery (a framework), are similar notions in many respects. In fact, the two ideas are so similar that both internal stable frameworks and operand transactional reservoirs may be grouped under the umbrella of system "resources." Resources, by this definition, also include relatively stable **non**-depreciable internal frameworks, such as policies and operating norms.

Sub-Classes of Framework Operations

Since one-shot frameworks have a life-span of only a single run of the functions which use them, they are not affected by framework operations. Longer-lasting frameworks, however, are influenced or maintained by special framework operations, whose sub-classifications are explained in the following.

Framework Influence Sub-Class

A framework is considered to be internal to a system to the extent that it may be maintained by system operations. A framework which may be **maintained** by system operations is probably wholly internal, while a framework which may only be **influenced** by system operations is most likely to be chiefly an external framework. For instance, the customers of a business cannot simply be maintained like a piece of machinery. At best, the business can maintain a certain relationship with the customer, with the rest beyond its control. This **relationship** is maintained, for example, by advertising campaigns, which are in the framework **influence** class of operations.

The drawing of the system boundary lines around functional elements largely affects the degree to which a framework is internal. If operations which maintain a framework used by a system have been assigned, in the analyst's view, to another system, then at best, all that a system can do is obtain the cooperation of that other system in maintaining the framework. In this case, the other system is contacted, but the framework cannot be directly maintained by the framework operations. To illustrate, the order entry system may require that a framework—namely, a customer account record—exist for every customer for whom an order is entered. This framework, consisting of static data in the customer master file, is maintained by the customer master file maintenance system, which is autonomous from the order entry subsystem. Therefore, should order entry detect that a customer who is requesting an order does not exist in the customer master file, set-up of this customer in this file may be requested by order entry control operations from the customer static data maintenance system.

Since only internal or partially internal frameworks may be maintained by a system's framework operations, the completely external frameworks are by definition added, changed, deleted, and upkept by framework operations belonging **to other systems**. Various ways exist in which a system may **influence** its totally external frameworks. As just seen, one way in which this might be done is by obtaining cooperation from other systems which are responsible for maintaining the external frameworks in question. To use another example, a business may affect the existence of one of its vendors, customers, or competitors by a strategy which makes it either easy or difficult for these to succeed.

Framework Add/Change/Delete and Upkeep Sub-Classes

Framework operations applicable to purely internal frameworks in a system are of either the "add/change/delete" or the "upkeep" type. Framework change, for instance, occurs in a business when a new set of operating procedures is installed. Framework upkeep, on the other hand, involves the restoration of frameworks, as in the case of machinery repair.

Framework upkeep may also perform the re-establishment of a state of equilibrium with an **external** system framework. Accounting operations, based on one view of the business organization (see Chapters

Two and Five), are framework upkeep operations, since they maintain the state of equilibrium between the business organization and external frameworks, such as customers (accounts receivable), suppliers (accounts payable), and personnel (payroll).

Planning/Framework-Ensuring Operations

Current Versus Non-Current Decision Making

The subject of frameworks leads directly to the subject of planning operations, defined as operations which make decisions affecting relatively stable internal or external frameworks. Framework changes of long-term duration may be considered to be the result of planning operations. On the other hand, very temporary changes to internal or partly internal long-term frameworks, and upkeep of long-term internal frameworks affecting only a single run of an objective-defined function, may be seen as being the result of decisions rendered by current operations. The grey areas between planning and current decision making often result from the fact that the definition of "long-term" may be dependent either on the actual results or on the intent with which the planning is done. Since it may be difficult to predict if the effects of decision making will be long-term, what is planning and what is current decision making is often just a question of interpretation.

Examples of planning operations in a business are provided in the following list:

- Make a plan for improving the quality of personnel.
- Decide to modernize machinery.
- Decide to purchase new filing cabinets.
- Work out a policy for evaluating suppliers.
- Formulate an advertising campaign.
- Decide upon the standards to be used in evaluating the performance of the sales force.
- Work out procedures to be used on the production line.
- Decide to borrow money to finance expansion plans.
- Decide to lobby the government to obtain more favourable legislation affecting the industry.
- Decide to hire X instead of Y.
- Decide to buy new computer hardware.
- Plan changes to computer programs.
- Decide to add new static data (e.g., customer names and account numbers) to the data base.
- Decide to set up a channel for collecting and distributing data.

- Decide to retrain operating staff in conjunction with the bringing in of new computer operating system software.

Link Between Planning and Framework Operations

Planning operations communicate decisions which may then be implemented by framework ones. Implementation via framework operations is needed only when mere communication of the decision by planning operations fails to result in the decision's being effected. For instance, achieving implementation of the decided upon policy may involve taking a labour union to court to allow the policy to take effect. In this case, framework influence operations are needed to take the union to court.

Conversely, planning operations are always needed before framework operations can occur. For example, the installation of a new piece of equipment may take place only after a decision to do this is made by planning operations.

The Defined Feedforward Role of Planning Operations

When planning operations initiate framework ones, as in Figure 9-1, they play the same type of role relative to framework operations as initiative coordination operations play in relation to baseline operations. The word "planning" is used here only in connection with feedforward roles, whereas the term "framework-ensuring" is used for external adaptation decision-making operations which validate the operations of framework or lower level planning functions. Figure 9-1 clearly distinguishes between the feedforward role of planning and the ensurance role of framework-ensuring. This is done by placing the framework-ensuring operations in the ensurance position on the page. Of course,

FIGURE 9-1: "Planning" versus "framework-ensuring."

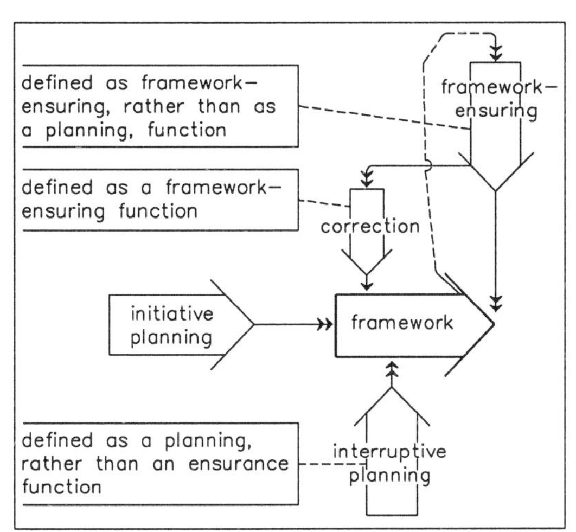

when these operations are viewed from the context of a larger structure-flow chart, **both** planning and framework-ensuring occur in the initiative position relative to the framework objective-defined function being controlled.

Current Adaptation Operations

Decision-making Roles

Like the planning/framework-ensuring (non-current decision-making) operations, the coordination/validation (current decision-making) operations detect problems or opportunities, decide what to do when problems or opportunities are spotted, and, then, based upon decisions made, emit data leading to execution of other operations.

In an ensurance time-orientation, this emitted data (communicating the decision) may be usedin relation to the base function to

- initiate corrections (either by redoing operations or by invoking decision-implementing functions) when errors are detected;
- to continue with the next step of the base function when no errors have been detected.

In a feedforward time-orientation, emitted data may be used in relation to the base function to

- initiate, and/or speed-up, and/or restart;
- and/or terminate, and/or slow-down, and/or suspend.

Roles of Decision-Implementing Operations

As in the case of non-current decision-making operations, when mere communication of the decision does not suffice to achieve implementation, decision-implementing operations may be needed to implement decisions made by coordination/validation operations. As in the link between non-current decision-making and decision-implementing operations, current decision-implementing operations must **always** be initiated by decision-making ones, whereas decision-making operations may or may not require decision-implementing operations to achieve decision implementation.

Decision-implementing current operations correct or prevent problems in accordance with decisions communicated by current decision-making operations. They may also procure something from the system environment when external contact is needed to implement a decision. Current decision-implementing operations implement solutions which do not add/change/delete/ stable frameworks, although temporary bypasses to framework use may be implemented by them.

Sub-Classes of Current Decision-Implementing Operations

As with non-current decision-implementing operations, the principles used in distinguishing the different sub-classes of current decision-implementing operations are those of time-orientation (ensurance versus feedforward) and focus (internal versus external effect). The associations involving each of the non-current and the current function classes with focus and time-orientation were documented in Table 3-4 (Chapter Three).

Procurative current decision-implementing, like framework influence, achieves its ends by going out to the environment and causing something to happen there which will have an intended effect on the system. Thus, to clear a roadblock foreseen at the start of a trip made by a truck, the driver physically lifts a number of barriers. This will, hopefully, clear the way in a **feedforward** direction, thereby enabling the journey to succeed. Alternatively, the roadblock is cleared, but the police, having found out about the incident, confiscate the transported goods. Procuration, in the form of legal action, is then needed to repossess the goods, thereby allowing the objective of the trip to be effectively completed in an **ensurance** time-orientation.

Correction, like framework A/C/D in the non-current operations, achieves its ends by actions taken **within** the system. Thus, if the legal action undertaken in the previous example is included somewhere within the boundaries of the system in which the error occurred, the situation is said to be remedied by **corrective**, rather than by **procurative**, decision-implementing operations.

Prevention, like correction, acts **within** the system, except that it is used only in a feed**forward** situation. For instance, the failure of a manufacturing run may be prevented by obtaining the availability of some needed resource.

MORE ABOUT FUNCTIONAL LEVELS

Scope of Effect Within Each Set of Levels

The analysis of functional levels begun in Chapter Three is now reviewed and expanded. In comparison to the simple triangular view of system control levels, the three-dimensional understandings obtained with the pyramid and the structure-flow charting method allow different sets of levels for each combination of time frame, time-orientation, and levelling dimension (external versus internal). Any **one** set of levels may be represented by a triangle.

Examples of Sets of Levels

Figures 9-2 and 9-3 provide two different examples of sets of control levels. Figure 9-2 shows control levels for the externally related non-current operations.

Figure 9-3, on the other hand, shows a set of external control levels without distinguishing different sets of levels for different classes of adaptation objective-defined functions or different time-orientations. In this figure, the normal situation is for upper level management to communicate with middle-level management, which in turn communicates with the supervisory staff. At the lowest level, only the supervisors normally interface directly with the baseline operations. However, in possible "pathological" cases of the company president operating a machine, the party with the broadest external scope of control finds himself directly controlling the baseline/operating core operations. An illustration of this idea is the line connecting the top module of the hierarchy to a bottom level in Figure 9-3.

Scope of Effect Within Any Set of Levels

All adaptation operations may be classified on a continuum, from very local (tactical) to very general (strategic), according to their scope of application within the system to either specific types of system components or to particular system functions. For example, the control operations which are specifically applicable to each baseline step are somewhat local in their scope of effect, while the overall administra-

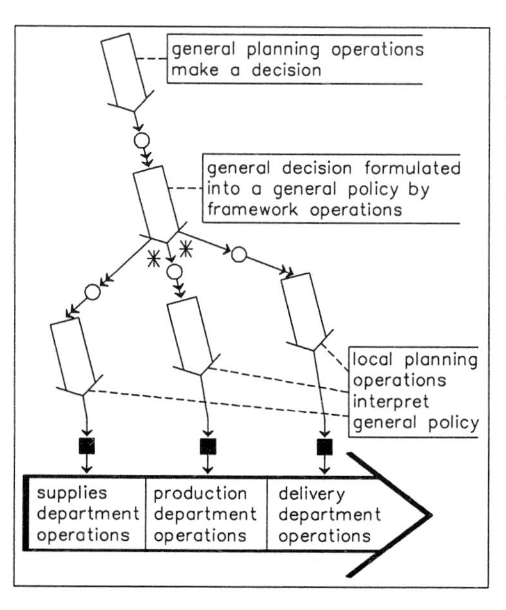

FIGURE 9-2: The spread-out effect of general decisions. *These decisions are interpreted at the more local level.*

tive operations are general in their pertinence to other operations in a system.

The discussion of locality or generality may also apply to the **type of system component** affected, e.g., machinery, personnel, computer programs, manual procedures, real estate, etc. Thus, a management policy (a methodological framework) is general in scope if it applies to many types of system components, but is local in scope if it affects only one type of component. Insofar as each component is treated by objective-defined functions uniquely associated with it, the locality patterns can be captured in the structure-flow chart.

Limitations to the Number of Levels

Obviously, very many levels of control operations may be gone through before the baseline operations are reached, as in a management hierarchy. The same applies to the framework and planning classes of adaptation operations. Theoretically, an unlimited number of possible degrees of removal from the baseline operations exists, in either the internal or external directions. However, in a simple system, such as one which has no adaptation operations except for inseparable control operations, the system control levels may be so close to being homologous (or flat) that one can all but forget the concept of hierarchies of local adaptation operations and degrees of distance from the baseline, at least in the external levelling dimension.

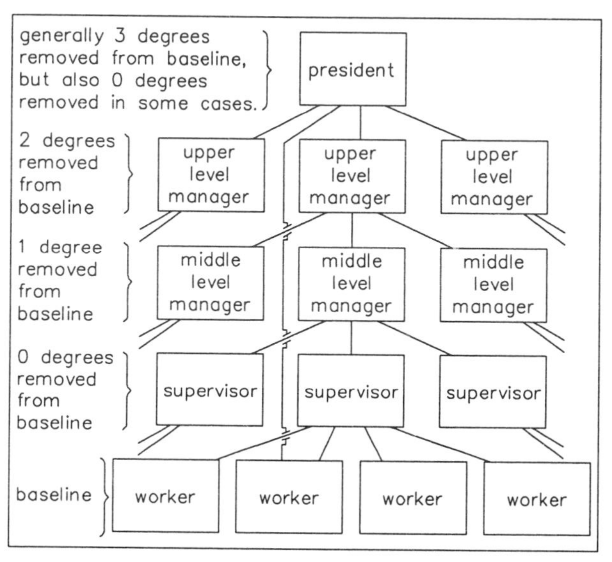

FIGURE 9-3: A conventional hierarchy of control levels. *This is the technique used by conventional levels-of-control charts, since different sets of levels are not used for different functional classes and time-orientations.*

Scope Versus Span

This degree of locality or generality refers to **scope**, rather than to **span**. Scope refers to all the levels of subordinates ultimately reporting to the decision-making module in question, while span just refers to how **many** directly reporting subordinates a module has in a levels-of-control chart. Thus, one manager may have many employees directly reporting to him, while another manager has fewer. However, the manager with the fewer subordinates reporting to him may in fact have a wider **scope** of control within the company, if his subordinates are more important and, for example, have many direct subordinates of their own. The scope of control is the main interest in reference to the locality or generality of adaptation operations.

The scope of adaptation operations may be to some extent represented by triangles of different shapes. The triangle has been used in many different forms, in various disciplines. A few of these are shown in Figures 9-4 a to d.

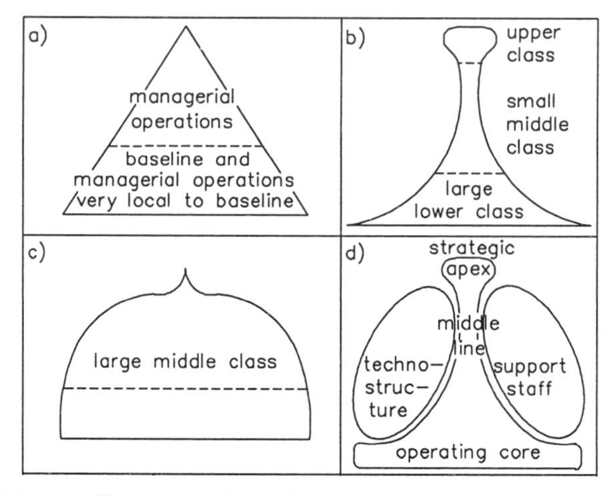

FIGURE 9-4: Some variations of the triangular idea.
a) the normal idea of a triangle, e.g., the Anthony triangle;
b) a representation of the class levels in developing countries;
c) a representation of the class levels in an industrialized country;
d) the Mintzberg form of the "triangle." From Henry Mintzberg, THE STRUCTURING OF ORGANIZATIONS, copyright 1979. Reprinted by permission of Prentice-Hall, Englewood Cliffs, N.J.

Types of Distance of Levels from the Operating Core

Idea of Degrees Removed from the Operating Core

The ideas of first-order interactions involving adaptation objective-defined functions was discussed in Chapter Five. It is, for example, illustrated in Figure 9-3. Both first-order and non-first order interaction can be discussed in terms of the degree of distance from the baseline or basic framework operations, or in other words, degree of distance from the system "operating core." Here, adaptation operations may be arranged on a continuum, from very directly applicable to very remotely applicable to the baseline or basic framework operations. In this context, the term "degrees removed" from the baseline or basic framework may be used; for example, one degree removed, two degrees removed, ten degrees removed, and so on. The term "degree" here is used in its figurative sense, since it has no precise, quantitative meaning. Figure 9-5 represents the idea of "degrees removed." Optionally included in this idea is that passage via the system environment on the way to the baseline adds an extra degree of removal. For example, the zero degrees removed function in Figure 9-5 has one direct connection with the baseline, and a second indirect connection via an environmental entity.

The "basic" framework operations referred to in the latter paragraph are those framework operations which are zero degrees removed from the baseline. Starting from these, the degree of removal may sometimes be measured in terms of distance from the basic **framework** operations,

FIGURE 9-5: Different senses of distance of operations from baseline.

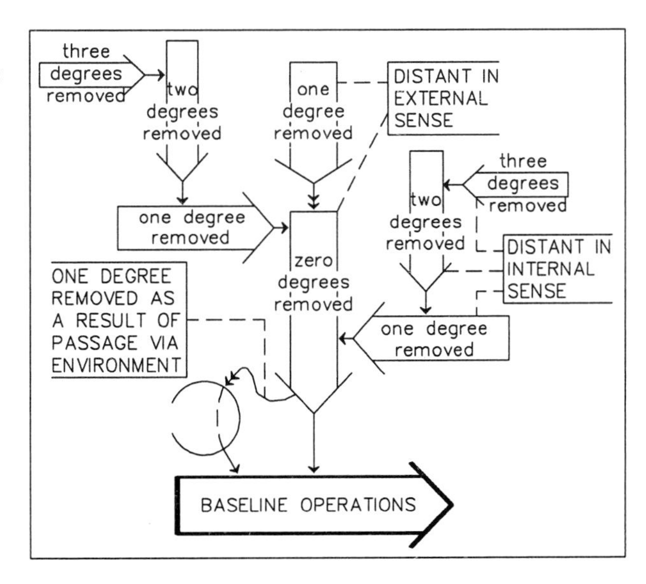

rather than of distance from the baseline operations, although an objective-defined function zero degrees removed from the basic framework operations is one degree removed from the baseline *per se*. Taken together with the baseline operations, these basic framework operations constitute the system's "operating core."*

Internal Versus External Directions of Distance from Operating Core

In the internal direction of distance from the operating core, the corporate planning process for establishing overall corporate strategy (see Figure 9-13 later) may be highly formalized via written procedures. The procedures-creation operations, which are unique to the corporate planning operations, are themselves, like the corporate planning operations, a form of planning operation distant from the baseline. They are, however, set up for use **only** in the corporate strategic planning operations. Consequently, they are many degrees removed from the baseline operations in a sense different from that of the corporate planning operations, because distance from the baseline in the case of the procedures-creation activity occurs in the internal levelling dimension (i.e., forms an internal level of control).

Tendency to Mix Up Degree of Distance and Scope Ideas

The two dimensions, degree of "internality/externality" and degree of distance from the operating core, both have values of a continuous nature. The first of these varies from very (inseparably) internal to very external, while the second varies on a continuum from very directly applicable, to very remotely applicable to the operating core. These two dimensions are often confused with one another in the sense that people tend to equate distance from the baseline and externality. However, as illustrated above, this may occur in two opposite directions. On the one hand (for instance, in a management hierarchy), the higher the level of management, the more external the scope of managerial control and the more distance there is from the baseline operations. On the other hand, the further removed from the baseline, the more internal may be the adaptation operations. This happens if a step is taken in the direction of the internal levelling dimension; for example, the establishment of procedures for the corporate planning process described above.

---

* This is a term borrowed from Mintzberg, and used by him for what Anthony roughly called the "operations level."

Management of, Versus Management by, Information

The popular "MIS" course subject, "management **by** information" versus "management **of** information," can be effectively reviewed in the light of the above discussion. Using the structure-flow chart, all levels of hierarchy of informational functions may be viewed as performing management **by** information, as long as these informational functions point towards the system base operations (or, alternatively, the operating core). In more precise words, management **by** information is involved as long as increase of distance from the baseline or basic system framework operations coincides with the increase of generality of scope in the external levelling dimension. As soon as a step is taken into the internal levelling dimension, informational functions begin to perform management **of** information, as illustrated in Figure 9-6. For example, information processing to initiate, interrupt, plan, or validate the base physical flow occurring in the business organization involves management **by** information. However, information processing to validate, etc. **other** information processing, such as the edit checking of data in a computer system, implies management **of** information. Thus, the data processing purchasing department performs management **by** information.

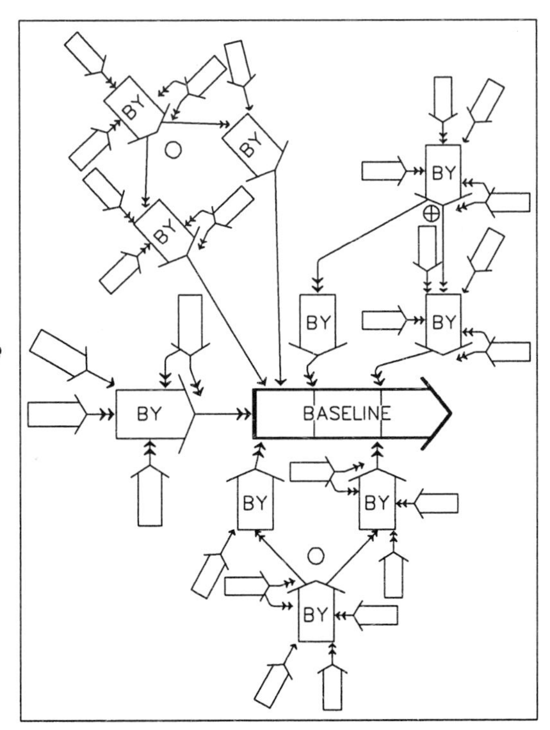

FIGURE 9-6: Management of, versus management by, information. *Here, the adaptation functions which are not labelled "BY" are in the internal levelling dimension, i.e., are management of information.*

RELATIVITY OF THE FUNCTIONAL CLASS/LEVEL SCHEMES

UNIVERSAL RE-APPLICATION OF THE SCHEMES

The Relativity of Functional Conceptualization

Meaning of the Relativity of Functional Conceptualization

It has been seen in earlier chapters, that the functional significance of an input/output item may change in a system, depending upon which function uses it. As well, the conceptualization of a functional element may change in that its class and/or level varies with the perspective from which it is viewed. This principle of "relativity of functional conceptualization" is the key to developing a paradigm of conceptual system structure which can be applied to different subsystems nested at different levels of the same parent system.

The functional analysis schemes for both classes and levels of objective-defined functions, as well as the scheme to be developed in Chapter Eleven for classifying and levelling flow functions, may be applied again and again, no matter what the type or level of system. Thus, each system and each subsystem of that system has its own baseline and adaptation functions. Reapplicability of the schemes means that the functional classification or level attributed to a given operation in a parent system may sometimes change when this same operation is viewed from the perspective of a subsystem. Figures 9-7 and 9-8 are designed to help with these ideas.

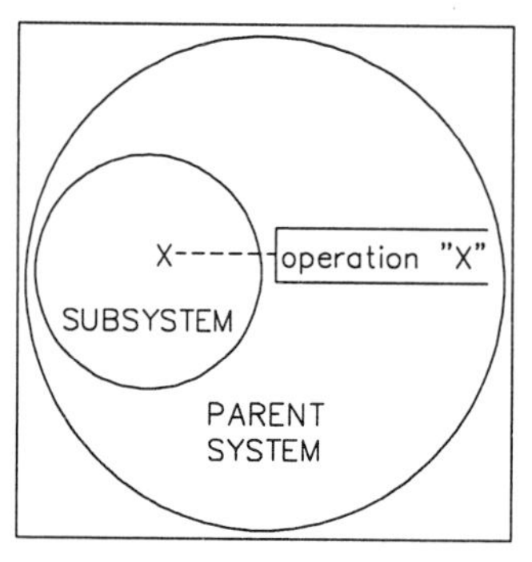

FIGURE 9-7: Idea of the relativity of functional conceptualization. *The functional conceptualization of operation "X" may not be the same when "X" is viewed from the differential perspectives of the subsystem and its parent system.*

The Four Different Dimensions of Analysis

Each type of classification or levelling scheme may be represented as a dimension of analysis, on its own mathematical axis in n-dimensional space. In the "space" being referred to (see Figure 9-9), there are four dimensions, or axes, for the following: 1) classes of objective-defined functions; 2) flow function classes; 3) distance from the baseline; and 4) internality/externality of effect. Since the levelling dimensions the depicted by the latter two axes are pertinent only to the description of adaptation operations, they are shown in the diagram as intersecting the classifications of objective-defined functions axis just at the point where the adaptation classes begin. For this reason, it may look as if these are just straight line functions in an ordinary two-dimensional space diagram. However, the diagram represents **four** dimensions. Thus, any operation in any system can be located according to a single set of four co-ordinate values, such as (a,b,c,d). Since the dimensions each represent independent variables, none of the four co-ordinate values is dependent on the values assigned to the other three co-ordinates, except that baseline operations cannot be analyzed in terms of either of the two levelling dimensions.

Although any operation in any one system can be represented by a single set of the four co-ordinate values, these values may change when the same operation is seen from the perspective of another system which contains it. To illustrate, an operation may be designated in the *parent system* as belonging to the **input** flow function of the current correction **decision-implementing** class, as being zero degrees **removed** from the baseline, and as being **internal** in effect. In contrast, it is designated in the *subsystem* as belonging to a **processing** flow function of the **baseline** class of operations, and in an **external** levelling dimension. The use of the same set of dimensions to analyze the operations in ques-

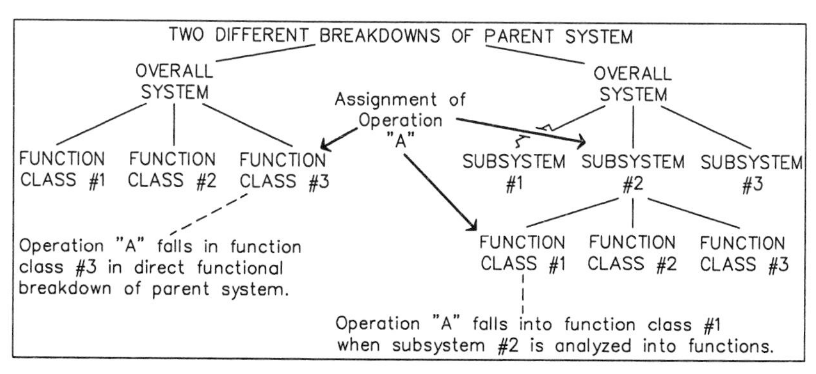

FIGURE 9-8: Functional class in parent system vs. that in subsystem. *Note: The subsystems in this figure may not themselves have been identified along functional lines.*

tion from the viewpoints of both levels of system is made possible by the principles of relativity of functional conceptualization, according to which the set of assigned values of the four dimensions changes from system to subsystem, or from system version to system version when the system essence-identity (boundaries/mission) is redefined.

Patterns of Change of Functional Conceptualization

Change from Adaptation to Baseline

In one type of change of functional conceptualization, adaptation operations become baseline ones. This occurs when the parent system adaptation operations are assigned to a given subsystem separately

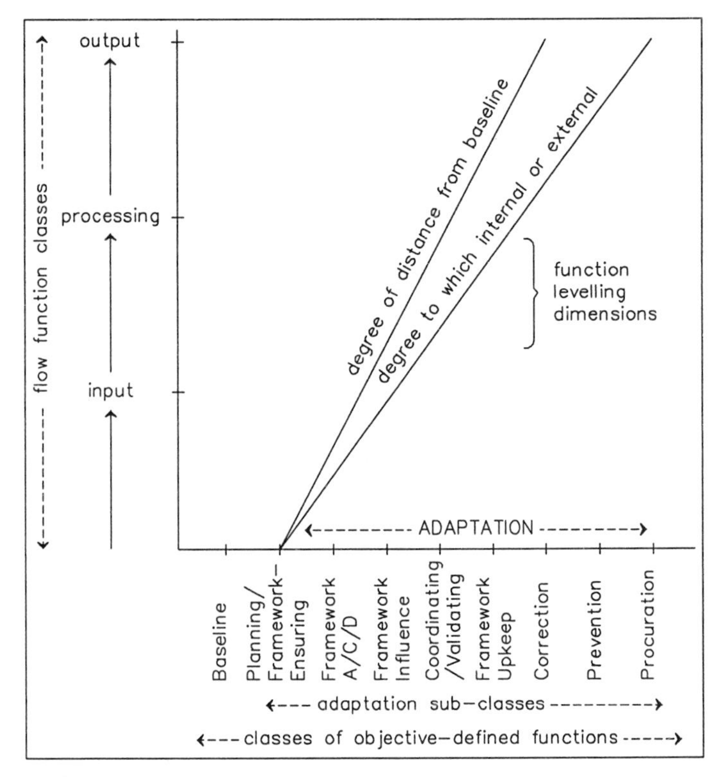

FIGURE 9-9:
Four-dimensional space representation of functional dimensions. *This is a set of mathematical axes, giving four dimensions for analyzing system functions. [Adapted from an article by Ed Baylin, "Logical System Structure," Journal of Systems Management.]*

from the baseline operations to which they apply, since, as they are the only operations in that subsystem, the adaptation operations no longer bear their adaptive type role relative to the subsystem in which they are located. An example is the government's baseline operations really being adaptation operations of the society in which the government exists. Similarly, the baseline operations of a managerial subsystem are themselves adaptation operations of the organization forming the parent system for these operations.

Actually, the above examples of changes in functional conceptualization involve the identification of information subsystems using the

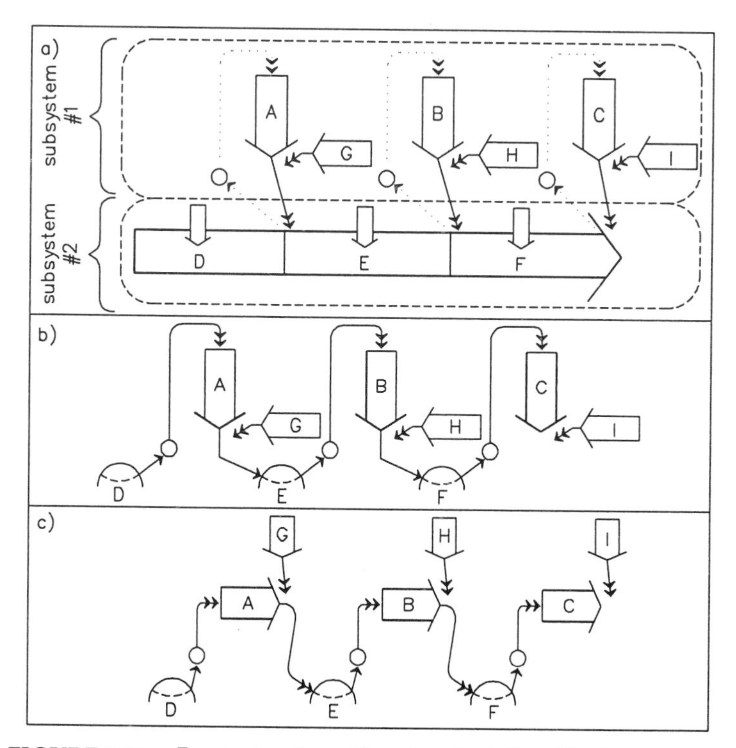

FIGURE 9-10: Reconstruction of functional relationships in a subsystem. *Here, the principles of relativity of functional conceptualization are used to construct the structure-flow chart of a subsystem identified by logical cohesion:*

a) parent system divided into two subsystems by logical cohesion;

b) subsystem #1 in "a" as seen from the perspective of the parent system;

c) same subsystem as in "b," but with the page alignments of the structure-flow chart seen from the perspective of ***this*** *system.*

logical cohesion method. Figure 9-10a shows a system divided into two subsystems, each circled by a non-continuous line. The top subsystem (the information system), identified by class of function, performs the control operations needed for subsystem #2 in the figure. Figure 9-10b then shows the information system of the previous figure (except for the inseparable internal adaptation operations, represented in the figure by embedded wideshafted arrows). However, the baseline functions in this system, namely, A, B and C, are not yet in the proper alignment for baseline functions in the structure-flow chart. Figure 9-10c gives the correct alignment for baseline A, B and C. As seen, A, B and C, which are current adaptation functions in the parent system, are now baseline functions in the subsystem. Also, the local current adaptation functions G, H and I in the parent system are now zero degrees removed from the baseline, rather than one degree removed as in the parent system perspective.

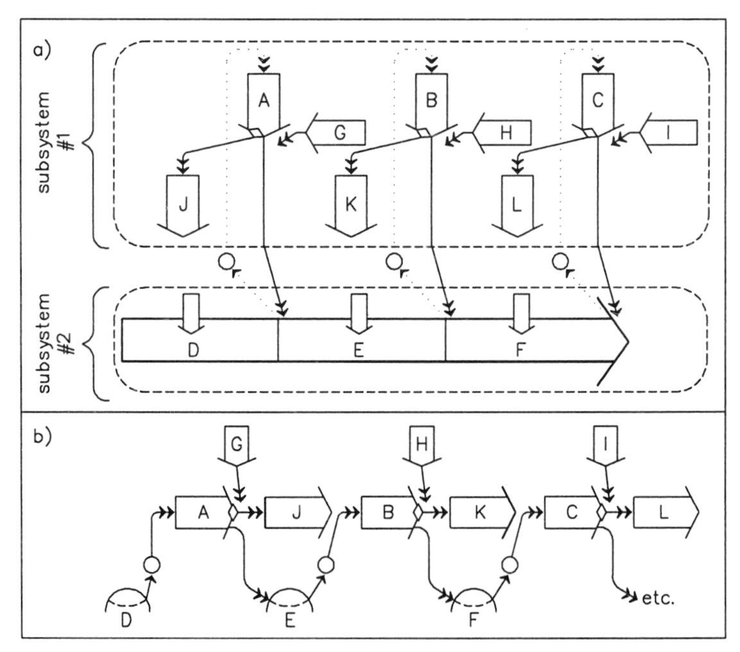

FIGURE 9-11: Reconstruction of functional relationships—example #2. *The examples shown here are similar to those in the preceding figure, but contain more functional elements and more complexity:*
a) circling of subsystems identified by logical cohesion;
b) subsystem #1 in "a" seen in the structure-flow chart from its own perspective.

There is more to this set of relationships than is shown in these figures, since the placements of coordination and decision-implementing current operations are not discussed. Decision-implementing operations

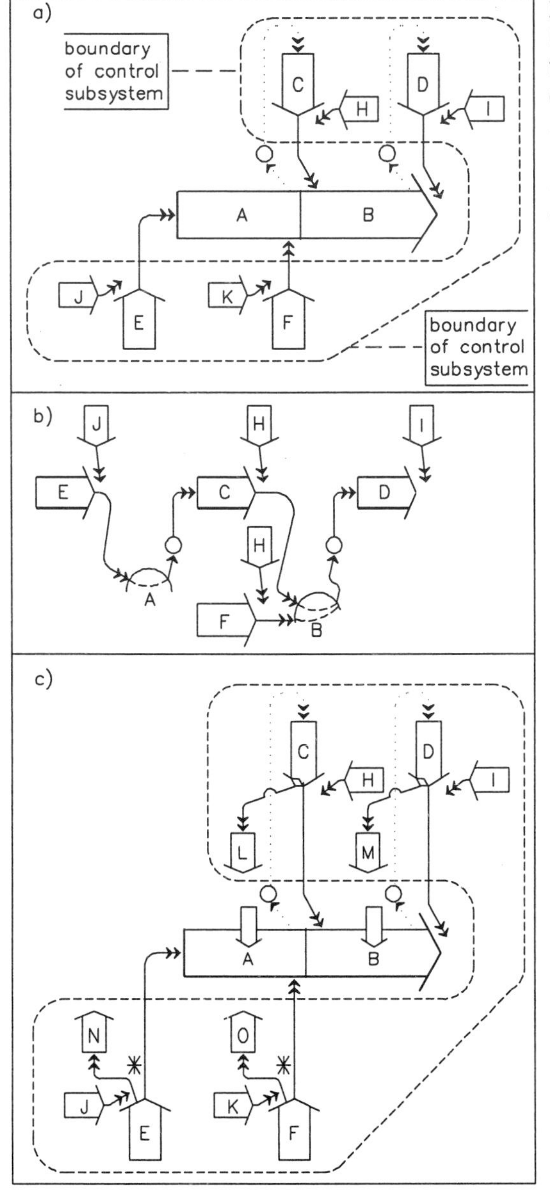

FIGURE 9-12: Reconstruction of functional relationships—example #3. *Here, the control subsystem identified contains decision-implementing functions, in both ensurance and feedforward time-orientations:*

a) the parent system with subsystems circled;

b) change in structure-flow chart page alignments as control subsystem of "a" is seen from its own perspective; B, C, D, and F are now seen as baseline functions;

c) an expanded view of the present example, to indicate further possible complexities in applying the principles of relativity of functional conceptualization. These would result from the addition of the decision-implementing functions L, M, N, and O.

and a pattern of analysis into subsystems are demonstrated in Figure 9-11a. Figure 9-11b then redisplays the functional relationships of the information subsystem (what was subsystem #1) of Figure 9-11a. The decision-implementing functions, not the coordination ones, are now the baseline operations relative to the information subsystem, while the coordination operations, which operate in an ensurance role in the parent system, are seen as initiative operations relative to each subsystem baseline.

Figure 9-12a shows the validating and coordinating operations in the parent system, along with internal validating functions. Figure 9-12b then redisplays the control subsystem from its own perspective. Finally, Figure 9-12c comprises a number of further ensurance and feed-forward current operations, by adding certain decision-implementing functions (labelled "L," "M," "N," and "O" in the figure). The complexities of the conversions of functional class and level increase accordingly. To keep things simple, the author has refrained from redisplaying the control subsystem corresponding to Figure 9-12c.

Change in Time Frame

A second type of change of class of objective-defined function exists when non-current adaptation operations are converted to current ones. This occurs when the baseline operations of a subsystem are the non-current operations of the parent system. What is non-current in the parent system becomes current relative to the subsystem; i.e., the base-

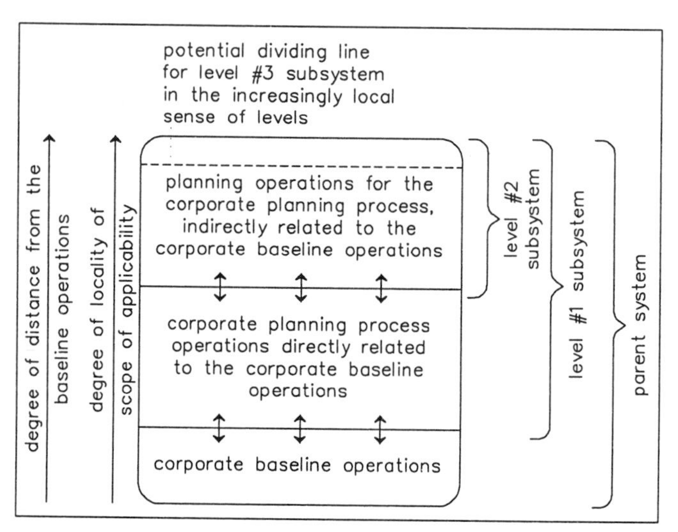

FIGURE 9-13:
Subsystems at different degrees of distance from baseline.

line of any system contains what are by definition current operations relative to that system.

Changes in Scope of Adaptation Function

The scope of adaptation operations increases as the degree of distance from the baseline decreases. As well, the decrease in distance from the baseline often causes a change of adaptation operations from the internal to the external levelling dimension. For example, in Figure 9-13 (the procedures-creation for corporate planning example used in the first section of this chapter), the level #2 subsystem operations are in direct contact with the level #1 subsystem operations. Assuming that the level #2 subsystem operations are internal to the level #1 subsystem operations, rather than being general planning operations for the corporation as a whole, the level #2 subsystem operations have a more external applicability in the level #1 subsystem than in the parent system.

Tendency for All Above Changes to Coincide

The following four types of change in the class/level of adaptation operations often tend to coincide with one another:

—change from adaptation to baseline;
—change in time frame, from non-current to current;
—decrease in degree of distance from the baseline;
—change from internal to external applicability.

For instance, Table 9-2 is the job description for a hospital quality assurance administrator/consultant. The essential baseline role of this person is to organize and assist those directly involved in managing the quality assurance activity, both with their planning functions and sometimes with their organizing and current controlling functions (using the Fayolian functional classes). Thus, the baseline activities of the quality assurance administrator (the job title) are strategic level external adaptation activities of the hospital as a whole, far from the baseline of hospital activities. The baseline activities of the quality assurance administrator are also mostly of a non-current nature with respect to the hospital's baseline, although they are current activities from the viewpoint of the job of the quality assurance administrator. This administrator also performs certain (external and internal) adaptation functions needed to administrate the function of the quality assurance programme itself, i.e.,those which are local to the quality assurance activity. With respect to the job of the quality assurance administrator, the latter activities are of various scopes and degrees of distance from the baseline and of varying internality, although they are all still far from the baseline in relation to the hospital system as a whole.

JOB DESCRIPTION: QUALITY ASSURANCE ADMINISTRATOR
SUMMARY: Under the immediate supervision of the Executive Director, the incumbent is responsible for organizing and assisting in the management of the QA (quality assurance) programme. Also, on invitation from the persons responsible for this management, the incumbent is responsible for organizing and controlling the carrying out of the QA activities.
DUTIES AND RESPONSIBILITIES *Organizing and Assisting in Management of the QA Programme*
The main objectives of the QA administrator are the following: 1 - to help department heads set and maintain standards, identify and investigate problems, prepare analysis reports, and determine actions; 2 - to help department heads, QA task forces, and the QA committee prepare agendas for meetings; 3 - to provide data for the above by actions such as the following: maintaining logs of problems and their follow-ups; collating review agendas and problem analysis reports; completion of summary monthly problem analysis reports; maintaining files. *Organizing and Controlling the Carrying Out of QA Activities*
On invitation from the department heads, the QA administrator may engage in activities such as the following: 1 - training of clinicians and support personnel in QA techniques; 2 - enforcing policies on the confidentiality of QA data. *Maintaining the QA Programme*
The perpetuation and effective maintenance of the QA programme require: 1 - participation in the preparation of the annual QA budget; 2 - periodic reporting to the executive committee and external agencies regarding QA activities; 3 - maintenance of accreditation status logs.

TABLE 9-2:
Job of a hospital quality assurance administrator/consultant.

DIAGRAMMATIC FORMS VERSUS FUNCTIONAL CONCEPTUALIZATION

That the same set of analytical schemes can be applied again and again to all system boundary definitions and levels is supported by the ability to represent the ideas in terms of a diagram such as the four-dimensional representation (Figure 9-9). While the triangle, and more so the pyramid, provide certain types of geometrical representations, they cannot conveniently capture **all** the understandings exhibited in the four-dimensional space diagram. Nor can any of the many other representations attempted by this author, with the exception of the structure-flow chart, portrayed in its full three spatial dimensions. As for the pyramid, at least three of the four dimensions in the four-dimensional space diagram can be well represented, with the exception of the flow dimension.

Limitations of the Triangular Representation

Some possible misinterpretations result from using the triangular form as opposed to the four-dimensional space diagram or structure-flow chart or the pyramid. A casual study of (for instance) the Anthony triangle might lead one to the following conclusions:

- that Anthony would not have separate axes for classifications of objective-defined functions and internality/externality, since the degree of the latter is assumed to vary with the degree of non-currency of the time frame; i.e., long-term operations are assumed to be external;
- that Anthony ignores the fact that degree of distance from the baseline and internality/externality are actually separate dimensions; i.e., Anthony's triangle assumes that externality always increases along with distance from the base of the triangle;
- that operations far from the baseline are always non-current.

The preceding conclusions are, however, probably not the types of simplistic assumptions that Anthony would make, but are based more on the impression which arises from the limiting nature of the triangular **form**. In other words, functional classes and functional levels are two separate, albeit highly interrelated, subjects.

Capturing Further Dimensions

This book does not deal with all dimensions for understanding systems. For instance, a fifth dimension is the evolution of the system structure over time. However, the structure-flow chart may possibly be able to capture this idea. Thus, the slant of the adaptation operations of all classes may also be exploited to capture the dynamics of system evolution. In a growing system, or a "one-shot" system like a project, the slant in the same direction as the operations in relation to which the adaptation operations play an adaptive role symbolizes a building state of a system. A slant of say, 45 degrees in relation to the base operations, shows a system undergoing moderate growth, while a slant of 90 degrees shows a system in a steady state. In contrast to the growing system, a slant in the contrary direction is symbolic of a dying system, or of a system being destroyed. A similar type of idea is to some extent demonstrated in Figure 9-14. Here, a framework is being taken down, namely, the de-allocation of file space on disk after all transactions have been processed by a computer program.

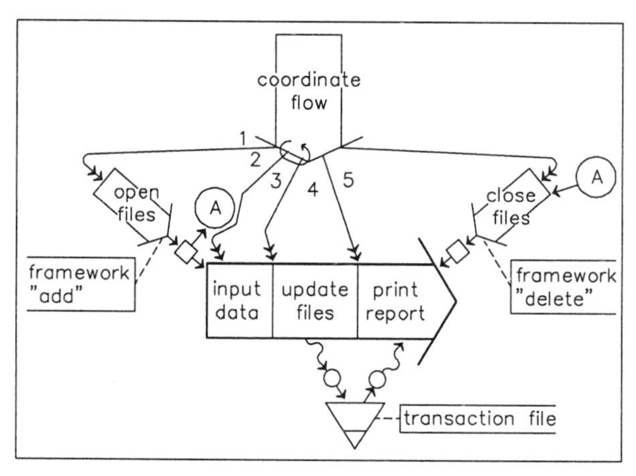

FIGURE 9-14: Showing system evolution dynamics in the structure-flow chart. *This uses the angles of adaptation function wideshafted arrows relative to the baseline wideshafted arrow to symbolize the dynamics of system evolution.*

CHAPTER TEN

MORE ABOUT SUBSYSTEM IDENTIFICATION METHODS

BACKGROUND ISSUES IN IDENTIFYING SUBSYSTEMS

PERCEPTION OF SUBSYSTEM ESSENCE-IDENTITY

Manner of Achievement As Part of the System Definition

The essence-identity of a system is its "common whole," or its "transcending purpose." This includes its "mission," ("end," "goal"), along with a broad definition of boundaries. The fulfillment or maintenance of its mission and essential boundaries provides the system with its identity, without which the system would simply cease to be what it is. This definition relies on the idea of "equifinality," by which the end result is all that matters, with the routes to this end being of no importance.

It is perhaps better to use the word "mission," rather than the word "objective," to describe the goal of the system, since the word mission has a more comprehensive ring to it. Specifically, "mission" may sometimes refer not only to objectives, but also to the **manners** in which objectives are achieved. For instance, a heating system's mission may be defined by one analyst as "to produce heat," and by another analyst as "to produce heat **electrically**." This is a very basic point, since many subsystems are identified in terms of some manner of operation within the parent system to which they belong. In fact, many of the non-functional cohesion (subsystem identification) techniques described in the computer literature refer to the *manner* of operation, in addition to the operational objective. For example, the procedural cohesion method may identify a subsystem carrying out those control operations of the parent system done by computer.

Characteristics Used to Define System Essence-Identity

Subsystems within any parent system may be identified in terms of sub-sets of the parent system characteristics. Different types of parent system characteristics are used in identifying a subsystem's essence-identity. These particular types of characteristic, or the mix of characteristic types, tells something about the type of approach taken to identifying a subsystem. In other words, these are *defining characteristics*. To quote: "Attributes may be one of two general kinds: defining and accompanying. **Defining** characteristics are those without which an entity would not be designated or defined as it is. **Accompanying** characteristics or attributes are those whose presence or absence would not make any difference with respect to the term describing it."* These characteristics may be conceptual ones, such as objectives or manners of achieving objectives, or they may be physical or formal ones, such as the territory occupied or the formal role identities of the physical system components.

To determine the parent system characteristics used to define the essence-identity of a subsystem is not always an easy task, as the following points out:

> By objectives of the system, Churchman means those goals or ends toward which the system tends. Hence, goal seek-

---

* Schoderbek, C.G., Schoderbek, P.P., and A.G. Kefalas, <u>Management Systems: Conceptual Considerations</u>, revised edition. Dallas, Texas: Business Publications, Inc.©, 1980, p. 21.

> ing or teleology is a characteristic of systems. With mechanical systems the determination of objectives is not really difficult since the objectives have been determined even before the mechanical system took shape. A watch is made to tell time, either in hours, minutes, seconds, or in days; it isn't supposed to cut grass or to slice tomatoes. The determination of objectives of human systems, however, can be a formidable task. One must beware of the distinction (often a real one) between the stated objective and the real objectives of the system. A student, to use Churchman's telling illustration, may give as his objective the attainment of knowledge while in fact his real objective may be the attainment of good scholastic grades. The test that Churchman proposes of distinguishing the real from the merely stated objective may be called the principle of primacy: will the system knowingly sacrifice other goals to obtain the stated objective? If the answer is yes, then the stated and real objectives are identical.*

The use of the word "objective" in the above quote should be noted. As mentioned in Chapter Eight, the word "objectives" is used in this book such that objectives do not involve *manners of achieving* objectives. However, as discussed above, the **mission** of a system could include a manner in which the objectives are achieved.

The identification of the essence-identity of a subsystem proceeds in two distinct steps. **First**, the type of approach must be selected, as in the decision about which of four variables in an algebraic equation to make the dependent one, that is, to associate with the algebraic equation's "essence-identity." **Second**, based on the general type of approach, specific parent system characteristics must be identified. To continue the analogy, by specifying the specific value, or value range, of the dependent variable in the algebraic equation, one is identifying a specific subsystem. Thus, if $W = 2X + 4Y/Z$, and if the desired value for W is 36, then any combination of the values of X, Y, and Z which causes W to be 36 will uphold the essence-identity of a particular subsystem.

Seeing The Essence-Identity of the Business Organization

A business organization, a subsystem of the economic system, can demonstrate the ideas just discussed. At least two distinct functional approaches identify what a business organization is all about. One

---

* Schoderbek, Schoderbek, and Kefalas, p. 9.

deals in terms of the clearly technical objectives of the business, e.g., to produce a certain product line. The choice of the product line, if not of the individual products within this line, is then tantamount to the choosing of the value of the mission part of the essence-identity of the business. The second approach deals in terms of profit (which may be considered either a technical objective, or, if the profit accrues to the business' operators, an agent-based objective). Where profit is the objective, the last step in achieving the objective is the payment of dividends, and the technical objectives of emitting certain product lines are seen as steps on the way to achieving the profit objective. Alternatively, the profit may be considered to be the primary objective, and the product line technical objectives may not only be seen as steps on the way to achieving the profit, but also as secondary elements in the definition of what the system is all about.

The latter example was discussed in Chapter Nine, in connection with the mutual contingency of two objectives where one objective is a step towards the other. This provides an example of using a functional approach in which a complex set of objectives is used to arrive at the mission part of the system essence-identity. In fact, the use of complex approaches seems to be the norm when dealing with socio-technical systems. Moreover, the approach to such systems also usually includes other types of agent-based objectives, such as the power and pleasure objectives of the managers who run the business. The inclusion of these other objectives types often makes it difficult, if not impossible, to perform functional analysis, since they badly distort the otherwise relatively simple functional relationships within the subsystem.

Question of Flexibility in Identifying Subsystems In An Actual System

Thus far, the question has not been discussed as to whether, at the one extreme, the subsystem essence-identities are fixed within their parent system, or whether, at the other extreme, an infinite number of ways exist in which any system can be carved into subsystems. The following quote seems to indicate that at least a limited number of possible essence-identities of the subsystems are inherent in the parent system:

> What we are concerned with in any system is relevant relatedness. When dealing with systems, one must be careful to acknowledge relatedness only when one is ready to declare relevancy. One can easily relate something in this world to almost anything else by reason of color, size, shape, density, distance, and so on. Many of these relationships may be spurious; they lack relevance. Perhaps this is why

> Beer states that there seem to be three stages in the recognition of a system. First, "we acknowledge particular relationships which are obtrusive, this turns a mere collection into something which may be called assemblage. Secondly, we detect a pattern in the set of relationships concerned; this turns an assemblage into a systematically arranged assemblage. Thirdly, we perceive a purpose served by this arrangement and there is a system."*

Regardless of the answer to the preceding question of whether or not the essence-identities of subsystems are inherent in their parent systems, it must remain clear that subsystems have essence-identities defined at their own levels. That is, the essence-identity of a subsystem should not usually be the same as that of its parent system. Conversely, one should obviously steer away from the error of defining the parent system's mission as that of one of its subsystems.

PLACEMENT OF SUBSYSTEM BOUNDARIES

Basics of Subsystem Boundary Placement

Defining the subsystem's mission is only part of defining its essence-identity; as well, its boundaries, or scope, must be identified.

A "subsystem" may be defined as a sub-set of the logical and physical components of a parent system which have a strong bearing on the subsystem essence-identity, albeit perhaps (see below) not subject to control by system operations. It is assumed here that every system is a subsystem of a larger, parent system (including any parent system itself). "Strong bearing" in this definition means a generally positive influence, although a negative influence may be tolerated in systems having contradictory elements, or "dysfunctionality," since disharmonious elements are blended together in its essence-identity. The influence of the logical and physical components having a strong bearing may be simple or complex, linear or non-linear, and so on, as long as it is "strong" enough.

The question of what elements of a parent system go inside the system boundary and which go outside has been discussed extensively in

---

* Schoderbek, Schoderbek, and Kefalas, Management Systems: Conceptual Considerations, pp. 23—24. The internal quote is from Stafford Beer, Decision and Control. New York: John Wiley & Sons©, 1967, p. 242.

general systems literature. One obvious point made is that the system boundaries are often not clearly definable.

In defining subsystem boundaries, the analyst must work within constraints related to the "inseparability" of certain logical or physical components; for example, if two different system operations are inseparable from one another, then either both must be placed inside the system boundaries, or both must be located in the system environment. To illustrate, the second-by-second scheduling of worker activity is inseparable from the operations performed by the workers, unlike hour-by-hour coordination, since the latter may be assigned to supervisors of the workers.

Thus, certain operations, given the definition of the system's essence-identity, must be included within the system. For the rest, the inclusion of operations within the system is at the discretion of the analyst who perceives the system. The distinctions between mandatory and optional, and between optional and necessarily excluded operations, and the grey areas between these categories, are represented in Figure 10-1. To quote:

> The boundary of a system is often arbitrarily drawn depending upon the particular variables under focus. One can adjust the boundary to determine whether certain variables are relevant or irrelevant within the environment or without. A system viewed from two different levels may have different boundaries. This arbitrariness is not necessarily undesirable, since researchers and organizational officials tend to view a particular system from their own intellectual perspectives much as managers tend to evaluate case study problems from the vantage point of their own specialities . . . different resolution levels call for different definitions of the system, different objectives of the investigation, different parameters, and different boundaries separating the system from its environment.*

Figure 10-2 shows the grey areas in defining the boundaries of the business organization. These are represented by the fact that the various environmental entities in this figure straddle the boundary.

Non-Overlapping Boundaries

To decide what is to go inside and what is to go outside the subsystem in ambiguous cases, the analyst should aim at creating a non-overlapping set of subsystem modules within the parent system. This forces

---

* Schoderbek, Schoderbek, and Kefalas, pp. 27—28.

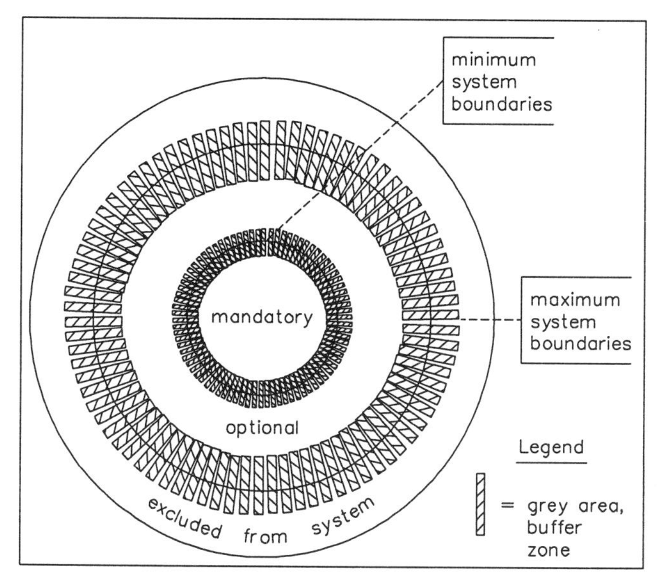

FIGURE 10-1: Latitude in system boundary placement. *This shows the extent of system boundaries, from minimum to maximum extents. Given the system mission, different operations may or may not be located within its boundaries. Moreover, system boundary lines are always grey areas, as indicated by the shading in the diagram.*

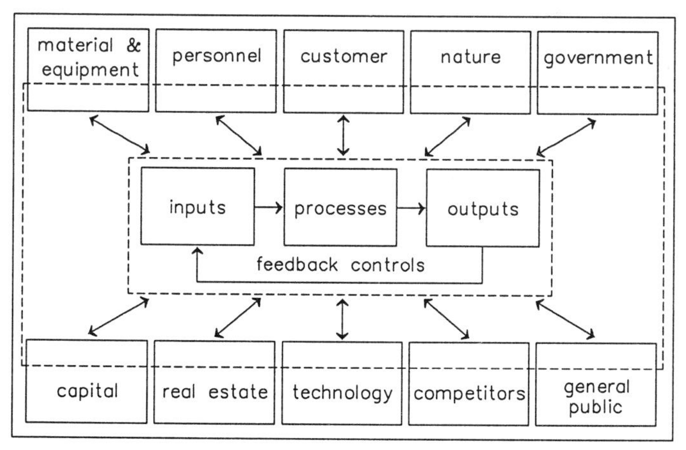

FIGURE 10-2:
Grey areas in the boundary of the business organization.

the analyst to make a choice in each case. Thus, each physical or logical component of the parent system should, ideally, be located in one subsystem only. Specific terms for discussing the subject of non-overlapping configuration modules of subsystems are:

- "root parent" system: the system at the highest level of hierarchy
- "sibling" subsystems: subsystems in the same branch of the hierarchy, (having the same parent) and also at the same level;
- "peer" subsystems: subsystems in different branches of the hierarchy (having different parents), but at the same level (provided that equating levels of different branches is possible).

These concepts are represented in Figure 10-3.

Further Questions About Boundary Placement

Two interesting and easy-to-answer questions related to the placement of the system boundaries are as follows:

—Do boundaries change depending upon a slight difference in interpretation of the mission (goals) of the system? Obviously, they may.

—Is it possible for different subsystems of the same parent system to have conflicting goals? Yes, they may.

The following covers two more difficult questions related to the subject of boundary placement.

Placement of Control Elements with the Elements they Control

Should control elements be placed within the same subsystem as the system elements which they control? Also, should system components

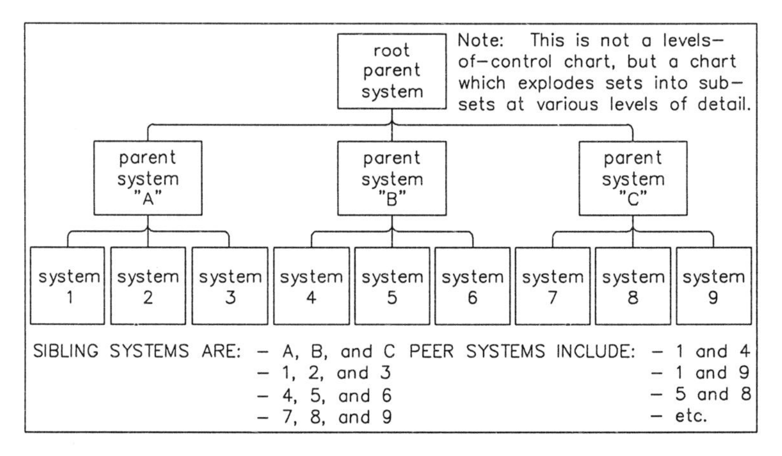

FIGURE 10-3: Sibling versus peer subsystems in a hierarchy. *This is a hierarchical functional decomposition chart.*

which are not controllable by other system components be placed outside the subsystem boundaries, where they would form environmental entities in direct or indirect interaction with the subsystem? It would seem that if this were so, the possibility would not exist of identifying subsystems by, say, the logical cohesion method, e.g., whereby general-level control operations might be separated into subsystems of their own, as in the case of the general managerial operations in a business organization being set apart as a subsystem of the business, apart from the worker operations which they manage. This is an example of some of the operations which control worker activity being located outside the worker system, and some of the operations which are controlled by the managerial system being located outside the managerial system.

It seems that the present answer to the questions being posed may not concur with some of the opinions held by general systems theorists, such as those which appear to be expressed in the following quote:

> . . . the environment includes all that lies outside the system's **control**. The system can do relatively little or nothing about the characteristics or behaviour of the environment. Because of this, the environment is often considered to be "fixed"—the "given" to be incorporated in any system's problem. Second, the environment must also include all that **determines**, at least in part, the manner of the system's performance. . . . Resources . . . are all the means available to the system for the execution of the activities necessary for goal realization; resources are inside the system; also, unlike the environment, they include all the things that the system can change and use to its own advantage.*

Goals and System Boundary Changes

Do system goals change as the system boundary is varied? The answer to this may not necessarily be affirmative, for the following two reasons:

First, as discussed above, not **everything** which a system may control or change, or which a system must control or change, needs to be included within its boundaries, as long as the controllable or controlling operations are separable, i.e., may be placed in another system. Goals do not change as further control elements are added to the system, (thereby expanding its boundaries).

---

* Schoderbek, Schoderbek, and Kefalas, p. 10.

Second, not all sections of flow in a functional sequence going towards the system mission need be included within the system. Only the operations corresponding to the **final step(s)** which achieve(s) the mission need be included. Thus, for example, steps 1, 2, 3, and 4 are needed to achieve a system objective in step 4. Step 4 and any combination of steps 1, 2, and 3 may be included within the system boundaries. For example, steps 2 and 3 may be located in the environment due to a particular analyst's manner of subsystem analysis, without this changing the system mission.

The preceding ideas may go somewhat against intuition. Thus, if four steps are involved and only the first and fourth are placed in the same system, it is very easy to see each of these first and fourth steps as being **separate** components of the system mission. However, the relationship between step one and step four is not lost, although it takes place through the intermediary of the other subsystems to which steps two and three were assigned during earlier subsystem analysis. This implies that step one is still seen as having an intermediate objective with respect to the subsystem to which it is assigned along with step four.

To restate this point, the mission of a system should not in some cases be expressed such that intermediate objectives are mentioned. For instance, one individual defines the mission of the supplies acquisition subsystem in a business as "to purchase and receive goods," while a second person defines it as "to obtain goods." This is more than a matter of semantics. In the first case, the purchasing step, which feeds forward to the next step, is seen as being a system objective, along with the receiving step; while, for the second person, only the receipt of goods is viewed as a system (or final) objective. The second person may then opt to include only the receiving step within the defined subsystem boundaries, on the assumption that the purchasing operations have been designated to another subsystem of the same parent system. The first individual has no such option, since the purchasing step is seen as part of the subsystem's very mission.

In conclusion, it may be said that **stating where to place the system boundaries is a matter separate from that of defining system mission.**

VARIOUS APPROACHES TO SUBSYSTEM IDENTIFICATION

Cohesion Methods

The many methods of dividing a system into subsystems may be classified into purposeful and coincidental categories. Coincidental ones involve subsystem identification according to the physical structure or the formal role identities of the parent system components. The purposeful methods, on the other hand, include a number of categories dealt with in the computer literature under the title of "cohesion methods."

Cohesion methods, a very popular subject in computer systems design and programming, evolved as part of the "structured design" of programs. A conglomeration of highly specialized, esoteric terminology has been developed to show computer systems specialists how to establish good modular breakdowns of programs. Perhaps the most important terms are "inter-modular coupling" and "modular cohesion." Computer systems analysts and programmers have been advised to use the functional cohesion method, i.e., to develop highly internally cohesive but loosely coupled modular sets in breaking down a system, such that each module performs a highly related series of tasks and does not require frequent interruption and branching of the sequence of execution to and from other modules during the course of executing a single run. On the subject of **how** to develop highly cohesive, loosely coupled modules, various types of cohesion are described.* From worst to best, these are as follows:

1. **Coincidental**—the lowest level; it means that the processing elements of a module are essentially unrelated by anything—no common function, data, or procedure.
2. **Logical**—the elements in a logically cohesive module are all oriented towards performing a certain logical **class** of functions; for example, handling all errors.
3. **Temporal**—the elements are related by time as well as logically; for example, initialization. They probably don't occur in a certain order or operate on the same data.
4. **Procedural**—the elements are all part of a procedure—a loop, an algorithm, a certain sequence of steps that **have to be done in a certain order.**
5. **Communicational**—the elements all operate on the same data; for example, a communicationally cohesive module might contain elements to perform anything that has to be done to a particular element of input data.

---

* Deltak.© The essence of these definitions comes from works such as those by Yourdon, Constantine, and Myers, e.g., Yourdon, E. and L. Constantine, *Structured Design: Fundamentals of a Discipline of Computer Program and System Design.* Englewood Cliffs, N.J.: Prentice-Hall©, 1979, pp. 108—130.

6. **Sequential**—the elements are related in that they perform different parts of a sequence of operations, where the output of one operation is input to the next.
7. **Functional**—each element is a necessary part of one single function, and essential to the function's performance.

Methods 2 to 7 above fall within the "purposeful" category of subsystem identification methods.

Sub-Groupings of Cohesion Methods

Some of the purposeful subsystem identification methods may be grouped into what might be called the "manner of operations" class. This includes all the above cohesion methods except for logical, sequential and functional ones. By "manner of operation," is meant such things as the following:

- timing—as in temporal cohesion;
- technology, instrument or methodological context—as in procedural cohesion;
- means through which contact is made with the system environment—as in communicational cohesion;
- other abstracts, such as task programmability, style of task execution, task duress, etc. Other cohesion methods might have been designed in the computer literature for these, but were not.

Mixtures of Different Cohesion Methods

When a subsystem within a particular parent system is inspected, difficulty may be encountered in understanding what cohesion method is used to identify it. This is often the case, since hybrids of the different cohesion methods are almost always present or unavoidable in practice. The real issue often seems to be not which cohesion methods are exclusively applied, but which method is the predominant one used to identify a particular subsystem.

When hybrids of different cohesion methods are used, the identification of the predominant method, if any, is weighted by the importance in the subsystem of the functional elements involved in different cohesion patterns. Where different cohesion patterns **coincide** with others with respect to the very same functional elements—for instance, two functional elements are related by **both** communicational and procedural cohesion—some might say that the functional elements are related by the so-called higher form of cohesion; e.g., the modules are communicationally, rather than merely procedurally cohesive, since communicational cohesion is the "higher" form of cohesion. The assumption that appears to be made here is that a communicationally cohesion module is also procedurally cohesive. The validity of the this assumption seems to be understood in the following statement: "The cohesion of a module is the highest level of cohesion which is applicable to all elements of processing in the module."*

(Based on the preceding quote, the reader may notice that the word "cohesion" may be used in two different senses, that is, in addition to cohesion **methods**, i.e., methods of identifying subsystems, it is also a **measure**, as the "degree of functional relatedness of processing elements within a single module."** This definition may be very confusing since cohesion methods such as the logical and temporal ones produce subsystems having very little internal functional relatedness. To get around this problem, the word "binding," referring to measuring internal functional relatedness, should be used instead of the word "cohesion" to describe the measure in question.)

The case of a less well evaluated cohesion method in a more important part of the system has to be dealt with. For instance, in a system whose major functions are related by logical cohesion (functional class), the presence of functional cohesion between internal (very local) adaptation functions should in no way lead to the conclusion that functional cohesion is used to identify the subsystem in question. Of the two methods, the one to choose must be logical cohesion.

SOME APPARENTLY NEW CONCEPTS BEHIND COHESION METHODS

Cohesion Versus Anti-Cohesion

Cohesion refers to the assigning of different functional elements to the same subsystem during analysis of a parent system into subsystems. In contrast, anti-cohesion refers to either the breaking up or the cloning of a single functional element into **more** than one subsystem during analysis of a parent system. Cohesion, as described below, is also called "grouping" in certain contexts, and "merge" in others; anti-cohesion may also be referred to as "segregation." Anti-cohesion (or segregation) will be discussed below in connection with the procedural and communicational cohesion methods. In general, each cohesion method has its own, corresponding form of anti-cohesion method.

---

* Yourdon, E. and L. Constantine, Structured Design: Fundamentals of a Discipline of Computer Program and System Design. Englewood Cliffs, N.J.: Prentice-Hall©, 1979, p. 135.

** Yourdon and Constantine, p. 447.

Grouping Versus Merging of Functional Elements

Grouping of different system elements, illustrated in Figure 10-4, means that each element retains its original identity. In contrast, the **merging** of different elements means that they are one and the same, or may be made one and the same by elimination of duplication.

When merge occurs, either the procedures involved must be duplicated for use in different functional contexts, or the procedures may be placed together such that they are in some way accessible from the different functional contexts. In the functional cohesion method, in order to avoid both duplication and subsystem conflicts, merged functional elements are placed into a separate, self-standing subsystem which is equally accessible from all contexts. As seen earlier, common accessibility of a subsystem module representing a non-duplicated resource may be represented by fan-in on levels-of-control charts.

Relationship Between Grouping and Fan-In on Levels-of-Control Charts

As just observed, merge and fan-in are related on the levels-of-control chart. Fan-in on such a chart does not, however, **necessarily** imply merge, or even partial merge (see Figure 10-4). This is because fan-in may represent grouping, or only partial merge, whenever communicational, procedural, temporal, logical, or coincidental cohesion methods are employed. In fact, the further the cohesion method is from the functional one, the greater the degree to which fan-in indicates grouping rather than merge. Thus, fan-in connected with communicational and procedural cohesion represents partial merge, while fan-in connected with logical, temporal, and coincidental cohesion increasingly symbolizes grouping rather than merge. Since fan-in on the levels-of-control chart may occur when any of the cohesion methods is used, it is not fan-in itself, but the degree to which fan-in represents the existence of merge, which points to the extent of subsystem configuration along functional lines.

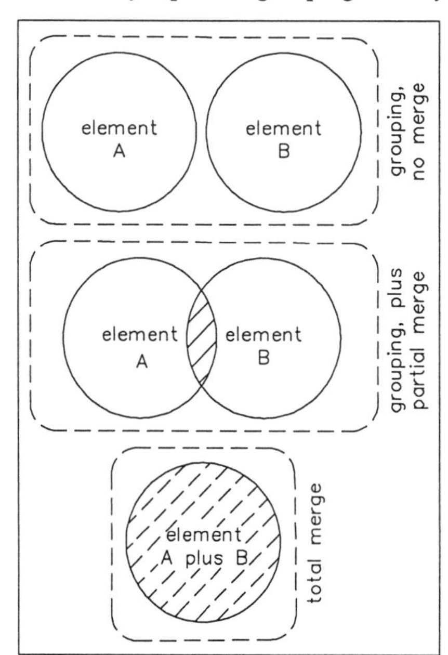

FIGURE 10-4: Grouping, partial merge, and total merge of elements.

The concepts of merge and grouping lay an important groundwork for comparing the different cohesion techniques and their relationship to the functional one. Whenever there is merge, it is likely that functional cohesion has been used as a method for perceiving subsystem structure, because a conscious attempt is being made to avoid subsystem conflict by taking a common resource and placing it in its own subsystem. However, partial merge may also appear in connection with the procedural and communicational cohesion methods, with the consequence that these cohesion methods tend to approach the functional one.

INDIVIDUAL METHODS OF IDENTIFYING SUBSYSTEMS

COINCIDENTAL METHODS

Two coincidental methods of subsystem identification are those by physical and formal structure of a parent system, namely, what are called here the "physical structure-interactions" and "formal identity" methods, respectively.

Physical Structure-Interactions Approach

In the physical structure-interactions approach, the **meaning** of the interactions among the agents observed in interaction within the given absolute or relative territory is of no importance. All that the analyst must observe are the physical (actual) patterns (e.g., frequency, path, timing, etc.) of interaction. A subsystem may be identified when the interaction is in some way intense enough between/among the various interacting agents. Here, as we are basically dealing with an organization, the physical structure-interactions approach is closely aligned to the idea of "work constellations."*

For example, the analyst wanders into an office building. The first thing that the analyst may notice in each office is the physical layout of desks, people in action (to be called "actors"), and office machinery. By

---

* Mintzberg, H., The Structuring of Organizations. Englewood Cliffs, N.J.: Prentice-Hall©, 1979, pp. 53—58.

observing this, and by staying long enough to observe and study the patterns of interaction involving these elements, the analyst should be in a position to form an idea of the subsystems that occur within each office. The essence-identities of these subsystems are specified in terms of agents (actors and physical catalysts), territories, and interaction patterns. Various levels of subsystems may be identified in this fashion, so as to create a consistent, non-overlapping hierarchy of subsystems in the analyst's imagination.

The lowest level of subsystems in this hierarchy may be composed of individual actors, as in Figure 10-5. Also, any individual actor may belong to a number of different subsystems. For example, an office worker in the building being studied may work in one room in the morning, and a different room in the afternoon, according to a regularized pattern. In each case, this office worker may be perceived by the analyst to belong to a different grouping of actors, i.e., to a different subsystem, in the morning as compared with the afternoon.

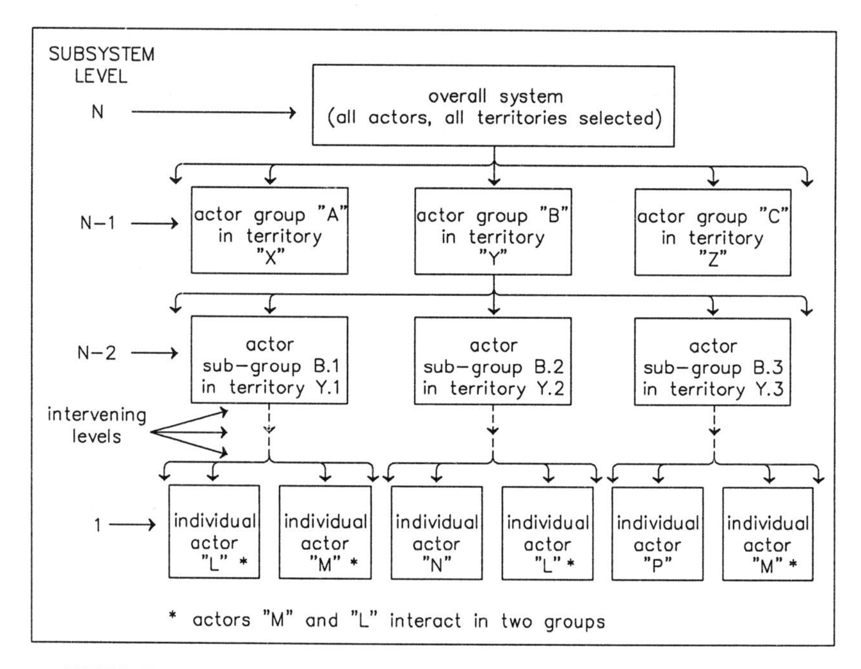

FIGURE 10-5:
Subsystems by the physical structure-interactions approach.
This is an hierarchical functional decomposition chart giving subsystem breakdown defined with the physical structure-interactions approach.

Formal Identity Approach

"Formal identity," the other coincidental approach, means identity in a legal or in a titular sense, e.g., a legal corporation, which is a "legal entity." Similarly, an organization chart depicts various titular entities, such as departments, units, and individual job positions, to each of which the analyst may identify a corresponding system or subsystem. Figure 10-6 is a hierarchy chart illustrating the formal identities which might exist in a hypothetical business organization. The asterisks in this figure show how the same job position may play roles corresponding to different formal identities, i.e., to different subsystems. The idea here (illustrated in Figure 10-6) is similar to that already shown in Figure 10-5.

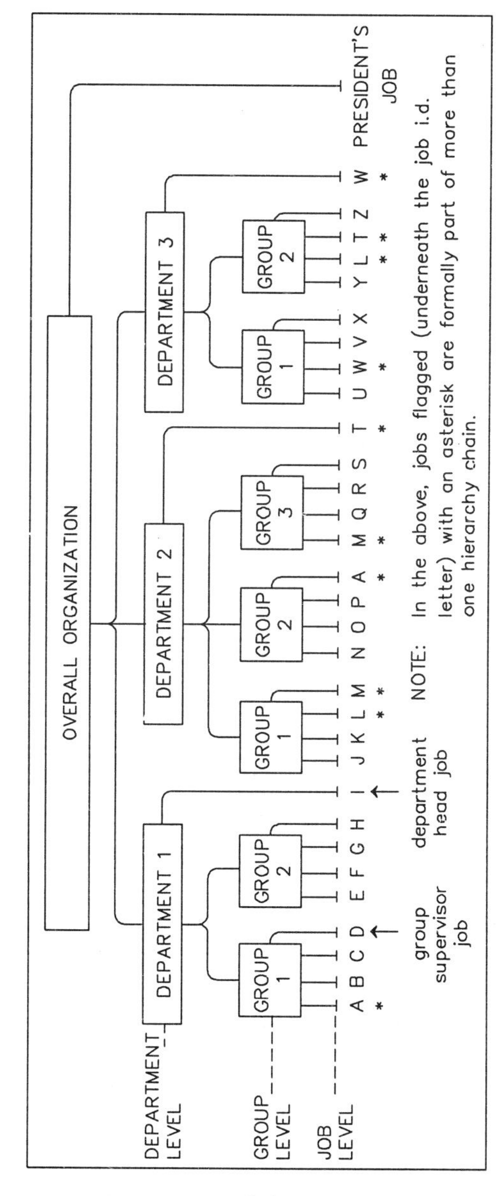

FIGURE 10-6: Subsystems by the formal identity approach.

Comparison of Two Approaches

To conclude, system components may be identified by their physical territories, characteristics, and patterns of interaction, or by their formal role titles. Identification of subsystems according to the first set of dimensions defines the physical structure-interactions approach type; while identification of subsystems by formal role titles constitutes the formal identity type of approach. The physical structure-interactions approach may also be referred to, if dealing in organizational theory, as the "work constellations" approach. Table 10-1 provides some examples of system essence-identities using the two coincidental approaches. In this table, a crude attempt is made to identify the same system by both types of approaches.

NON-FUNCTIONAL PURPOSEFUL METHODS

Logical Cohesion

Definition

Logical cohesion refers to the grouping of parent system functions into subsystems according to some scheme of classification in the mind of the perceiver. Frequently, logical cohesion is applied using the functional classification scheme, e.g., according to the classification schemes put forth in this book for objective-defined and flow functions. As has been noted several times in this book, the identification of the information subsystem(s) within a parent system relies on logical cohesion because all informational (decision-making) operations form a functional class.

Illogical Results of Logical Cohesion

The logical cohesion method can produce some fairly "illogical" results. **First**, unrelated elements may be placed together. For instance, a car maker and a manufacturer of household metal products merge.

PHYSICAL STRUCTURE-INTERACTIONS	FORMAL IDENTITY
interacting actors in a given territory	nation-state
interacting actors using telephone lines	telephone company clients
interacting actors in the same house over many years of living, in which some actors are older than others by more than twenty years	family
people occupying the same office	company office workers

TABLE 10-1: Examples of subsystems by the coincidental methods.

Using logical cohesion, the assembly line for car production is then combined with that for household objects, since both are in the baseline class of functions in the new, combined system. One can imagine the confusion.

Second, highly coupled functions related to a common objective may be separated into different subsystems; e.g., the informational functions are separated from the functions which they control. This is the reverse of the example of the above manufacturer example, which shows the grouping into the same subsystem of functions which are not even coupled.

To clarify an issue which might arise from the preceding example, it may make sense in terms of the functional, rather than the logical, cohesion method to integrate, say, all managerial operations in a system. This occurs when the control operations are separated into their own subsystem in order to avoid subsystem conflict. Thus, if a system consists of five persons, one individual may be given charge of all general decision-making duties.

Since logical cohesion often results in either the destruction of close functional relationships or the artificial grouping of unrelated functional elements, representation on levels-of control charts of subsystem modules identified by logical cohesion destroys many of the left to right

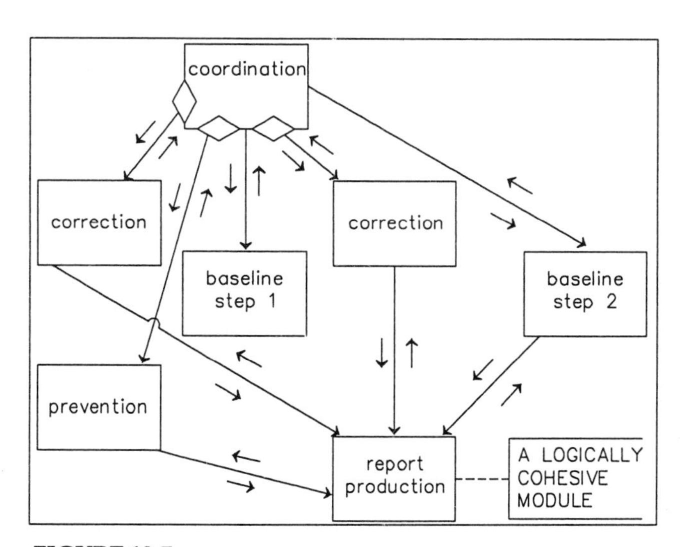

FIGURE 10-7:
Logical cohesion and fan-in on the levels-of-control chart.
The report production module produces different messages, in different formats, for each of its superordinates.

sequencing patterns. Consequently, modules in the chart which are identified by logical cohesion will be "fanned into" from the other modules. To illustrate, in Figure 10-7, all functional elements in the "report printing" class are identifiedby logical cohesion. For example, in a computer system, both user reports and run control reports, intended for use in controlling the program run itself, are grouped together. Moreover, the run control report may contain messages on errors occurring in a number of different functional sections of the program.

Temporal Cohesion

Temporal cohesion refers to the grouping of functional elements according to their schedule or timing. Thus, two weakly coupled or completely uncoupled functional elements may be placed in the same module, because they are performed, say, hourly, or at month-end, or according to a random schedule. The uncoupled functional elements may be parallel operations within the same objective-defined function; for instance, the opening of files and initialization of variable values near the start of executable statements in a computer program are often located in the same sub-program, since both are done near the start of the program.

Figure 10-8a exhibits exclusive use of the temporal cohesion method in the design of all of the subsystems of a parent system. In contrast, in Figure 10-8b only one of the subsystem modules is identified by temporal cohesion. Finally, Figure 10-8c provides an example of fan-in on the levels-of-control chart because of temporal cohesion. Temporal cohesion may also often be represented as a step in the transform-centered design in the levels-of-control chart.

As a design method, temporal cohesion tends to suffer from the same defects as logical cohesion, i.e., it may result in the grouping of uncoupled functional elements and/or the separation of highly coupled ones.

Procedural Cohesion and Anti-Cohesion

Procedural cohesion refers to the use of a similar methodology to perform different functional elements; for example, all operations done by computer, all highly routine operations, all operations requiring a blow-gun, all operations done by the same worker, all operations involving a chemical treatment, etc. For instance, a standard form letter of a business for sending different messages, with the particular message being ticked off from a number of options, has an aspect of procedural cohesion, since it involves a common method, which is varied for specific circumstances.

The identification of procedural cohesion may be a difficult task, since the similarity of methods used by different functions is sometimes a matter of interpretation. For instance, items are picked from warehouse shelves during the operations of two different objective-defined functions. If different types of shelving and/or different picking equipment is/are used in each case, the analyst may or may not decide that the two operations are procedurally cohesive, depending upon how the similarities lead to use of variations of a common procedure.

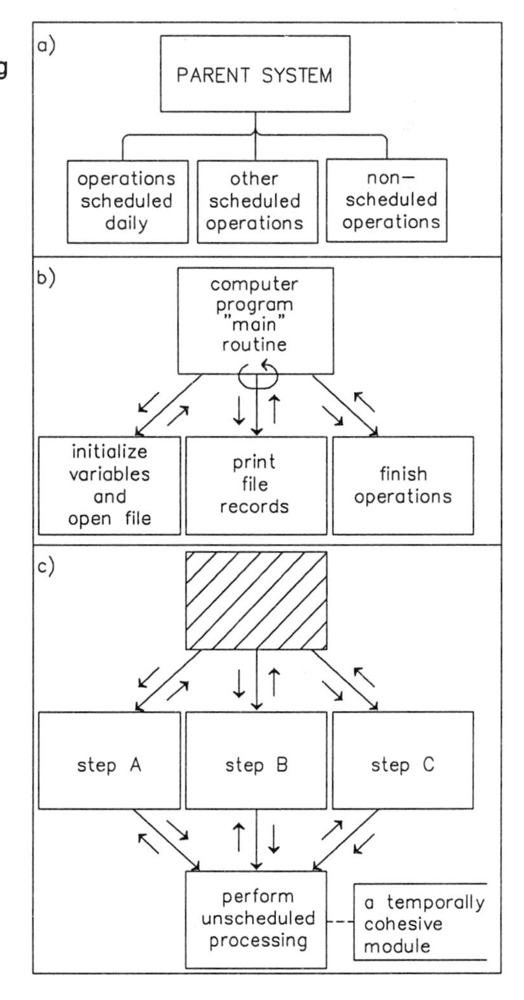

FIGURE 10-8:
Figures for demonstrating temporal cohesion.
a) hierarchical functional decomposition chart showing subsystems identified by temporal cohesion;
b) temporal cohesion in one module of a levels-of-control chart. Initialization of variables and opening of files may take place in parallel, as the output of one is not the input of the other. However, because of temporal cohesion (both occurring near the start of the program), they are located in the same module;
c) temporal cohesion involving fan-in on the levels-of-control chart.

Procedural cohesion may involve some merge of different functions, in addition to the grouping of them, i.e., it may involve partial merge. For example, a business makes use of the computer for only its accounts receivables and sales reporting functions, and so these are procedurally cohesed into the same subsystem in the organization chart, as in Figure 10-9a. Figure 10-9b shows the equivalent in terms of a levels-of-control chart, i.e., with direction of flow indicated. Within each of these sets of computer operations, certain standard program utilities are used to sort and back-up data, i.e., the procedures are identical in certain parts of these two different computer applications. Thus, the fan-in on the levels-of-control chart to show procedural cohesion indicates some merge along with the grouping of the different functions. The merge implies a certain existence of functional cohesion within the basically procedurally cohesive pattern.

Following is another example of procedural cohesion. The identification of computer information systems within information systems having both computerized and non-computerized aspects was introduced earlier in this text. Figure 7-19 (Chapter Seven) illustrates a control break reporting computer program which may happen when proce-

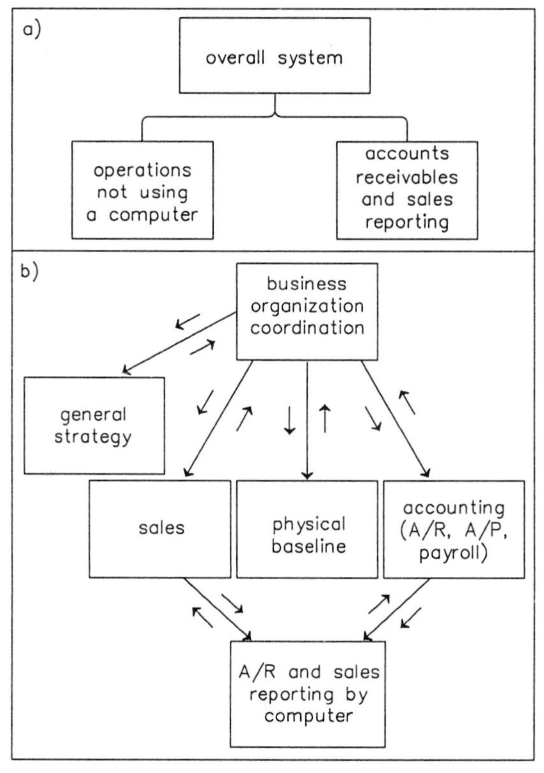

FIGURE 10-9: Hierarchy charts to demonstrate procedural cohesion.
a) procedural cohesion of the accounts receivable and sales reporting functions, demonstrated in a hierarchical functional decomposition chart;
b) procedural cohesion and fan-in on the levels-of-control chart.

dural cohesion is applied. In this diagram, all rectangles are connected planning functions, based on their role within the larger, parent information system; i.e., they represent parts of certain problem or opportunity detection flow functions within planning functions. Figure 10-10, in an impressionistic way, demonstrates this parent information system, in which the approximate locations of the functional blocks of previous Figure 7-19 are indicated by text. Thus, the attempt is made to reconstruct the parent system from which the computer information subsystems are derived.

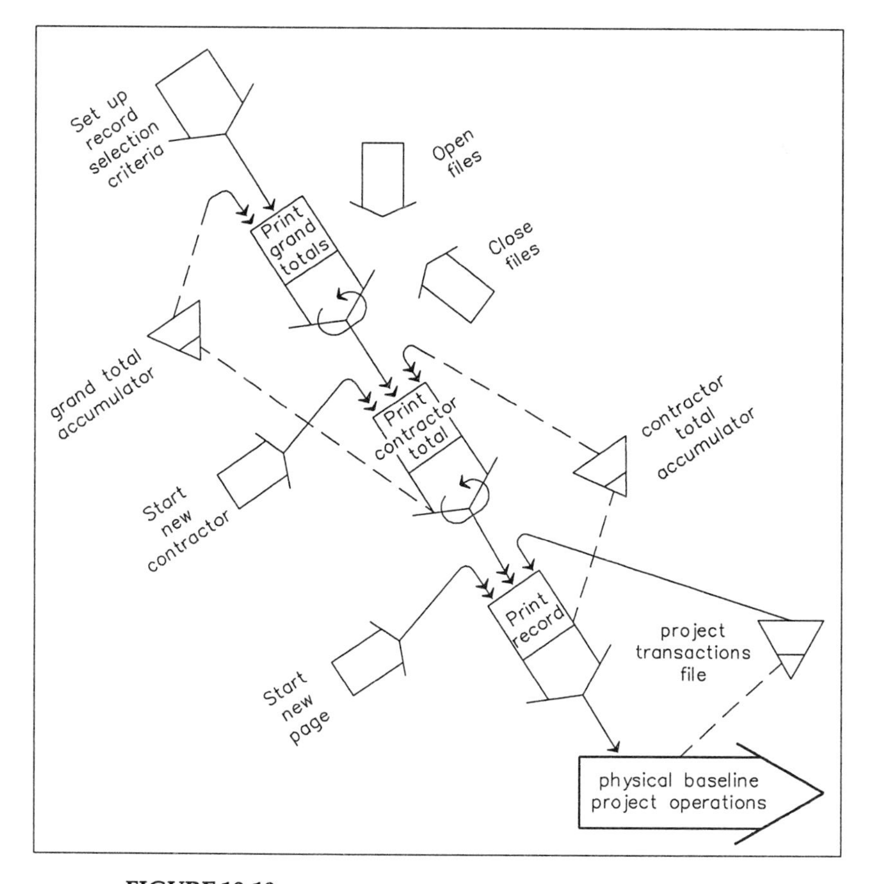

FIGURE 10-10:
Procedural cohesion from which Figure 7-19 was derived.
In this chart, the non-computer operations are unlabelled, but occur as parts of the different wideshafted arrows.

Procedural Segregation

The subject of procedural **anti**-cohesion, or segregation, refers to different procedural means used in various runs of the same objective-defined function. For example, as in Figure 10-11a, the identical boat may be fabricated either by hand or by machine. In consequence, the objective-defined function is broken up into different subsystems, as in Figure 10-11b. The diagramming of both procedural segregation and parallel operations within an objective-defined function is demonstrated in Figure 10-11c.

Procedural segregation of a different variety from that just illustrated is often connected with a mutually contingent set of functional objec-

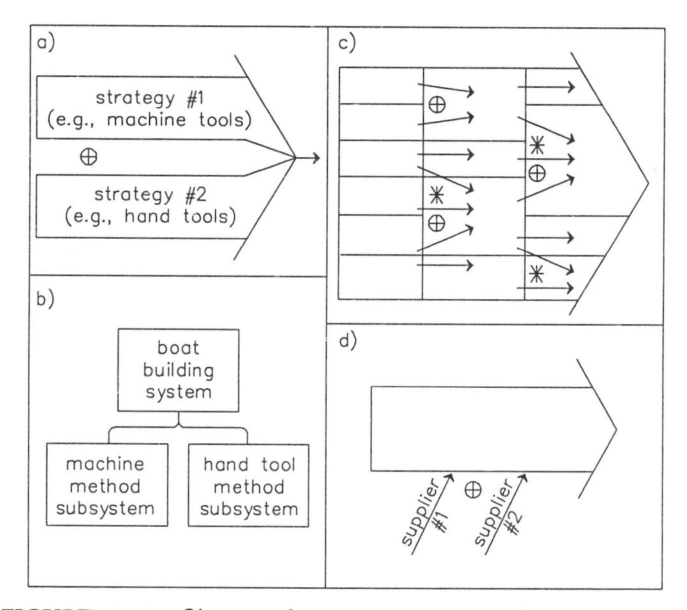

FIGURE 10-11: Charts to demonstrate procedural segregation.

a) procedural segregation (or anti-cohesion), whereby the objective is achieved using different strategies in different runs of the objective-defined function; procedurally segregated functional elements are separated by the Boolean EXCLUSIVE OR symbol;

b) hierarchical functional decomposition chart giving subsystems derived from "a" using procedural segregation;

c) parallel operations (con-routines) and alternative paths in the structure-flow chart within the wideshafted arrow, using Boolean gating logic symbols;

d) procedural segregation resulting from alternative sources of supply.

tives. Here, mutual contingency results from a mutually exclusive set of options, rather than from different ways of achieving the same objective. It is also often caused by the possibility of different sources of input to an objective-defined function, as diagrammed in Figure 10-11d. To illustrate, a business requires different receiving procedures to quality check goods, based upon which supplier provides its raw materials, since one supplier is known to be more reliable than the other.

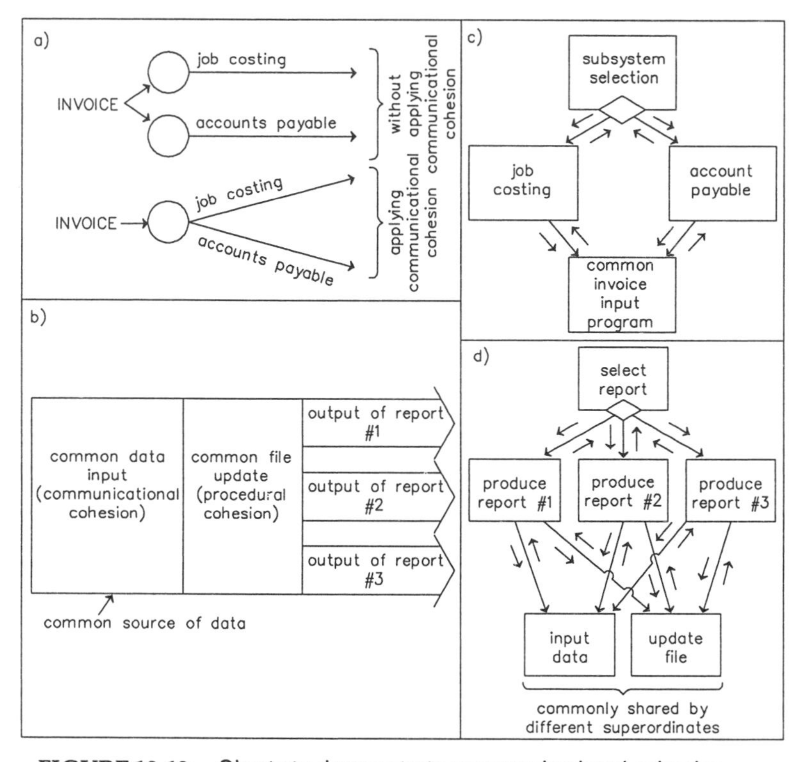

FIGURE 10-12: Charts to demonstrate communicational cohesion.
a) communicational cohesion based on a common source of input;
b) communicational cohesion in data input step (with procedural cohesion in the file update step), shown using wideshafted arrow;
c) communicational cohesion shown using a levels-of-control chart, with the same example as in "a";
d) communicational cohesion shown using a levels-of-control chart, with the same example as in "b."

Communicational Cohesion and Anti-Cohesion

Definition and Examples

In communicational cohesion, different functional elements may be grouped together because the same entity is communicated with.

First, for example, a man visits a bridge engineer for advice on a bridge-building problem. This bridge engineer also happens to be a medical doctor. During the discussion with the bridge engineer, the man is seized with a severe chest pain. Now, the man would normally go to the clinic, ten miles away, to see a doctor. However, there is no need in this case, since a doctor, who is also the bridge engineer, is already on site.

Second, both a job costing and an accounts payable computer system derive their information from supplier invoices. Without communicational cohesion, one program would input invoices to the job costing system, and a rather similar second program would input invoices to the accounts payable. However, because of communicational cohesion, only one program is written and used, thus eliminating duplication, as diagrammed in Figure 10-12a.

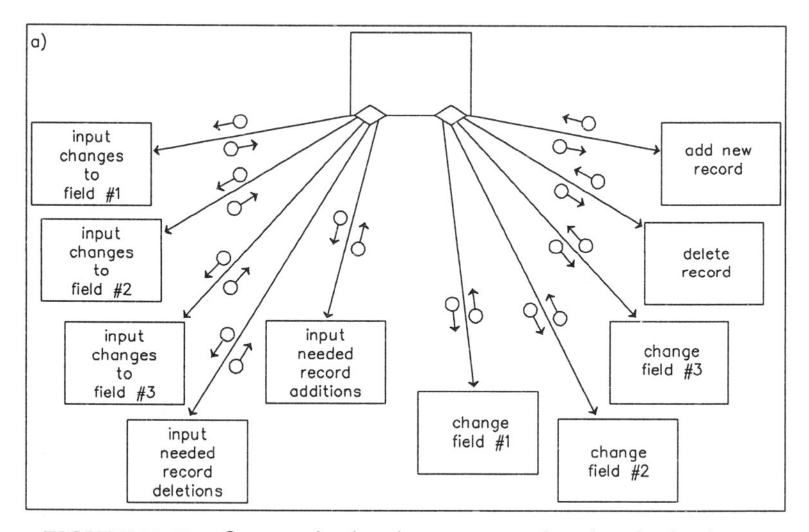

FIGURE 10-13: Communicational versus more functional cohesion.

a) ***(above)*** *example of levels-of-control chart in which functional elements, represented by modules, may be communicationally cohesed;*

b) ***(opposite)***

c) ***(opposite)***

Third, using a variation of the wideshafted arrow, Figure 10-12b demonstrates a similar type of situation, in which different reports, each with somewhat different data fields, are produced from the same file and input. Here, a number of different objective-defined functions are involved, if it is assumed that the productions of the different reports are not mutually contingent objectives.

Figure 10-12c and 10-12d show fan-in on the levels-of-control chart to demonstrate the same things as in Figures 10-12a and 10-12b, respectively.

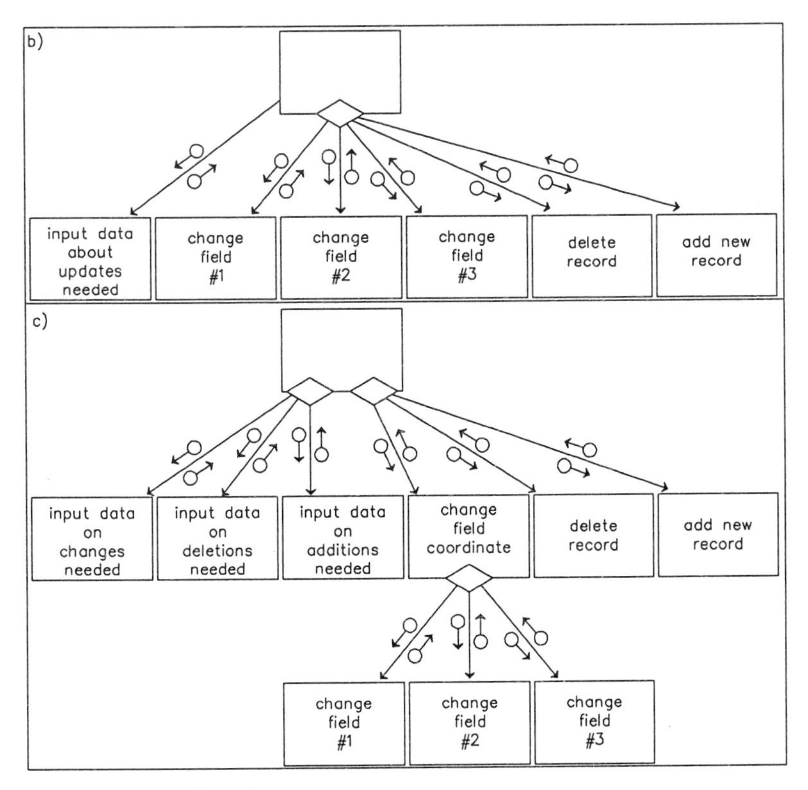

FIGURE 10-13b and 13c

b) ***(above)*** *chart in "a," but with input functional elements communicationally cohesed;*

c) ***(above)*** *the same example, but with a degree of communicational cohesion in between that in "a" and "b." (Also seen is a new module to coordinate change of the functional elements for changing fields. The latter has been included to complete the parallel to Figures 8-5 in Chapter Eight.)*

Fourth, a different example of communicational cohesion is given by Figures 10-13. These figures are equivalent to Figures 8-5 in Chapter Eight, which use the structure-flow chart.

Communicational Cohesion and Merge

As with procedural cohesion, communicational cohesion often implies some merge along with the grouping of different functions. For example, the input to the accounts payable and job costing computer applications includes nine fields from each supplier invoice. Three of these are relevant uniquely to job costing, three are pertinent only to accounts payable, and three are applicable to both job costing and accounts payable. The three common fields imply an extent of merge represented by the fan-in on the levels-of-control chart in Figure 10-12c.

Procedural cohesion often coincides with communicational cohesion, since both techniques may involve merge, and hence may be used to reduce redundancy.

Communicational Segregation

Communicational segregation, an anti-cohesion method, is the inverse of communicational cohesion. It involves the **cloning** of different subsystems performing the same function for different communicational regions. To illustrate, a bank opens three branches in three different cities, each branch performing the same functions as the other two branches. Even though the functions are identical, three branches must be established to communicate in the three different geographical areas. Communicational segregation may be diagrammed as in Figure 10-14.

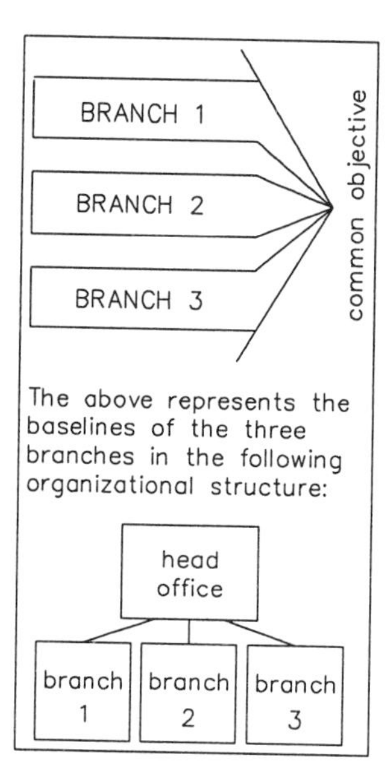

FIGURE 10-14: Communicational segregation (anti-cohesion).

Sequential Cohesion

Definition

Sequential cohesion refers to the grouping of functional elements into the same subsystem because the output of one functional element serves as input to the next. This formula obviously results in the grouping of the different flow functions in the same objective-defined function, as well as of sequentially or reciprocally interacting objective-defined functions. Moreover, if certain steps in the sequence are allowed to execute less frequently than others and/or to be performed on an optional basis (such that, for example, return loops and optional branches are permitted to occur in the sequence), the results may be almost or completely identical to those achieved by the functional cohesion method. This is because all medium or highly coupled functional elements related to the same objective are thereby grouped together.

In fact, applied to a parent system which was itself already identified by the functional method, the sequential cohesion method produces no further decomposition of the parent system, since all functional elements in a functionally cohesive cluster will already be sequentially related (or related by reciprocal coupling, which is, in effect, two-way sequential coupling). It is only when dividing a parent system which is not itself functionally cohesive that different sequences become discernible at the level of perception used. Even here, more than one sequence will only exist if unrelated elements are present.

Thus, it might be thought that when applied to a functionally cohesive parent system, the sequential cohesion method is the same as the functional one. However, two important differences are present. **First**, the avoidance of subsystem conflict criterion of the functional cohesion method is not fulfilled by the sequential cohesion one. **Second**, the option of separating sub-clusters of functional elements whose **only** bond is sequential does not exist, as it does in the functional cohesion method. Thus, while another level of functionally cohesive modules may be identified by the functional method within a functionally cohesive parent system, a further level of sequentially cohesive modules is not possible to identify within a parent system itself identified by the sequential cohesion method.

Places Where Sequential Bonds Occur

Where do sequential relationships occur? **First**, sequential cohesion occurs typically when error control steps are inserted into a sequence to which the control steps apply, as in Figure 10-15a. For instance, the essential baseline steps in a student registration system are:

—BASELINE #1— Enter the student for the semester;

—BASELINE #2— Record the particular courses selected.

If acceptance of the student and optional correction of errors are inserted between steps 1 and 2, and if a last step to authorize the courses selected is added at the end of the sequence, then control steps are added to the baseline sequence.

Second, sequential coupling occurs when framework operations are joined into the same subsystem as the base objective-defined function to which they apply. Since framework add/change/delete operations usually precede or follow the operations in the base objective-defined function, they may be aligned together with it, as in Figure 10-15b. That the framework operations execute less frequently is indicated in the diagram by iteration symbols, similar to the way the less frequent flow functions are shown. Specifically, in this example, the relative frequencies of execution of the modules are such that:

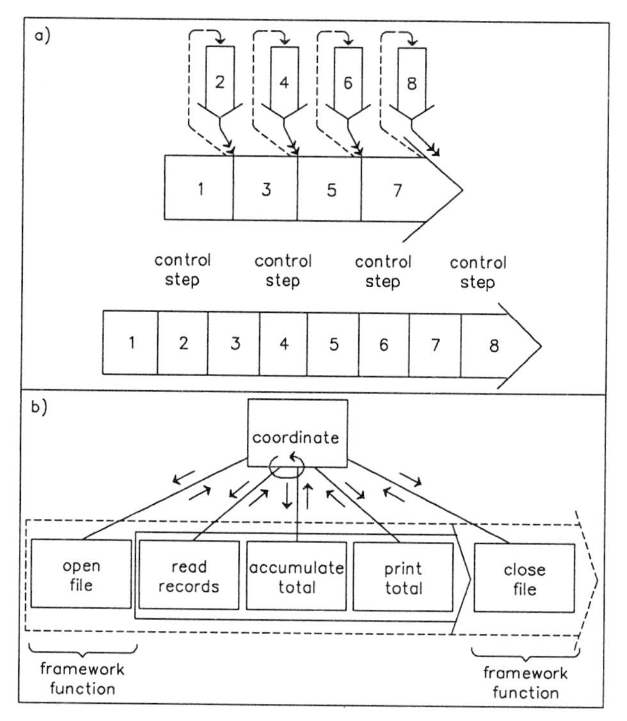

FIGURE 10-15: Charts to demonstrate sequential cohesion.

a) *The basic order of steps is given by the numbers. If the control steps were to be merged into the same wideshafted arrow as the steps which they control, then the lower half of this figure would represent the flow. Unfortunately, the lower half of the figure is not consistent with the use of the wideshafted arrow in this book, as it contains both directly and indirectly related functional elements.*

b) *The alignment of the framework and baseline functions is intended to format this levels-of-control chart to highlight the sequential relationships over the purely functional ones (as established in this book).*

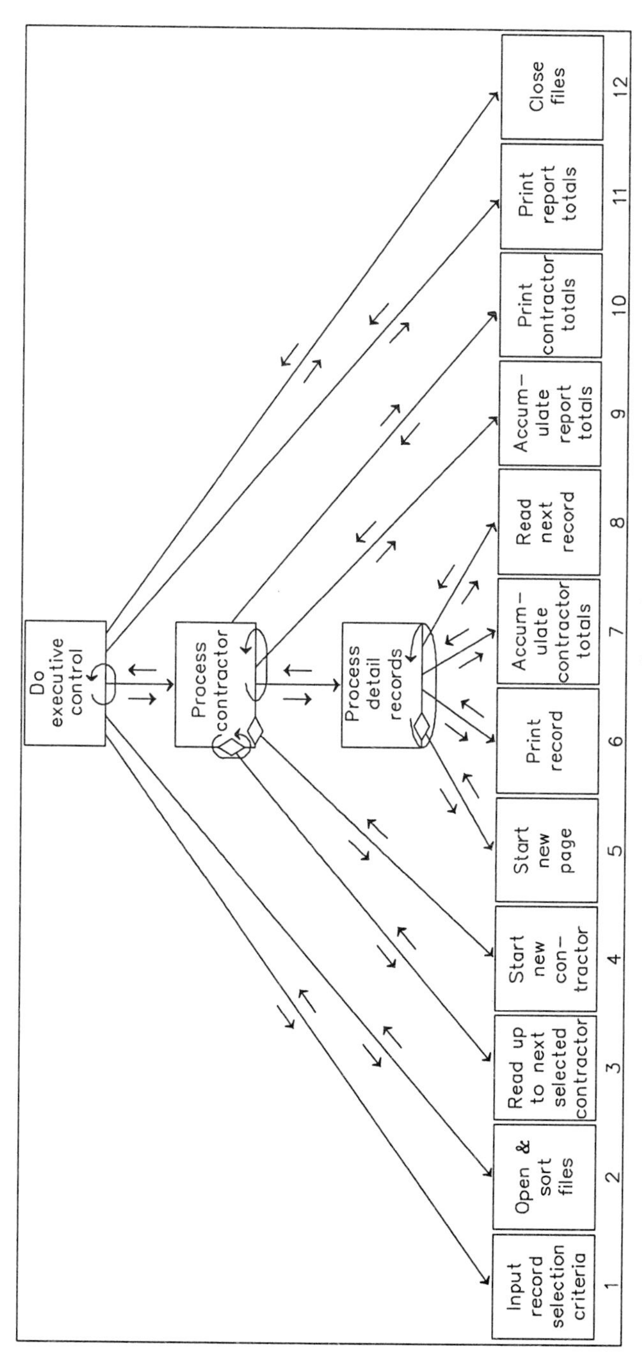

FIGURE 10-16: Formatting the levels-of-control chart to highlight sequence. *This may be contrasted with the same levels-of-control chart in Chapter Seven, namely, Figure 7-19 which was specially sub-formatted to highlight the functional cohesion method, rather than the sequential one.*

- open and close files, which are framework operations, have a frequency of 1;
- print total, a baseline flow function, also has a frequency of 1;
- read records and accumulate total have a frequency of greater than 1.

A **third**, more complex example of sequential coupling is exhibited in Figure 10-16. This is another version of Figure 7-19 (Chapter Seven), and of the equivalent computer operations in Figure 10-10 (above). As seen, highlighting sequential relationships between functional elements involves aligning all the bottommost modules in all branches of the hierarchy, as opposed to keeping them at the more meaningful levels seen in Figure 7-19 (Chapter Seven). Drawing a wideshafted arrow around them is a temptation, although this use of the wide-shafted arrow would not be consistent with previous concepts, since both direct and indirect functional elements are present.

THE FUNCTIONAL PURPOSEFUL METHOD

Effect of the Method

Where subsystems are identified within a parent system using the functional cohesion method, the different subsystems become as independent of one another (loosely coupled) as possible. Thus, minimum interaction occurs among them, such that changes can be made to one subsystem while minimizing the changes needed to other subsystems as a result of changes to the first.

Latitude in Applying the Method

The functional cohesion method groups together into the same subsystem functional elements according to the heuristic rules outlined in Chapter Two. As is discussed in the following, a high degree of latitude exists as to how the functional cohesion method may be applied and as to what does or does not constitute functional cohesion.

Types of Coupling

High relationship to a common objective does not automatically imply high coupling of functions. For example, two sets of parallel operations may be directly related to the same objective, as in the payroll function example in Figure 10-17. This is known as "zero coupling," since the parallel operations are related **only** by a common objective. Conversely, high coupling of functions does not necessarily mean di-

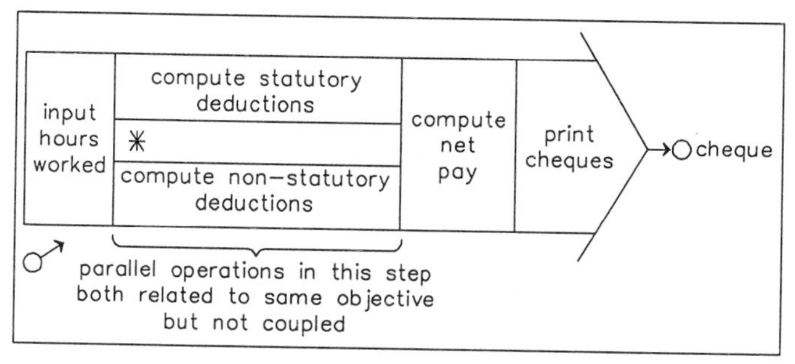

FIGURE 10-17: Zero coupling between operations in parallel branches. *This is demonstrated using the wideshafted arrow.*

rect relationship to a common objective, although it usually implies at least an indirect relationship to this. What then is involved in high coupling? To answer this, Figure 10-18 demonstrates three basic coupling patterns identified in Chapter Two. Mintzberg describes this diagram as follows:

> Thompson . . . distinguishes three ways in which . . . work can be coupled. First is **pooled coupling**, where members share common resources but are otherwise independent . . .. In **sequential coupling**, members work in series, as in a relay race where the baton passes from runner to runner. . .. In **reciprocal coupling**, the members feed their work back and forth among themselves; in effect each receives inputs

FIGURE 10-18: Three types of coupling. From Thompson, J.D., Organizations in Action (New York: McGraw-Hill, 1967), via Mintzberg, The Structuring of Organizations, p. 230.
a) pooled coupling;
b) sequential coupling;
c) reciprocal coupling.

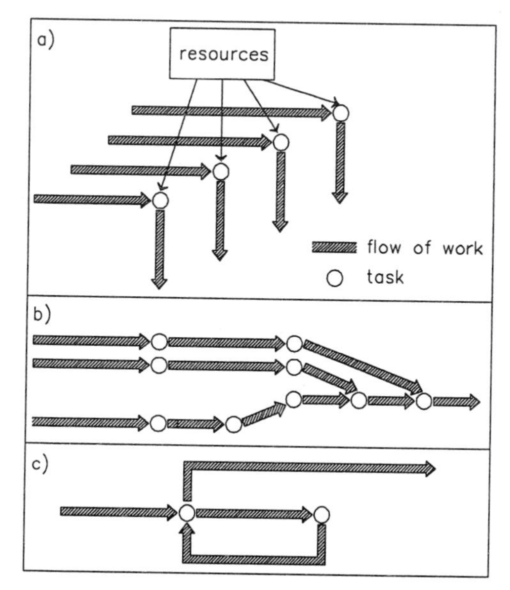

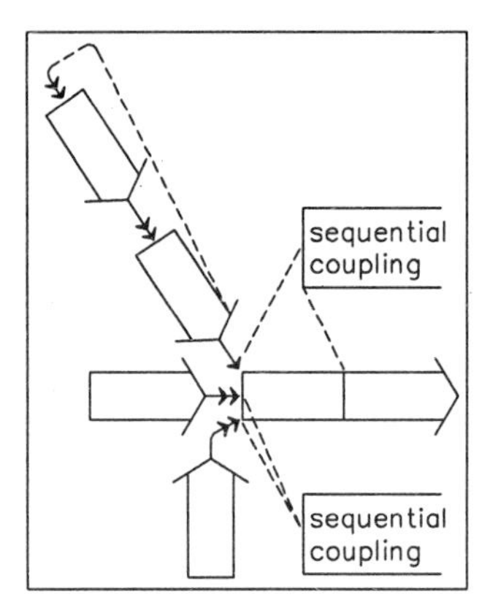

FIGURE 10-19:
Examples of sequential coupling in the structure-flow chart.

form and provides outputs to the others . . ..Clearly, pooled coupling involves the least amount of interdependence among members. Anyone can be plucked out; and, as long as there is no great change in the resources available, the others can continue to work uninterrupted. Pulling out a member of a sequentially coupled organization, however, is like breaking a link in a chain—the whole activity must cease to function. Reciprocal coupling is, of course, more interdependent still, since a change in one task affects not only those farther along but also those behind.*

Reciprocal coupling between functions would occur in the event of feedback loops and two-way symbiotic (or offshoot, or competitive) relationships of functions. As in Figure 10-19, sequential coupling would take place between successive flow functions of the same objective-defined function, or between different objective-defined functions which are in sequential interaction with one another. Sequential coupling is always found within reciprocal coupling. Moreover, pooled coupling may be found within either of these.

Pooled coupling, for instance, occurs in three different ways in Figure 10-20a. Here, the two baseline objective-defined functions parallel to one another are related by:

—a common initiative coordination function;
—usage of a common framework;
—a common validation function

Pooled coupling does **not** necessarily occur between objective-defined functions which can in some way be cohesed into the same sequence, as can, for example, the validation functions in Figure 10-20b; i.e., common relationship to an objective, or to another function, in no way implies even pooled coupling, since no coupling at all may be the case.

* Mintzberg, pp. 53—58.

Pooled Coupling and Fan-In on the Levels-of-Control Chart

Fan-in on levels-of-control charts may represent merge and consequent avoidance of duplication. In other words, it may be used for demonstrating that form of pooled coupling resulting from common usage of a resource. Figure 10-20c shows a "fan-in," whereby a resource used by two different flow functions is made equally accessible to both. As a result, for example, in a computer program, a sub-routine can be invoked from different parts of the program. Thus, a square root calculation sub-routine could be called during two different sets of program calculations.

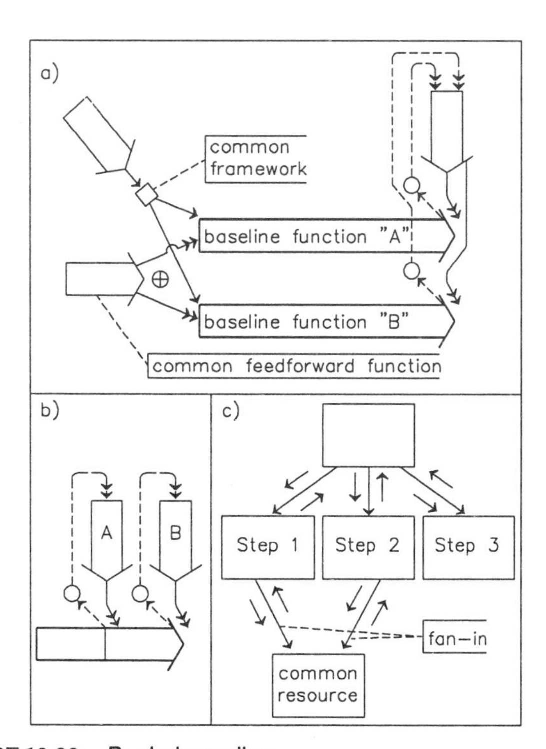

FIGURE 10-20: Pooled coupling.

a) *three examples of pooled coupling in the structure-flow chart;*

b) *an example to highlight the difference between zero coupling and pooled coupling. The two validation functions are* not *related by pooled coupling, since function A executes earlier than function B, and relates to a different area of the base objective-defined function;*

c) *an example of fan-in on the levels-of-control chart involving pooled coupling.*

Examples of Subsystem Divisions Satisfying Criteria of Functional Method

Figures 10-21a and 10-22a provide some examples of subsystem identification patterns which satisfy the criteria of functional cohesion. In each of these figures (also see their structure chart equivalents in parts "b"), the various subsystems are encircled by dashed lines. Each subsystem is identified by a letter, which is also the identification used for it in the structure chart. The dashed lines may circle just a single objective-defined function, in which case just the baseline of the subsystem is indicated in the diagram. As well, the dashed lines may enclose more than one objective-defined function, and/or part of a baseline objective-defined function plus the ensurance adaptation operations which

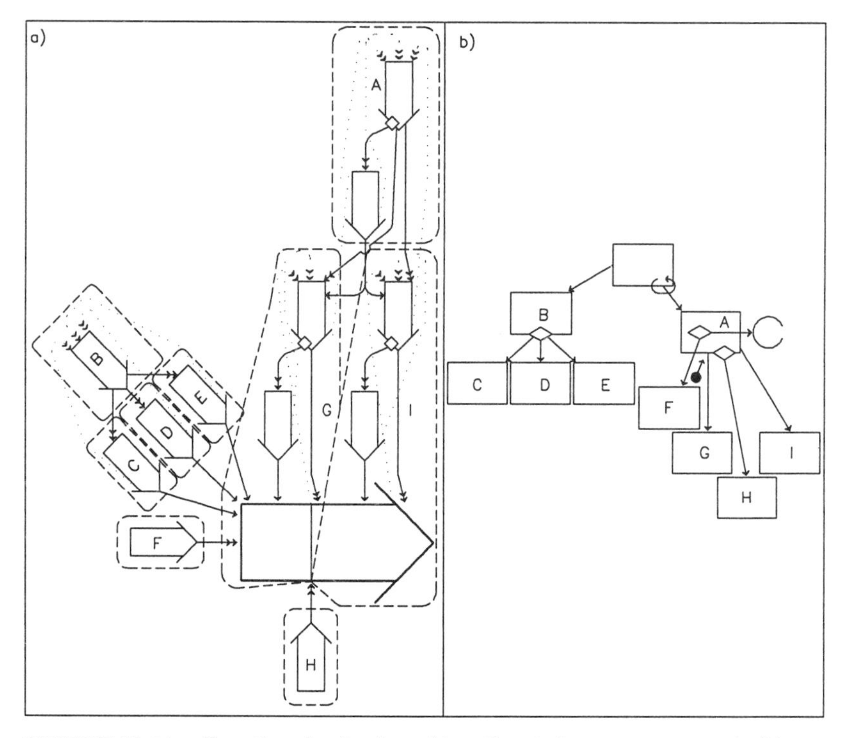

FIGURE 10-21: Functional cohesion of functional elements—example #1. *The shown structure-flow chart, and its equivalent structure chart, give a pattern of division of functional elements into subsystems such that the requirements of the functional cohesion method are satisfied. Here, subsystems A and B are identified as separate subsystems in order to avoid conflict among their respective subordinate subsystems. The two sides of the figure are:*
a) structure-flow chart; *b) structure chart equivalent.*

locally apply to that part of the baseline objective-defined function. For example, in Figure 10-21, subsystems G and I each consist of part of the overall system baseline, plus the applicable ensurance functions, while subsystem A consists of the ensurance functions at the top of the overall system hierarchy of these. Each of the subsystems, G, I, and A, contains functional elements which are reciprocally coupled to one another.

Unlike the preceding figures, Figure 10-23a (also see Figure 10-23b), shows subsystem identification which does **not** satisfy all the criteria of functional cohesion, since subsystem conflict is present between subsystems C and E, as the executive controller in E is also the executive controller of C. It should be noted that the criteria of functional cohesion are satisfied within subsystem A in Figure 10-23a, since each lower

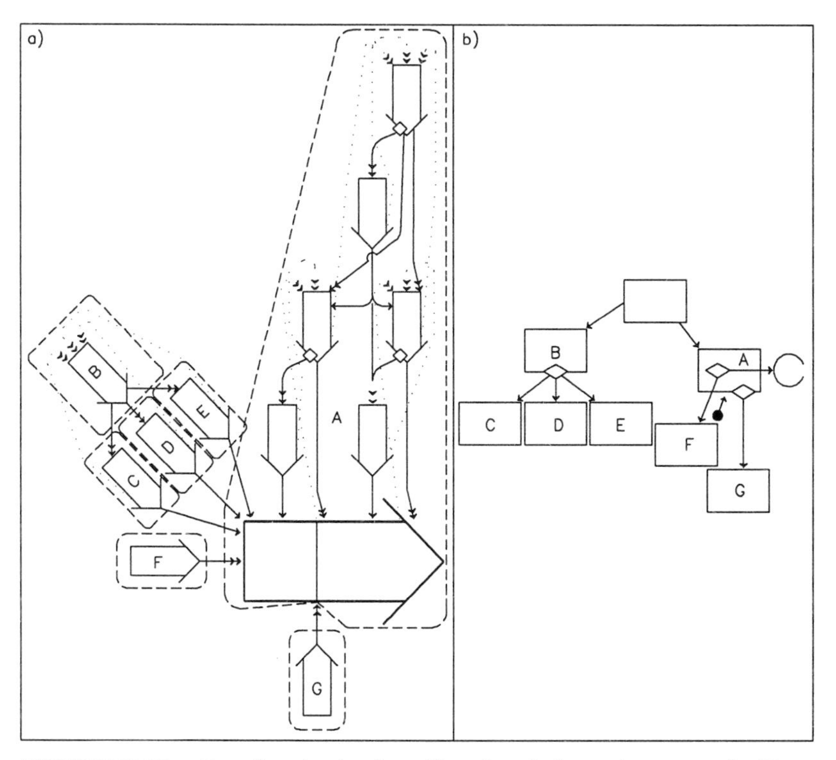

FIGURE 10-22: Functional cohesion of functional elements—example #2. *This gives a second possible pattern of dividing functional elements into functionally cohesive subsystems. This is like the preceding charts (Figures 10-21), except that all baseline and ensurance functions are joined into a single subsystem, namely, subsystem A.*

level function in the hierarchy is reciprocally coupled to the higher level one.

Grey Areas in Using the Functional Method

As is evident from these examples, many acceptable patterns of subsystem identification according to the functional cohesion method are possible within any given system. This variety is largely due to the large grey area between tight and loose coupling whenever coupling is sequential. As well, many other situations may exist in which the extent of coupling is debatable, e.g., when symbiotic relationships occur between two different decision-making functions, as in Figure 10-24.

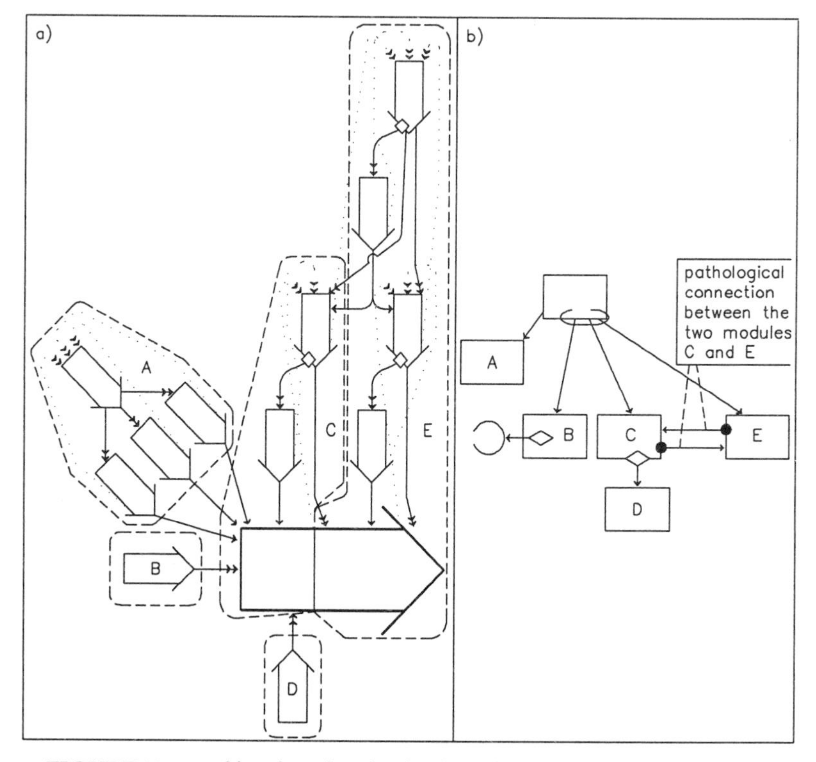

FIGURE 10-23: Non-functional cohesion of functional elements. *This gives an example of division of structure-flow chart of preceding figures which does not satisfy the criteria of the functional cohesion method, since C needs to invade E to gain access to the top level current adaptation functions; i.e., subsystem conflict is present:*
a) structure-flow chart;
b) a possible structure chart equivalent.

Also, the variety of possible patterns satisfying the functional cohesion criteria is increased by using varying levels of perception of functional elements.

Another ambiguous type of case concerns the separation into subsystems of parallel functional elements within the same objective-defined function. Such a separation might be desirable even when the functional elements are viewed from a macro-perspective (see discussion in Chapter Two of micro versus macro views of functional elements). In Figure 10-25, the validation function at the top of the ensurance hierarchy may be split down the center, into sections applicable to each of the two baseline steps. However, because of the nature of validation operations, it may be that symbiotic relationships exist between the two split sections. Intuitively (although perhaps not logically), this type of symbiotic relationship **within** an objective-defined function becomes increasingly likely the more general the level of decision-making function. Thus, the higher the managerial level in a business organization,

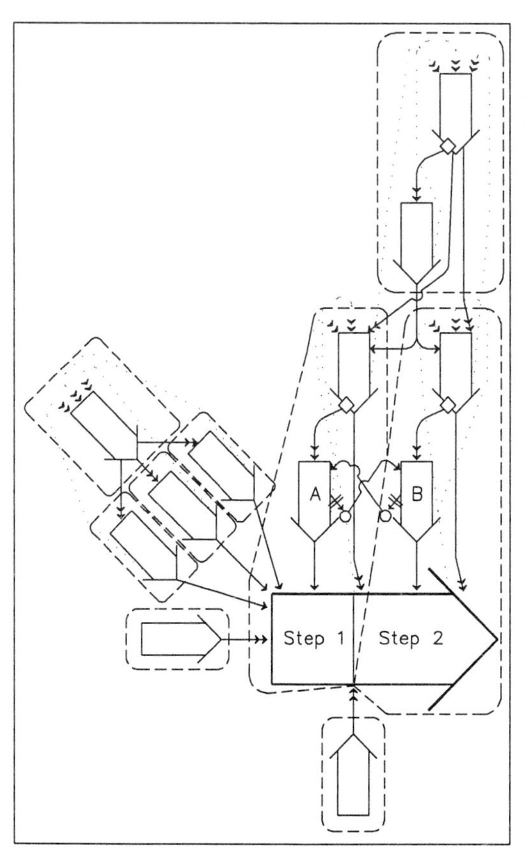

FIGURE 10-24: Ambiguity as to whether cohesion is functional—case #1.
This is an example of a division into subsystems where an ambiguity still exists as to whether or not the criteria of functional cohesion have been satisfied. Specifically, functions A and B are located in different subsystems although they are reciprocally coupled by symbiotic interchanges.

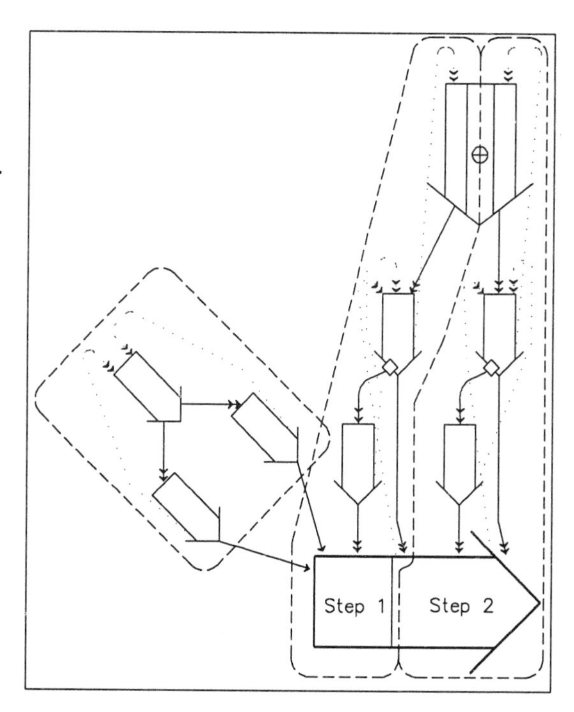

FIGURE 10-25: Ambiguity as to whether cohesion is functional—case #2. *This is a second example of a grey area in applying the criteria of functional cohesion. Here, to properly satisfy the requirements of the method, the top coordinate is split into parts belonging to each subsystem.*

the more likely it is that cross-communication occurs between the managers of different functional subsystems, since problems are more highly interrelated. The latter notion is presented in Figure 10-26, using the Anthony triangle.

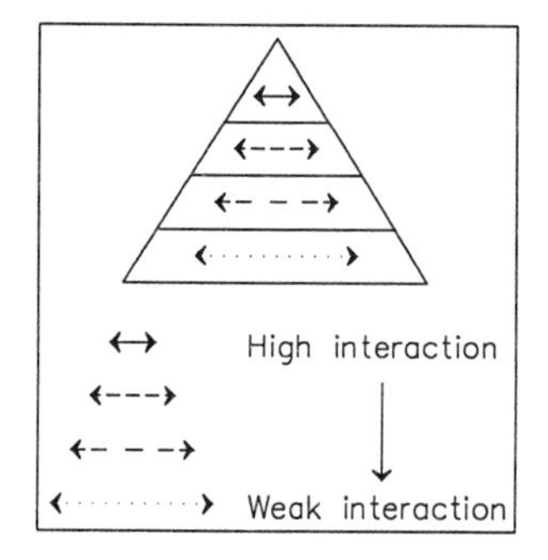

FIGURE 10-26: Cross-communication between basically parallel elements. *This is an illustration of the heuristic, that the higher the level of the function, the greater the cross-communication between otherwise basically parallel functional elements.*

COMBINATIONS AND COMPARISONS OF THE DIFFERENT METHODS

COMBINATIONS

Gestalt Mixtures

Different cohesion methods may apply to the description of the relationship between different pairs of functional elements in a system. Moreover, more than one cohesion type may apply to a given pair of functional elements; e.g., both communicational and functional cohesion may relate functional elements A and B in a system. Thus, different subsystems in a parent system may be defined using different cohesion methods and any given subsystem may be identified by a combination of cohesion methods and/or coincidental approaches. The mixture of subsystem identification methods provides a "gestalt" approach.

The following examples of gestalt approaches represent a few of the virtually infinite number of different possible combinations of the various approach types outlined above. Each of these is assigned a name.

Effective Organization

The effective organization point of view associates any purposeful approach type with any coincidental approach. The aim is to arrive at a very concrete and current definition of the way responsibilities for execution of operations of the larger system have been assigned to subsystems first identified by the coincidental approaches. For example, the effective organization approach can show the operational responsibilities of the system agents (human and non-human).

Qualified Technical Functional Objectives

The qualified technical functional objectives view bases the continuance of the subsystem mainly on its achievement of stated overall technical functional objectives. In addition to technical objectives, certain other criteria may be added, such as the ability to uphold its formal identity in the legal system. The emphasis here is on technical objectives, rather than on other aspects.

Justified Formal Identity

The justified formal identity approach is allied to the purely formal identity type. It tacks certain extra conditions onto the purely formal approach's definition of subsystem essence-identity. The subsystem retains its "right" to its formal identity as long as there are no other "fundamental" changes in certain other of its chief properties, such as its physical structure, its functional objectives, its manner of operation, and so on. The properties to be associated with the formal identity, and what constitutes a fundamental change in these are up to the analyst who defines the essence-identity. For example, the law of a nation may state certain types of limitations in the Charter granted to a corporation. If the corporation fails to meet the conditions specified in the law, its Charter may be dissolved. To illustrate, the law may state that the corporation in question

—be of a certain approximate size in terms of number of employees and operate within certain territorial confines (physical structure-interactions criterion);

—remain in a certain very generally defined type of industry (technical functional objectives criterion);

—continue to be financially profitable (profit functional objectives criterion);

—continue to express support for the political regime in existence (attitudes, or subjective manner, criterion);

—use certain types of machinery supplied by certain companies of the same nation (manner of carrying out technical objectives criterion).

Certain other conditions may also be stated in connection with the justified formal identity view, by which the subsystem continues to exist despite "artificial" changes in its formal identity. For example, if the corporation simply changes its name, but effectively remains the same organization doing the same things in the same way, this would be considered to be a **non**-change according to the justified formal identity approach.

Methods in Different Levels and Branches of Hierarchy

The computer analyst's view of the business organization in Chapter Two is explained such that only a single cohesion method is used to create each new level of subsystems within a given parent system. However, it might have been possible to jump levels, say from level two to four, by using a combination of approach types in a single division into subsystems. For example, both logical and procedural cohesion could be applied in a single division into subsystems. In general, however, to obtain a hierarchy whose component subsystem modules have non-overlapping boundaries, it may be good practice to use only a single, pure approach type **per division** of a system into the next level of subsystems. This rule of thumb simplifies the ensuring that each division of a system into a non-overlapping hierarchy of subsystems is internally consistent. For example, if one subsystem resulting from a given division is created in terms of a schedule of events, then all other subsystems resulting from that same division should also have their essence-identities specified in terms of a schedule of events. If, for example, one tried to assign all "daily scheduled tasks" to one subsystem, and "routine activities" to a second subsystem in that same division, the contents of the two subsystems would overlap.

Figure 10-27 gives some examples of invalid mixing of different approaches resulting from the use of inconsistent division criteria for each of the shown systems/subsystems. These approaches are inconsistent, not because different approaches are used in different **levels or legs** of the hierarchy, but because each sibling subsystem (i.e., subsystem of the **same** parent system) is identified according to a different approach. In contrast, in Figure 10-28, no inconsistency exists, as the same approach is used wherever there are sibling subsystems (and because each approach has been used in an internally consistent way).

Thus, while different approaches may be used to identify subsystems in a hierarchy of subsystems, the analyst must be careful to achieve consistency between siblings of the same parent system.

COMPARATIVE EVALUATION OF THE DIFFERENT METHODS

Effect of Criteria on Ranking Results

The comparative ranking of the different purposeful and coincidental subsystem identification methods may vary, depending upon the criteria used for evaluation. Overall, there can be no doubt about the supremacy of the functional method, since it is by objectives, and because it creates an optimally independent set of subsystems. This independence, or loose coupling, except where subsystem conflict must be avoided, allows changes to be made to one module while minimizing the extent of changes needed to other modules as a result of changes to the first module. The functional method, since it is by objectives, is also the easiest to implement and communicate. If applied again and again, at each new level of subsystem, it creates an easily comprehensible subsystem pattern, in which there is little problem in seeing the precise boundaries between subsystems.

Comparative Ranking

The easiest way of ranking the various purposeful methods is to compare them to the functional one. Here is a summary of conclusions already arrived at above:

- The logical and temporal cohesion methods are very unlike the functional one, in that they may result in dissociation of tightly coupled functions, on the one hand, and grouping of functions with little or no coupling, on the other.
- The procedural and communicational cohesion methods tend to apply the avoidance of both duplication and of subsystem conflict criteria of the functional cohesion method, so that common proce-

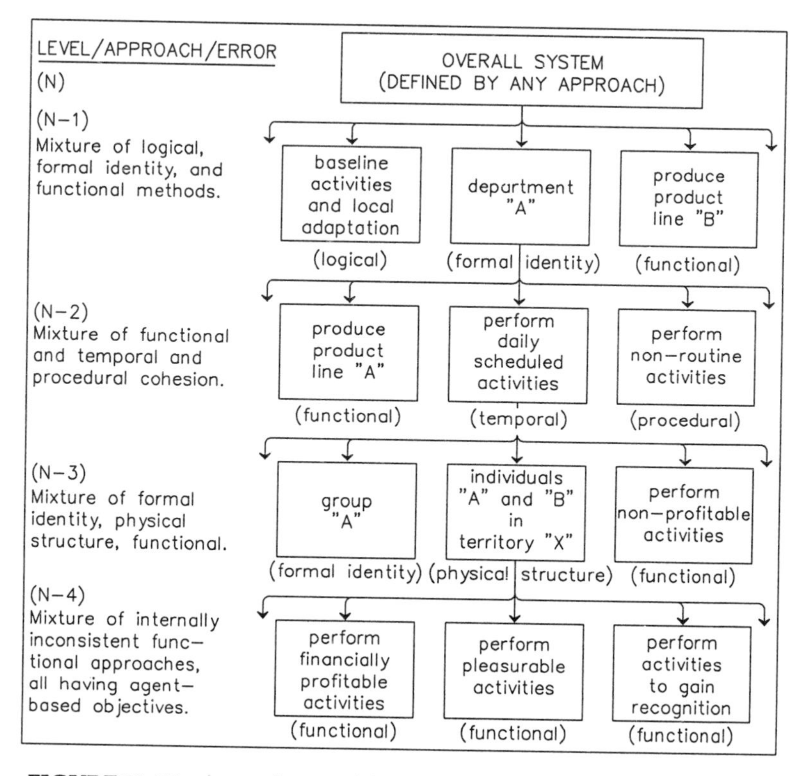

FIGURE 10-27: Inconsistent mixing of subsystem identification approaches. *In this, inconsistency is found at given levels of subsystem hierarchy, since different approaches may always be used for divisions at different levels of hierarchy.*

dures tend to be designed for different functions and then merged into the same subsystem.

- The sequential cohesion method tends to apply both the relationship to a common objective and sufficiently tight coupling criteria of the functional method, and thus is very close to the functional

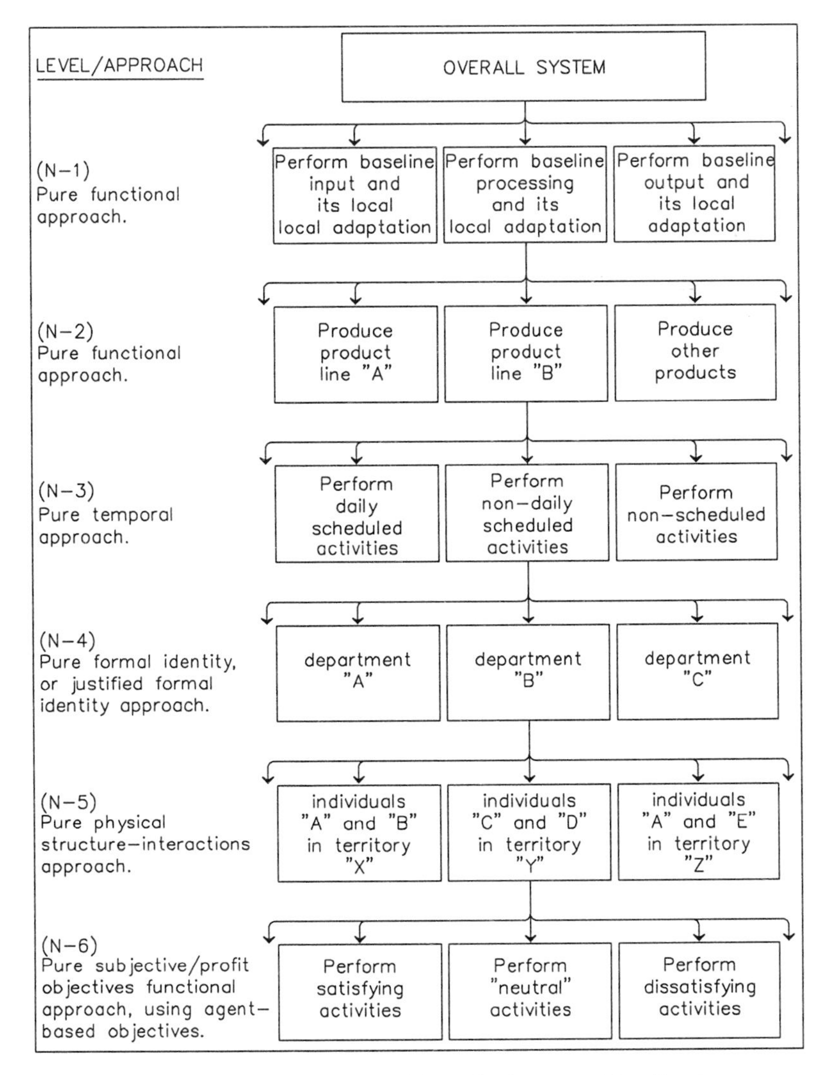

FIGURE 10-28: Consistent mixing of subsystem identification approaches. *Here, each level uses a single type of approach.*

method. However, the following differences from the functional method are present:

—When the coupling sought between element clusters is sequential, using the functional method may involve placing these element clusters in different subsystems, since sequential coupling occupies a grey area between tight and loose coupling.

—The avoidance of subsystem conflict criterion is not applied by the sequential cohesion method.

—The sequential method provides no further decomposition into subsystems of a parent system which is sequentially cohesive to begin with, while the functional method is able to further sub-divide a functionally cohesive parent system.

Limitations in the Use of the Functional Method

The functional method has certain limitations. **First**, it cannot always be applied, for any one or combination of the following reasons:

- *If the objectives of the function cannot be defined in a specific enough manner*— For instance, the objective is to produce pleasure, for either those who carry out the function or for some other recipient. What exactly does this mean?
- *If the mutually contingent functional objectives are in conflict, i.e., are dysfunctional to one another*— For example getting the job done and personal pleasure are both identified as being objectives of the same function. In this case, different functional subsystems may have to be defined corresponding to each objective.

The **second** limitation of the functional method, even if it can be successfully applied, is that it may not conform to either human (social) or physical reality. To illustrate, the essence-identity of a business organization cannot "really" be understood in terms of pure functional analysis, since a gestalt method, such as justified formal identity, provides an intuitively understandable representation of how a business organization "really" fits into the human world. The same applies with physical phenomena in nature, since nature does not necessarily employ the functional method in evolving its characteristics; i.e., to assume that nature does this might be to presuppose a teleological view of life, e.g., one based on a a particular type of God who assigns a purpose to everything in the universe.

CHAPTER ELEVEN

FLOW FUNCTIONS

The operations in each objective-defined function flowing towards the functional objective may be decomposed into functional elements, some of which are called "flow functions." Flow functions are the basic conceptual groupings into which the various basic flow phases may be divided within each wideshafted arrow. They exist at various levels of perception, and, as in the case of the objective-defined functions of which they are a part, certain principles of relativity of functional perception apply to their classification.

Like objective-defined functions, flow functions have certain patterns of interaction. The supplement to this chapter, "Patterns of Functional Interaction," really applies to both objective-defined and flow functions. Its materials have been delayed until the end of the present chapter, since the complete understanding of the subject requires both the concepts of objective-defined and of flow functions.

FLOW FUNCTION CLASSES

Flow functions are classified into three different categories, which may be referred to by the following alternative, equivalent sets of terms:

—input, processing, and output
—afferent, transform, and efferent;
—entry, production, and delivery.

To logically define each class, various matters need to be resolved over the remaining sections of this chapter, since a certain amount of counter-intuitive thinking is involved. The matters to be resolved in this section include the following:

- the difference between inputs and outputs, on the one hand, and input and output **operations**, on the other;
- inputs and outputs internal to the objective-defined function, versus those external to it;
- input and output versus interaction with the system environment;
- the necessity that any objective-defined function have at least one of each of the input, processing, and output flow function types;
- variances in the meanings of input, or processing, or output operations in relation to the class of **objective-defined** function.

In any objective-defined functional sequence, three different classes exist, as follows:

1. **Input flow function**: This obtains the transactional inputs which initiate and which are to be operated upon in the run of the objective-defined function. These may not be the only inputs to the input flow function. For instance, framework and transactional input to be used internally *within the input flow function itself* may be acquired by the input flow function.
2. **Processing flow function**: This operates upon (i.e., transforms or further transmits) the operand transactional input obtained by the input flow function, thereby obtaining a product ready for delivery. In addition to the transactional input from the input flow function, stimuli, transactional, and framework inputs to be used internally *within the processing flow function itself* may be input into the processing flow function from outside the objective-defined function.
3. **Output flow function**: This delivers the product obtained from the processing flow function. As well, stimuli, framework and transactional input to be used internally *within the output flow function itself* may be input from outside the objective-defined function by the output flow function.

Whether or not the output flow function delivers all those outputs which correspond to the objectives of the objective-defined function is a tricky matter. Such an output would only be delivered by an **output** flow function if this output reaches the output flow function from the preceding processing flow function. To ensure that it does reach this far, arrows may be drawn **within** the wideshafted arrow, as in Figure 11-1. The matter being discussed in this paragraph is related to the perception of an intermediate flow function objective as an objective of the objective-defined function as a whole (see mutual contingency concept in Chapter Eight). This is not the same matter as the existence of more than one output flow function within the sequence, as explained in the second section of this chapter, in connection with the "intertwine complexity."

Inputs and Outputs Versus Input and Output Operations

In any objective-defined function, different types of incoming (input, afferent) and outgoing (output, efferent) phenomena affect the flow at various points. Incoming and outgoing phenomena means **what goes into and out of** the objective-defined function, rather than the **types of operations** which occur during the flow; that is, there are input and **operations** as well as inputs and outputs.

Further, input and output operations may both receive input from outside, and send output to outside the objective-defined function. To

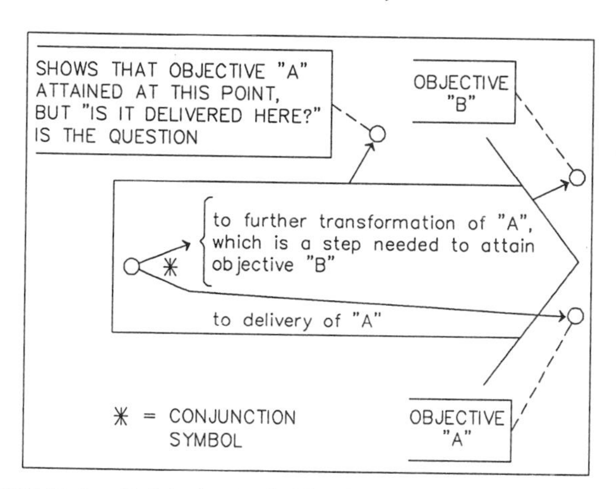

FIGURE 11-1: Point of exit of objective output from wideshafted arrow. *This concerns the question of identifying the exit point in leaving the wideshafted arrow when the objective output is attained earlier than the end of the last processing flow function, i.e., when a flow function objective is also seen as an objective of the objective-defined function as a whole.*

demonstrate that the phenomenon of output does not necessarily imply the phenomenon of achievement of objectives, an input flow function may output feedback on how it is doing.

Inputs and Outputs Internal to the Objective-Defined Function

Since the output of one step becomes the input to the next step in a sequence, inputs and outputs also exist **within** objective-defined functions.

Moreover, different instruments used to carry out operations within the objective-defined functions have their inputs and outputs. Thus, when operations are done by computer, it is incorrect to say that **any** operations which write or print data, such as those done on a computer printer, belong to output flow functions. These constraints, if applied, would straitjacket thinking.

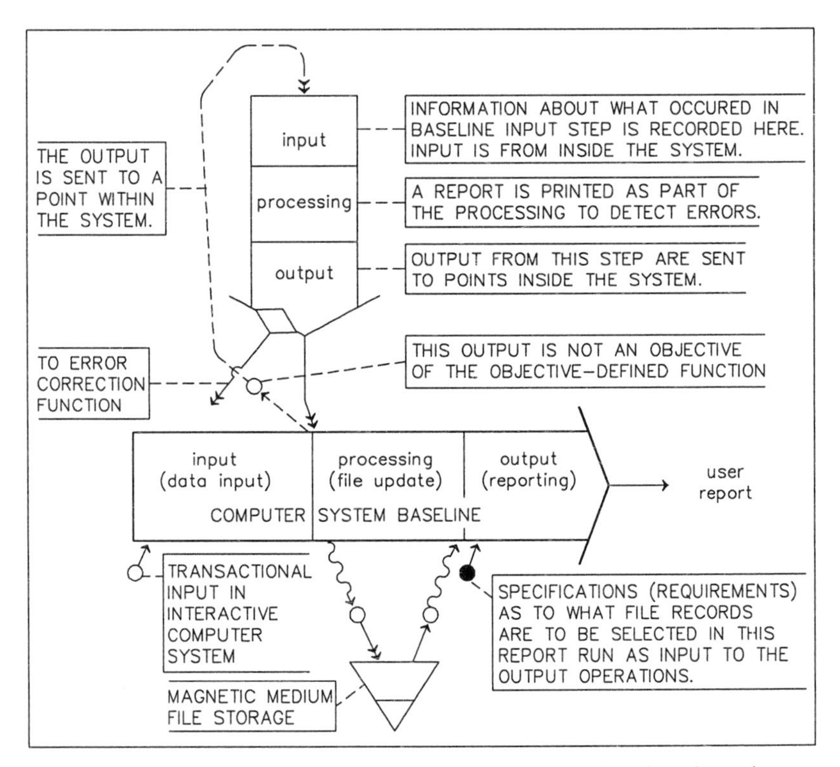

FIGURE 11-2: Misconceptions about input and output flow function roles. *These examples are designed to support the discussion of misconceptions about what are input and output flow functions.*

Input/Output Operations and Relationship to System Environment

Input operations may input their transactional inputs entirely from within the system, while output operations may deliver their objective outputs to destinations within the system. That is, an environmental interface does not necessarily exist. If such were the case, it would, for instance, become impossible to have a closed system, i.e., a system which consumes its own output, having no interactions with its environment. Also, it would be impossible to have interactions between different objective-defined functions of a system.

Summary

The above several paragraphs are intended to clear up various misconceptions about flow functions and how they may be categorized. An attempt has been made in Figure 11-2 to demonstrate areas in which some of these misconceptions might arise.

Question of Absence of Given Flow Function Types from an Objective-Defined Function

So far, the possibility has not been discussed that an objective-defined function may entirely lack a given flow function type in all runs of its existence; for example, it may have input and output flow functions, but no processing flow function. In relation to the input-processing-output model of flow, it has been said that

> . . . we are faced with the possibility of an application for which the model is inappropriate—such as one with no processing and degenerate output, or with a purely mathematical computation.*

In this book, it is considered that at least one of each of the input, processing, and output flow function types occurs in every objective-defined function. This opinion is based on the assumption that new levels of input-processing-output relationships can be perceived within preceding levels. The latter matter will be discussed in the third section of this chapter.

Class of Objective-Defined, Vs. Class of Flow Function

Flow function categories are defined above in a general way, without regard to the class or level of the objective-defined function in which

* Yourdon, E. and L. Constantine, Structured Design: Fundamentals of a Discipline of Computer Program and System Design. Englewood Cliffs, N.J.: Prentice-Hall©, 1979, pp. 148—149.

the flow function occurs. These definitions may be tailored to provide certain generalizations about these flow function types when they occur in objective-defined functions of particular classes. Much work has been done, for instance, to describe the internal steps (flow sub-functions) of the processing flow function type of decision-making classes of objective-defined functions. Simon, for instance, identifies three simple stages in this decision-making process:

1- intelligence: detect problem or opportunity; and collect data;
2- design: outline alternative solutions or actions to be taken;
3- choice: select the best alternative.

Alternatively, Mintzberg has the following to say about the decision-making process:

> What is a "decision?" It may be defined as a commitment to action, usually a commitment of resources. In other words, a decision signals an explicit intention to act.... And how about a decision process? One thing it is not is just the selection of a course of action. Our research (Mintzberg, Raisinghani, and Theoret, 1976) indicates that selection is often the icing on the cake, one of a series of steps leading to a decision, not necessarily the most important. A decision process encompasses all those steps taken from the time a stimulus for an action is perceived until the time the commitment to the action is made. This research suggests that those steps draw on seven fundamentally different kinds of activities, or "routines." Two take place in the identification phase of decision making: the recognition routine, wherein the need to initiate a decision process is perceived, and the diagnosis routine, where the decision situation is assessed. Two routines are associated with the phase of development of solutions: the search routine, to find ready-made solutions, and the design routine, to develop custom-made ones. The selection phase includes three routines: the screening of ready-made solutions, the evaluation-choice of one solution, and the authorization of this by people not otherwise involved in the decision process. A single decision process can encompass any and all of these routines, each in fact executed a number of times.*

---

* Mintzberg, H., The Structuring of Organizations. Englewood Cliffs, N.J.: Prentice-Hall©, 1979, p. 58.

As seen by the present author, the stages of decision making, and how they may be classified and sub-classified, are as follows:

1. **Input flow function**: Detect opportunity or problem.
2. **Processing flow function**:
   a) Collect further data needed to investigate detected situation.
   b) Study alternatives.
   c) Make choice of appropriate alternative.
3. **Output flow function**: Communicate (notify, record, report) decision details.

In the above, it is understood that both input and output flow functions, in that order, usually have the highest degree of structure (repetitiveness according to pre-established rules). The decision-making **process** is defined in a more restrictive sense that used by either Simon or Mintzberg.

POSSIBLE COMPLEXITIES OF FLOW

INTRODUCTION

The ideas of reservoirs and the intertwine complexity will now be dealt with. It might appear from the above simple input-processing-output model of flow function classes that the objective-defined function completes all input, then commences and finishes all processing, and finally does all the output. However, it is quite possible that the flow in any given run of an objective-defined function does not unfold in this simple pattern. One, or any combination of the following may occur during the flow of an objective-defined function:

- A flow function type may not be executed in certain runs of the objective-defined function; e.g., input is performed less often than processing.
- Different parts of a flow function may not execute in direct succession; e.g., they may be broken up by the execution of other flow function types in the same sequence.
- A particular flow function type may be executed more than once during the run of an objective-defined function, in the sense of being repeated, rather than of being broken up, e.g., input, followed by

processing, followed by a repeat of the **same** processing operations, etc.

- Different flow function types may execute in parallel during a given run of an objective-defined function; e.g., the first processing phase occurs at the same time as a second input phase. In fact, a single flow function type, executed more than once in a given run of an objective-defined function, may execute in parallel with itself, where more than one occurrence of this flow function type exists along the line of flow.

The possibility of all, or any combination of the preceding phenomena occurring during a particular run of an objective-defined function complicates the study of flow functions. The task here is to understand these phenomena, both in terms of why they occur, and of the patterns in which they occur. To this end, two concepts are needed, namely, that of reservoirs (previously introduced), and that of the intertwine complexity. The idea of reservoirs explains how given flow function types may be absent in some runs of an objective-defined function, while the intertwine complexity concept explains all the other phenomena mentioned above.

RESERVOIRS

The Concept

A reservoir may be defined as an accumulation (aggregation, batch) of framework and/or operand transactional inputs built up or drawn down during the run of an objective-defined function, or during the interaction of different objective-defined functions. This aggregate may be stored in a physical storage location, such as a warehouse, or a file of data. However, the reservoir is not defined by the storage location, but by the conceptual aggregate.

A reservoir may contain physical and/or data items. In many cases, data reservoirs are reflections of the contents of physical reservoirs (i.e., reservoirs which contain materially tangible objects). For instance, changes in personnel (physical objects) are reflected not only in the personnel reservoir, but also in the personnel master file reservoir, which contains *data* about the personnel.

Effect of Reservoir on Flow Scheduling

A reservoir of operand transactional inputs (which may also be referred to as "a pool of depletable resources") may be built up either gradually or "in a shot," and may be depleted in either of these two ways. The depletion occurs during a later run of either the objective-defined function which established it, or by the operation of another objective-defined function.

Since the reservoir results in unequal frequencies of executions of flow functions, it is connected with the phenomenon of iteration, or looping. To illustrate, a sequence may consist of three steps. Steps one and two execute, in that order, ten times, in straight repetition, before each execution of step three.

Reservoirs play roles in on-going systems as long as an objective-defined function in that system involved in drawing down the reservoir has a life-span of more than one run. In other words, reservoirs are made usable in that the objective-defined function which uses them is not a one-shot affair. In other words, repeat runs of the using objective-defined function are needed to establish the possibility of run-to-run relationships. (However, this does not imply that **all** run-to-run relationships are connected with reservoirs.) In general, the irregular—or regular but periodic—execution of a flow function in the run of an objective-defined function, may be viewed as being the result of accumulations/depletions, i.e., reservoirs.

Example of Reservoir of Transactional Inputs Built-Up All At Once

In a transactional reservoir input within a sequence, an objective-defined function is executed to make a product called "A," ordered by a customer. To produce "A," the input operations obtain the necessary raw materials. These are ordered in a quantity sufficient not only to cover the present customer order, but also to fill many future ones. As a result, the input operations need not be executed during the runs of the objective-defined function to fill the second, third, fourth, etc. customer orders for product "A." They can be omitted, since supplies are already on hand. Until the reservoir of raw materials (see Figure 11-3a), set-up during the handling of the first customer order, is sufficiently drawn down, the input flow function's execution may be omitted during functional runs. Similarly, should, for instance, the manufacturing operations executed in response to the first customer order for product "A" create enough of product "A" to provide shelf stock in addition to the amount of product "A" required to satisfy the first customer order, the omission of the processing operations in subsequent runs of the objective-defined function thus becomes possible, until the shelf stock is depleted.

In contrast to the accumulation/depletion pattern Figure 11-3a, that in Figure 11-3b shows gradual accumulation followed by depletion all at once.

Structure and Structure-Flow Charting Reservoir Flow

Showing the Effects of Reservoirs in Structure Charts

Loops and/or optional steps may be used in structure charts to symbolize unequal and/or irregular flow frequencies. These may result from execution errors, or from accumulations/depletions (i.e., reservoirs), or from differences in time frames (between current and non-current operations). In Figure 11-4a, the loop represents the greater relative frequency of the current operations. The optional step symbol within the loop arrangement could represent the effect of a reservoir, depending upon how it is interpreted. If this indeed represents the effect of a reservoir, then it is one which is both built up and drawn down gradually, since both current steps 1 and 3 execute more frequently than step

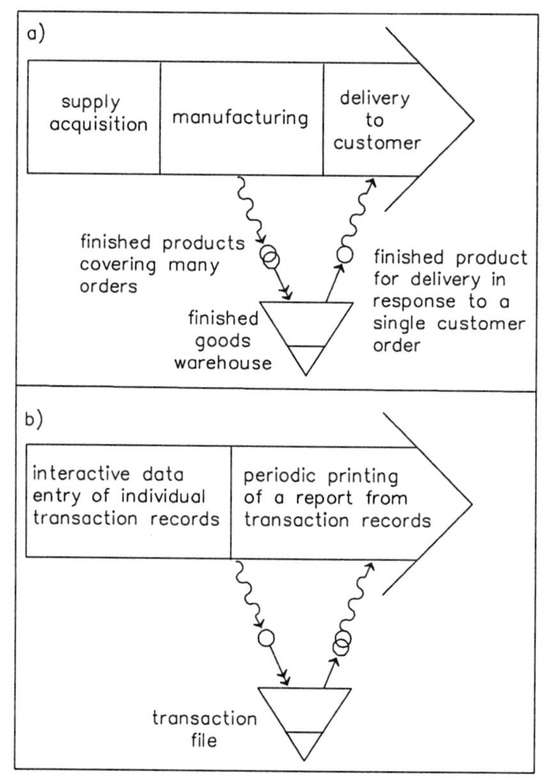

FIGURE 11-3: Reservoirs along the line of flow.
a) a flow pattern in which the finished goods warehouse stocks are built up all at once, and then depleted gradually in response to individual customer orders over time;
b) a flow pattern in which the reservoir is gradually accumulated, but drawn down in a shot.

2. Another way of showing this would have been to place loops around each of steps 1 and 3, and to make step 2 mandatory along with steps 1 and 3. (These loops would be internal to the loop already shown.)

The Effect of Reservoirs Shown in the Structure-Flow Chart

In the structure-flow chart, the effect of a reservoir on relative flow function execution frequency is diagrammed, as in Figure 11-4b, using both loop and reservoir (inverted triangle) symbols. The two loop symbols indicate that the reservoir is both built up and drawn down bit by bit, as each transactional input item is obtained from the reservoir. The first loop symbol in this example occurs in connection with initiation of the

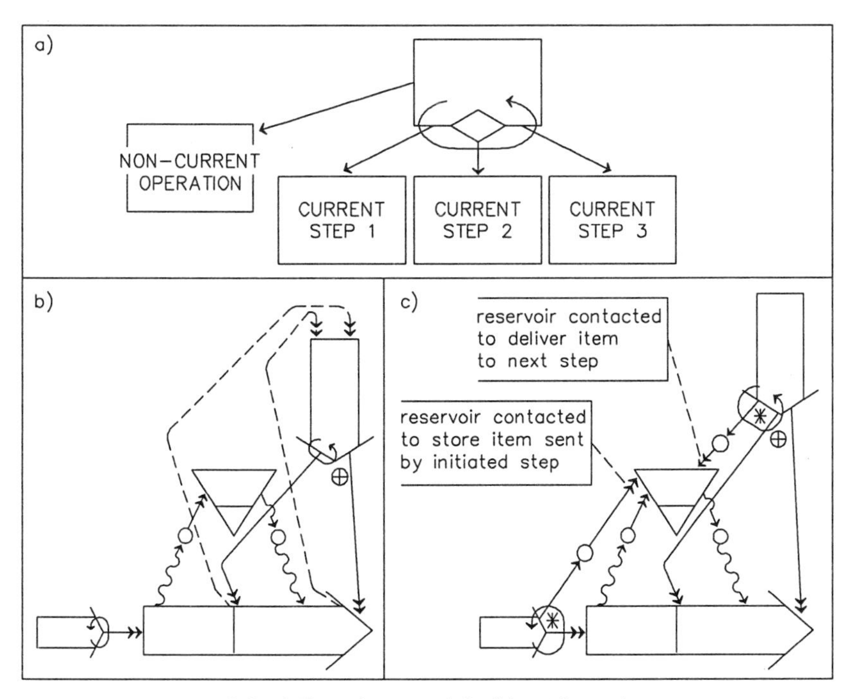

FIGURE 11-4: Scheduling of reservoir build-up/draw-down.

a) one way of indicating in a structure chart irregular execution frequency in current operations resulting from a reservoir along the line of flow;

b) gradual building up and drawing down of the reservoir indicated by the presence of loops in structure-flow charts to drive accumulations/depletions;

c) same as "b," except that instructions sent to the reservoir operating system are explicitly diagrammed, rather than assumed.

functional run, while the second one occurs in connection with continuation after validation of the results of the previous step.

Figure 11-4c is the same as Figure 11-4b, except that two new outputs are added. These show that, in conjunction with initiating (or continuing with) the next step, the reservoir system must be contacted and requested to send/receive items. As was indicated in Chapters Five and Six, this type of flow is not always shown in the structure-flow chart, since it would make it unnecessarily complex. Thus, it is often assumed that the input from outside a function arrives when it is needed, it being understood that the outside sources have been contacted by the function which initiates the next step.

Effect of Reservoirs on Control Flows

Any output of a function may be stored in a reservoir before it is used by another function or functional run. Thus, reservoirs may be involved in relationships between different flow functions of the same run of an objective-defined function, different objective-defined functions, and different runs of the very same objective-defined function. Accumulation/depletion in reservoirs may occur in relation to any type of functional output, such as feedback, symbiotic, objective, etc.

An interesting facet of storing feedback output used to ensure current run effectiveness in reservoirs is how this affects the timing of control of the current run. Obviously, if the feedback is stored in a reservoir, it may not be used to validate a given run of the objective-defined function until enough runs have been completed to begin drawing down the reservoir for use in the validation function. To illustrate, if, in an accounts payable system, the feedback output concerning bank withdrawals is accumulated in a reservoir, the effectiveness of a withdrawal may not be reviewed until such time as the reservoir contents are accessed. Thus, for example, bank withdrawals may not be validated until month-end, when bank statements are received by the business, although bank withdrawals occur throughout the month. To handle this problem in the structure-flow charting method, the analyst must assume that the run of the objective-defined function does not complete until validation of its results has been completed. Thus, while the money withdrawn from the bank may go to the supplier before the bank withdrawal is validated, successful completion of the withdrawal does not occur from the viewpoint of the business until the withdrawal has been checked and acknowledged by the business bank reconciliation function.

Operations Performed by Reservoir Systems

Reservoir Operations

The existence or reservoirs involved in the flow of an objective-defined function implies functions to operate the reservoir. In computer systems, for instance, the reservoir operating system is the set of operating system routines called upon by application programs to perform input/output. The inverted triangle symbol, chosen to represent the reservoir, implies the existence of a reservoir operating subsystem. Structure-flow charts of this subsystem were provided in Chapters Five and Six.

Functions Performed and Not Performed by Reservoir Operating Systems

Figure 11-5 shows which functions are or are not performed by the reservoir operating system. The following paragraphs discuss the conclusions demonstrated in this figure:

- **First**, although the items in a reservoir are accumulated in the reservoir, they are not originally **acquired** by the reservoir system; e.g., the business raw material inputs are obtained by the supply acquisition system, whence they go to the supplies warehouse, i.e., to the physical supplies reservoir operating system.
- **Second**, the items in a reservoir are neither used nor changed by the reservoir operating system. That is, they may only be used or changed after removal from the reservoir, following which they may possibly be returned to the reservoir where they are received by the reservoir operating system. For instance, to retrain personnel, employees are "removed" from the employee "pool" (reservoir), and then "restored" after retraining by the personnel system. Although removal and storage functions are done by the reservoir

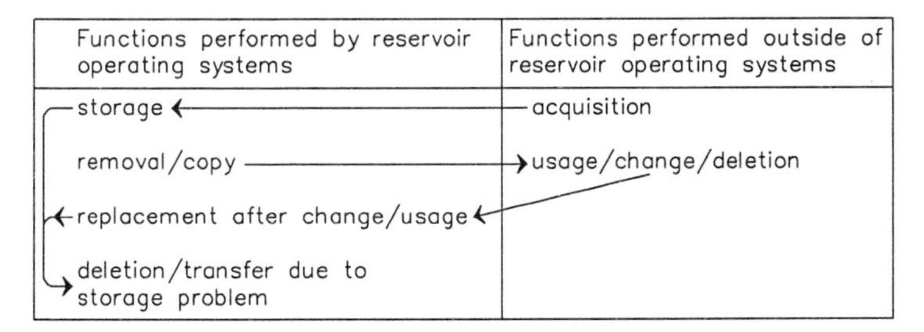

FIGURE 11-5: Functions performed by reservoir operating systems. *This figure shows both the functions performed and those not performed. The directional arrows indicate the sequential relationships which order the execution of these different functions.*

system (which is a buffer subsystem of the business, external to the personnel system), retraining is done by the personnel system.

Similarly, in a computer static data master file organized as a random access file, records to be changed are first queried, by obtaining a copy of the record via the copy function of the data reservoir operating system. Then the record copy may be changed in main memory of the computer by a computer **applications** program. Finally, the changed record copy is returned to its original location on the random access storage device, and replaced in this location by the replace function of the reservoir operating system. These read/write reservoir operations used to update a file are carried out by a group of operating system (as opposed to "application") programs, referred to as the "IOCS" (Input/Output Control System). The IOCS is, in effect, the reservoir operating system.

- **Third**, the reservoir operating system handles deletions only in cases of storage problems, e.g., spoiled, damaged, outdated, or stolen supplies in the supplies warehouse may be deleted by the reservoir operating system. Other deletions, like changes, are performed outside the system, after removal from the reservoir by its operating system functions. (Examples of deletion were given above in connection with the supplies warehouse.) An interesting point, however, arises in connection with deletion of items from, say, a random access data file. While the records may be deleted outside the reservoir operating system, it must be remembered that only a copy of the data item is thus deleted. In addition, the "delete flag" may be placed on the original data item, in the file, by the reservoir operating system.
- **Fourth**, the reservoir operating system handles transfers of storage location **within** the reservoir, as in moving items from one bin location in the warehouse to the other.

Factors Which Determine Functions Performed

The baseline functions of a reservoir system may be used to add, replace (after change), remove (for use), delete (if data, or if storage problem), and copy (if data) stored items. The question of which combination of these functions is performed by a given reservoir system involves factors such as the following:

- whether the reservoir contains data or physical items; e.g., where the reservoir contains data, copy functions may be present in the reservoir operating system;
- whether the items stored in a reservoir are relatively permanent, as in the distinction between temporary transaction and master files;

e.g., where the reservoir contains either very temporary items, replace functions may not be present in the reservoir operating system, as the item is completely consumed after removal;
- whether the reservoir items are so permanent that they cannot be changed at all; e.g., ROM (Read-Only-Memory) in a computer system means that replace functions to rewrite the data are not possible.

Thus, depending upon factors such as the above, different sets of baseline functions may be present in different reservoir operating systems.

INTERTWINE OF FLOW FUNCTION CLASSES IN ACTUAL FLOW

Concepts

Flow Problems Accounted For By Intertwine

On the one hand, the concept of **reservoirs** explains how the same flow function type, consisting of the same operations, may execute more than once in relation to another flow function of an objective-defined functional sequence. This "looping," or iterative, phenomenon involves multiple, successive executions of the same segment of operations for every one execution of another segment of the operations of the same sequence.

On the other hand, the **intertwine complexity** concept explains single run, multiple, but **non**-successive, executions of the same segment of operations, where the repetitions are dispersed in between other segments of operation in the sequence. The intertwine complexity also explains the breaking up of a contiguous segment of operations in a sequence into a number of non-contiguous sub-segments, dispersed in their execution between other parts of the sequence. Perhaps the best illustration of the intertwine complexity is the decision-making process, whose generic steps may be repeated many times before a final decision is reached.

Non-Successive Repetition Versus Non-Successive Continuation

To explain further, the non-successive **repetition** of operations during a single run of an objective-defined function occurs in the following succession of phases:

1. first input of two identical executions;
2. first processing of two identical executions;

3. second input of two identical executions (identical to #1);
4. second processing of two identical executions (identical to #2);
5. first output and only execution.

In contrast to the non-successive **repetition** of flow segments, the non-successive execution of different sub-segments of operations in a flow function is contained in the following operations:

1. input, first sub-segment;
2. processing, first sub-segment;
3. input, second sub-segment;
4. processing, second sub-segment;
5. first and only execution.

The latter case illustrates the breaking up of a single flow function into non-successive **continuations** (as opposed to repetitions) of the flow function operations; that is, the operations in a flow function continue after interruption by other operations belonging to other flow functions of the same objective-defined function. Unlike the cases in which the **same** operations are repeated after interruption by operations in another flow function type, a little of one flow function type is executed, then a little of another flow function type, then the remainder of the first flow function type, then a third flow function type, and so on, in an infinite variety of possible forms of switching back and forth. In the words of the computer systems literature, this switching back and forth occurs along "an incrementally changing line" of flow between "**co**-routines" (not to be confused with **con**routines) represented in Figures 11-6a and 11-6b. Figure 11-6c illustrates the contrasting idea of non-successive **repetition**. Finally, Figure 11-6d shows how simple things would look without any intertwine.

It must be emphasized that the intertwine complexity is very easy to confuse with the phenomena arising from the existence of the relativity of flow function type conceptualization, as discussed in the next section, based on which many different occurrences of a given type of flow function may appear along the line of flow of an objective-defined function. However, the latter perception depends on differences of **level** at which the sequential relationships are understood, rather than on the intertwining of flow function types understood at the **same** level of detail viewing.

Flow Sequencing Possibilities With Intertwine

The possibility of multiple, non-successive execution of a given flow function type, either in the sense of continuation or of repetition, opens the door to the further possibility of **parallel** execution of different flow function types in the course of a single run of an objective-defined function, e.g., a second input phase executed in parallel with the first processing phase. In fact, the configuration of flow function type sequencing in an objective-defined function's run can vary widely from the configuration found in another run of this same objective-defined function. An almost infinite variety of patterns is possible, subject to the following constraints:

—Each run of an objective-defined functional sequence must start with the execution of an input flow function

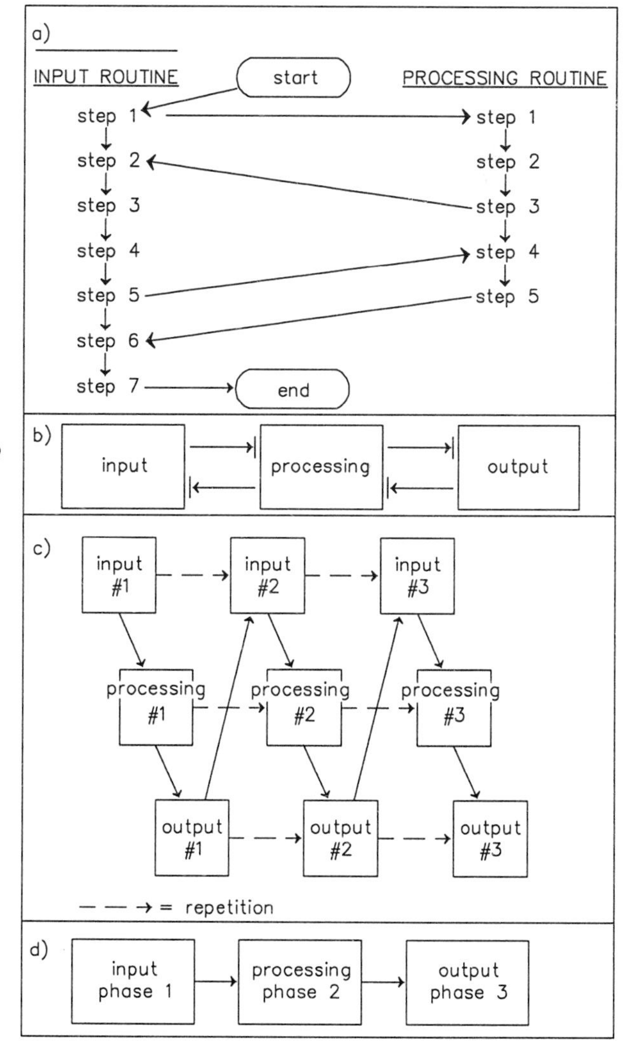

FIGURE 11-6: Charts showing the effects of intertwine of flow functions.
a) coroutines for intertwine involving non-successive continuation of given flow functions;
b) representation of coroutines in structure charts;
c) non-successive repetition of flow functions with intertwine;
d) simplicity of flow function sequencing without intertwine.

type, and must end with the performance of an output type.
—The last input operations must occur before the last processing phase; and the final processing operations must take place earlier than the last output phase.
—The first input operations must occur prior to the initial processing phase; and the first processing phase must be enacted before the initial output operations.

These constraints are reflected in the following examples.

Examples of Intertwine of Phases

Example One

A sequence is executed to manufacture a durable item consisting of ten component parts, to be assembled by the manufacturing process. To fill customer orders, the following events take place:
—Five of the ten component parts are acquired by input operations.
—The five component parts are assembled, by the first of two processing operations phases.
—The five remaining component parts are acquired, by a second input phase.
—The remainder of the product assembly is performed by a second processing phase.
—The finished product is delivered to the customer by output operations.

Example Two

A customer order is received for the manufacture of three items of a certain finished product. The following sequence of operations is enacted:
—The materials needed to manufacture all three items are acquired by input operations.
—Two of the three ordered items are manufactured, by the first of two processing phases.
—The first two completed items are delivered to the customer by output operations.
—The third item is manufactured by the second set of processing operations.
—The third item is delivered to the customer.

Example Three

The very large number of processing-output phase "couplets" of a repetitive nature is a very interesting type of phase intertwine. For all effective purposes, the "processing" is inseparable from the "output."

A woman receives a facial in a beauty parlour. Giving the facial massage involves positioning of the fingers and then pressing. Each finger positioning act is synonymous with "processing," while the final act of pressing is synonymous with "output." In the course of providing the face massage, there may be about ten thousand acts in which positioning of the fingers is followed by pressing; that is, about ten thousand processing-output couplets.

Example Four

The following sequence of operations occurs:

—A customer orders product "X," but does not fully specify the modalities of creating this product.

—The transactional inputs are acquired by the input operations for producing the ordered product "X."

—The first of two processing phases is performed, to produce "X" in its preliminary, "draft" form.

—The customer reviews the results of the first processing phase, and completes the requirements input by providing the remainder of the specifications on how to produce "X." (These customer activities are coordination operations, and not part of the direct line of flow.)

—The transactional inputs are acquired by input operations for the final processing phase.

—The final processing phase is executed, and the product completed.

—The product is delivered by output operations to the customer.

Example Five

The following flow function types execute in the order indicated:

STEP #			
1	input #1 — first input		
2		processing #1 — first processing	
3	input #2		
4		processing #2	
5			output #1 — first output
6		processing #3	
7	input #3		
8		processing #4	
9	input #4		
10		processing #5	
11	input #5		
12		processing #6	
13			output #2
14	input #6		
15			output #3
16			output #4
17		processing #7	
18	input #7		
19			output #5
20		processing #8	
21			output #6
22	input #8 — last input		
23		processing #9	
24			output #7
25		processing #10 — last processing	
26			output #8 — final output

DIFFERENT LEVELS OF INPUT-PROCESSING-OUTPUT RELATIONSHIPS

THE BASIC IDEA

Different levels of input-processing-output relationships exist within the operations of an objective-defined function. Any segment of flow within an objective-defined function, no matter how small, may be conceptually structured into these relationships. In Figure 11-7, the upper part of the diagram depicts the macro view, that is, the view of input-processing-output seen from the level of the overall system. The bottom part shows how each box in the upper part may be blown up into a similar set of boxes representing input-processing-output relationships. By applying this type of logic, a recursive perception of similar generic types of flow functions within any level of flow function may be obtained.

For example, the objective of the output flow function of a baseline objective-defined function is to deliver goods produced by the business to customers who order them. The delivery has its own internal steps, including receiving the order to make the delivery, obtaining the delivery vehicle for the trip, obtaining the fuel and drivers for the delivery vehicle, making the delivery trip, and finally, handing the goods over to the customer. In effect, the delivery phase with respect to the baseline objective-defined function of the overall business has its own level of input, processing, and output.

Table 11-1 shows how the following steps of the baseline objective-defined function of a distribution company may be seen at two different levels of hierarchy:

1. **Input:** Receive inventory supplies from manufacturers, and send this to the warehouse for storage;
2. **Processing:** Pick and pack goods needed to fill customer orders;
3. **Output:** Ship goods to customers.

Showing Flow Within the Wideshafted Arrow

At levels of hierarchy lower than that of the overall objective-defined function, input and output operations may reference points of receipt and delivery which are not only inside the system, but also inside the objective-defined function, e.g., items D and E in Figure 11-8a. As seen in this figure, inputs and outputs can take place in the input operations of the second level of perception of flow function elements, just as they do in the elements of the first level of detail.

Relativity Principles In Designating Type of Flow Function

The relativity of flow function conceptualization is yet another aspect of the idea of the relativity of functional conceptualization, besides those which affect classifications and levels of objective-defined func-

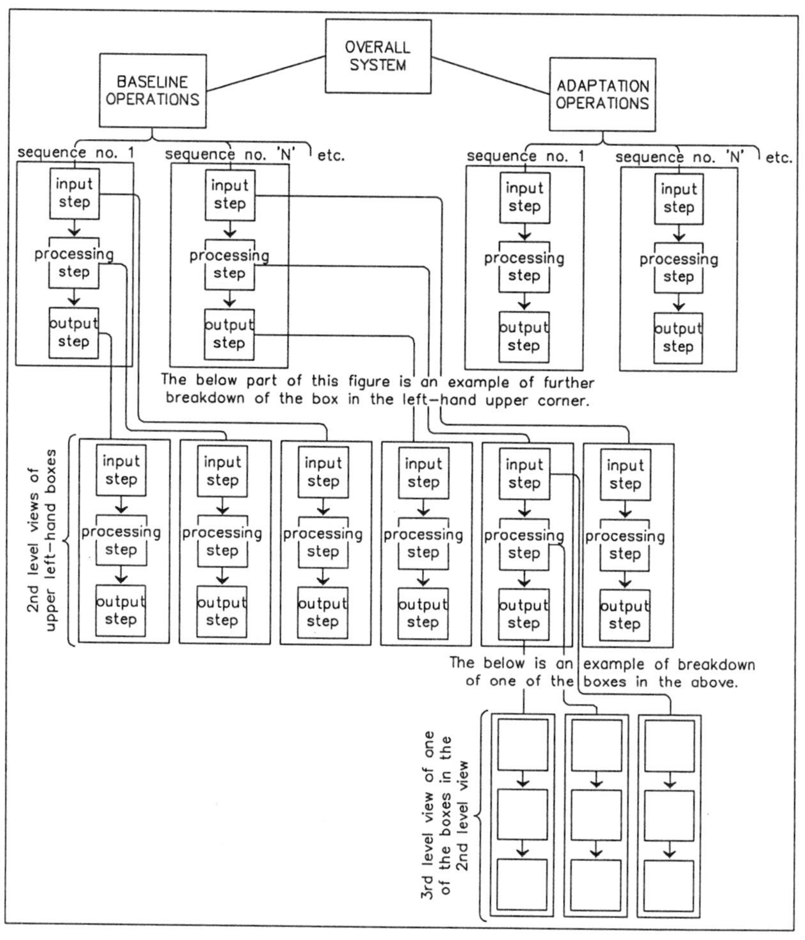

FIGURE 11-7: Hierarchical decomposition of flow function segments. *This produces new levels of input-processing-output relationships.*

	OVERALL FLOW		
	INPUT RECEIVE INVENTORY	**PROCESSING** PICK AND PACK GOODS	**OUTPUT** SHIP GOODS
INPUT	Become aware of shipment of goods from manufacturer. (1)	Receive workorder to ship goods to customer. (4)	Acquire products for delivery from step 6, and send to step 8. (7)
PRO-CESSING	Receive shipment of goods. (2)	Pick goods from warehouse shelves. (5)	Transport products to destinations. (8)
OUTPUT	Send goods for warehousing. (3)	Pack goods for delivery. (6)	Hand products over to customers. (9)

TABLE 11-1: Different levels of input-processing-output relationships. *Flow goes in the sequence indicated by the numbers in the lower right-hand corners of the cells of this matrix.*

tions (see Chapter Nine). This principle, which affects flow function classification, is applied during subsystem analysis when only parts of the flow of an objective-defined function of the parent system are assigned to a subsystem. In this case, it is necessary, as in Figure 11-8b, to blow up the assigned part(s) of the objective-defined function into a "full-fledged" one, i.e., one having a first level of input-processing-output in relation to the subsystem in question.

To illustrate, a business organization extracting oil from the earth processes this oil to a stage of refinement needed in the manufacture of plastics, and then manufactures plastic objects; i.e., there is a high degree of what is called in the science of economics "vertical integration." However, since the government dislikes the extent of this vertical integration, it breaks up the company into three separate companies, one for mining, a second company for refining, and a third for plastics manufacture. Before disintegration of the company, the processing flow function consisted of refining and plastics manufacture. After break-up, the mining operation becomes the **processing** flow function of the first company; the refinement the processing flow function of the second company; and the plastics manufacture the processing flow function of the third company.

PRINCIPLES OF CHANGE OF FLOW FUNCTION CONCEPTUALIZATION

The Problem of Finding Principles

In order to analyze a single flow function so as to create a new set of input-processing-output relationships at the next level of detail perception, it is necessary to "blow-up" ("explode," or "stretch") the single flow function phase being studied. In fact, input-processing-output relationships can also be perceived in a fractional part of a flow function and/or in more than one series of flow functions within the sequence of operations in an objective-defined function.

Searching for a way of explaining the intuitively understandable idea that new sets of input-processing-output relationships can be seen as the level of detail increases, it is possible to use the following analogy

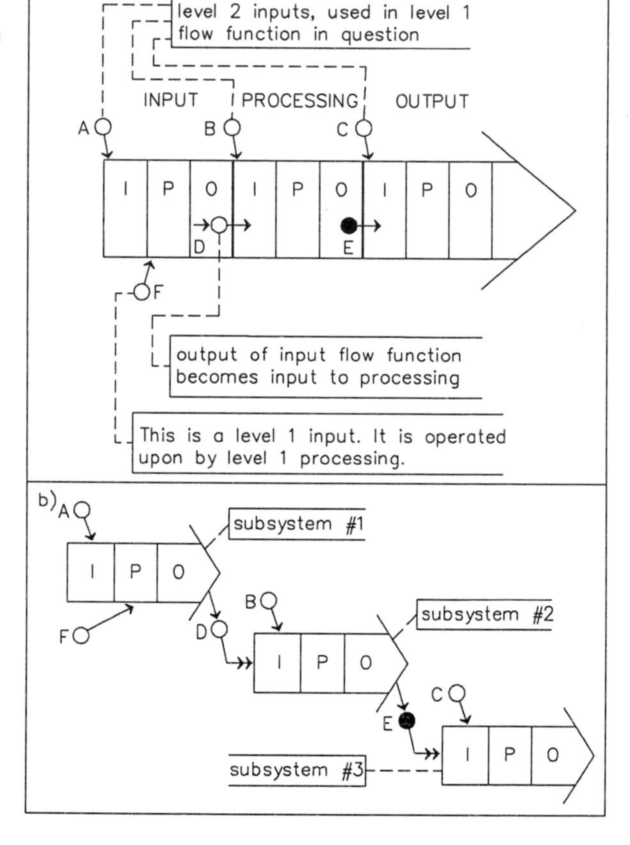

FIGURE 11-8: Inputs/outputs vs. level of flow function perception.
a) *two levels, 1 (overall) and 2 (internal) of input-processing-output relationships in a wide-shafted arrow;*
b) *equivalent of "a," except that each first level flow function is now a baseline function in its own subsystem.*

of an elastic band (hence, the word "stretch" as a synonym for "blow up" or "explosion"). This analogy is admittedly very contrived and probably defies the laws of physics. Moreover, understanding the principles stated may require an excessive amount of attention from most readers. However, it does represent an original attempt at a logical explanation.

In this analogy, imaginary elastic bands are placed in the various columns in Figure 11-9. Each such band, when fully extended, may reach no further than the entire column length, from the start of phase one to the end of the nine phases shown. As seen, various cases exist of elastic bands which, when at rest, are shorter than the entire column length, and so represent a series of operations which are shorter in extent than the entire length of the objective-defined functional sequence. The elastics at rest are indicated by the continuous lines forming sausage-shaped loops. The non-continuous lines forming loop projections, i.e., extensions of the elastics at rest, portray the type of stretch which must be performed to extend a segment of a sequence to a full sequence as seen from the viewpoint of the segment of the objective-defined function represented by the solid (unstretched) elastic.

Statement of Principles

The principles underlying the above example are as follows. Figure 11-9 contains arrows pointing into each elastic from either side of the loop. These represent, say, clamps, which, when the elastic is stretched, limit the stretch to the part of the elastic above the clamp (if the elastic is stretched upward) or below it (if the elastic is stretched downward). When each elastic is stretched, each sub-section of rubber in the part of the loop being stretched is extended an equal amount. Based on these (undoubtedly unscientific) assumptions, the following rules apply to the placement of clamps:

1. The clamp must always be placed somewhere within the **last** integral phase prior to the end of the elastic at rest, in the direction in which it is to be extended. If the elastic tip at rest does not end **exactly** at the start or end of an integral phase, further operations, as shown in number 4 below, must be enacted before the clamp is placed for the operation described here.
2. The exact position of the clamp within the last integral phase is at the start or end of the **sub**-phase which is the equivalent of the integral phase in question; e.g., sub-phase 3.1 within phase 3 is the equivalent of phase 1 within the more macro view. If the elastic is to be elongated upward, the clamp is placed at the end of the sub-phase; otherwise, it is placed at the start.

FIGURE 11-9: The phase stretch principle.
Loops portrayed by continuous lines represent elastics at rest, while non-continuous lines forming loops are for projected extensions of the elastic. (The rest of this figure is on the facing page.)

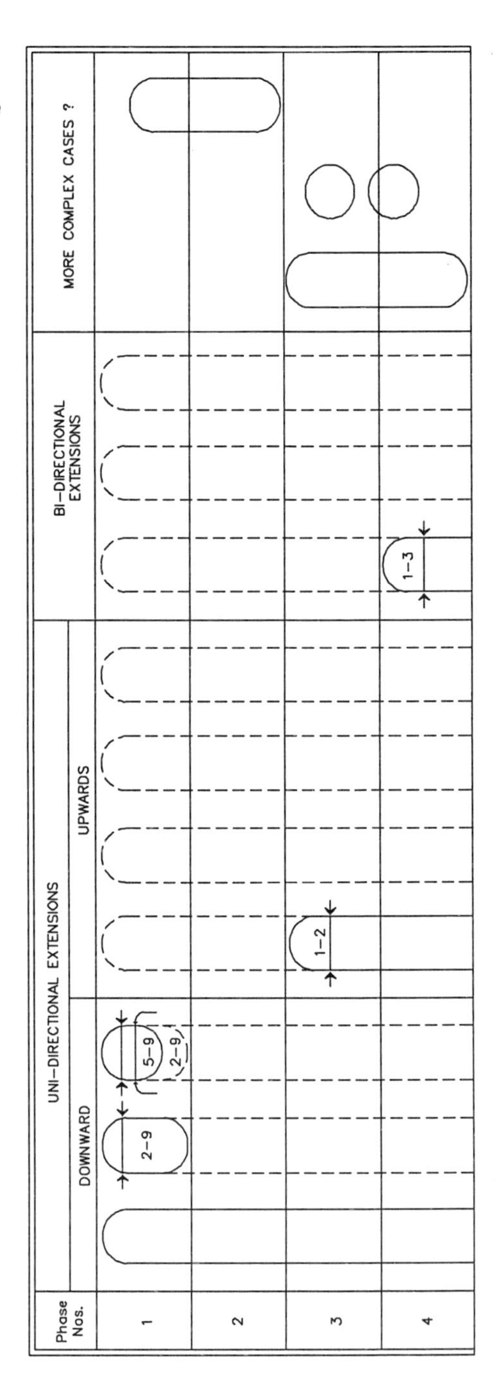

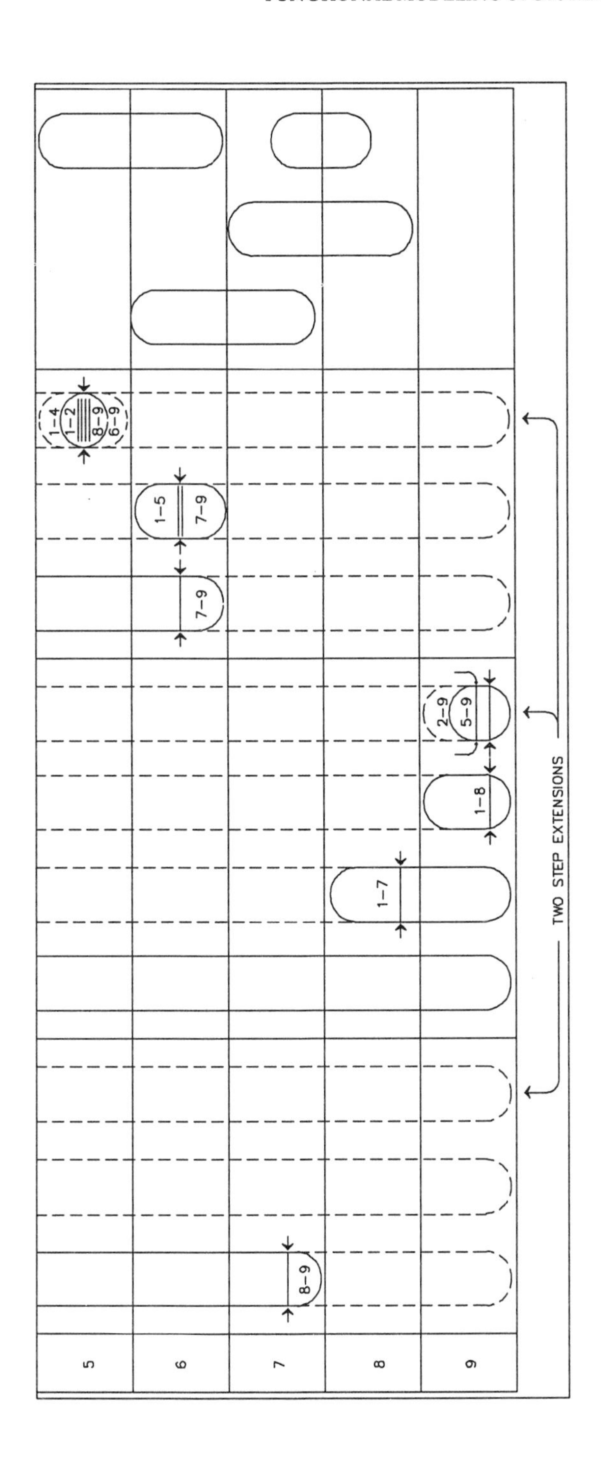
1-4
1-2
8-9
6-9
1-5
7-9
7-9
2-9
5-9
1-8
1-7
8-9
TWO STEP EXTENSIONS
5
6
7
8
9

3. If the elastic is to be stretched in **both** directions, the preceding numbers one and two are applied so as to create stretch at both ends of the elastic. However, this can only be done where the elastic at rest covers two or more integral phases; otherwise, further preliminary operations are needed, as in the following.
4a. If the elastic tip, in either direction, does not coincide exactly with the point of start or end of an integral phase in the direction in which it is to be stretched, numbers 1 to 3 above are first applied in relation to the sub-phase within the overall phase to be stretched. This extends the elastic within the phase to achieve stretch covering the whole phase. Once this is completed, the whole elastic can be stretched, using number 1 to 3 above, as if the starting position were at the beginning or end of an integral phase.
4b. However, 4a cannot be executed unless the tip of the elastic at rest coincides exactly with a **sub**-phase start or end point. If this is not the case, 4a cannot be executed until still further preliminary operations are applied to the elastic. These apply 1 to 3 above in relation to the sub-sub-phase within the sub-phase within the overall phase to be stretched.
4c. Along the same lines, 4b cannot be executed if the second level sub-phase in question is such that the elastic at rest fails to have a tip whose placement coincides exactly with a sub-sub-phase start or end. In this case, 1 to 3 will first have to be executed on the sub-sub-sub-phase. Then 4b, followed by 4a, may be executed, to complete the preliminary operations.
4d. The preceding logic continues until the end point of the elastic, in the direction of stretch, coincides with the start or end of a sub-phase identified at some level of hierarchy, i.e., with the needed number of "sub-" prefixes before it.

Example of Results of Applying Principles

The following illustration of the results of applying the above principles may be useful. An elastic at rest extends from the start of phase three to the end of phase nine, e.g., as in the fourth elastic from the left of Figure 11-9. Assuming that phase three can be decomposed into sub-phases, such that by chance, there are nine sub-phases, labelled 3.1, 3.2, 3.3, . . . , 3.9, all extensions of the elastic are now made so that the elastic is stretched upward in the following manner:

1. Phase 3.1 is displaced from its original position to occupy overall phase 1, and phase 3.2 is displaced such as to cover overall phase 2.

2. Phase 3.1, since it has been moved from its original position, must be replaced. To do this, assuming nine sub-phases in any phase at any level of detail, phase 3.3.1 is moved to phase 3.1. Similarly, phase 3.3.2 replaces phase 3.2.
3. Along the same lines, phase 3.3.3.1 and 3.3.3.2 replace phases 3.3.1 and 3.3.2, respectively.
4. This process of displacement-replacement continues and continues until the displacements-replacements become infinitesimally small, i.e., with a limit of infinity.

Simplifying Assumptions Needed

Of course, all the above logic can only be performed if simplifying assumptions are made. In particular, it is assumed at all levels of detail that the phase or sub-phase to which stretch is applied can be divided into the same number, types, and order of phases as the previous level phase which is being further divided. These simplifying assumptions are, of course, far-fetched, since, for example, variations of the effects of reservoirs and of the intertwine complexity are ignored at every step of the logic.

Furthermore, what if, as represented by Figure 11-9, columns with **more** than one elastic, two or more segments of the sequence in the parent system are assigned to the subsystem to which the principles of conversion of sequential conceptualization are applied? To handle this situation, does one simply assume that a number of objective-defined functions perceived within the subsystem equals the number of elastics in the column, whereby each elastic is expanded as if it were the **only** elastic in the column? This subject of different flow segments of an objective-defined function being assigned to the same subsystem was brought up in Chapter Eight, in connection with the subject of "functional elements appearing to belong to different objective-defined functions but which belong to the same one." Based on that discussion, stretching cannot be handled by treating each elastic as if it were the only elastic in the column. To handle such cases, the elementary principles outlined obviously need further elaboration, even when all other simplifying assumptions have been made.

Conclusion

In conclusion, the above principles are intended to provide an idea of how a new set of input-processing-output relationships can be perceived in any flow segment, thereby demonstrating the workings of another aspect of the relativity of the functional classification schemes.

SUPPLEMENT—PATTERNS OF FUNCTIONAL INTERACTION

This supplement is intended to expand the discussion of interactions between functions in different classes begun in Chapter Five. Both Chapters Five and Eight, and ideally Chapter Eleven as well, should be completed before reading this supplement.

Interactions can be defined as relationships involving inputs and outputs flowing between:

- one flow function and another, or one run and another, of the same objective-defined function;
- or, an objective-defined function and the system environment;
- or, two different objective-defined functions of the same system.

Interactions in any of the above senses may take place indirectly, through the system environment, and/or via reservoirs. To illustrate, interaction between the initiative coordination and baseline functions of the supply acquisition system occurs via the supplier, making this an environment-linked interaction pattern. As in Figure 11-10, whether interactions are internal or external may change, depending upon where the analyst draws the system boundary lines.

Points of Interaction

In interactions, the following may, for instance, occur:

- One objective-defined function sends framework inputs of some sort to another objective-defined function at some point along the line of flow of the receiving objective-defined function, e.g., guidelines in order to modify the modalities of execution of the receiving objective-defined function.
- One objective-defined function sends stimuli to another objective-defined function, to initiate, encourage, discourage, continue, or cancel operations in the receiving objective-defined function, e.g., due to a problem spotted.
- An objective-defined function provides another such function with transactional inputs during the course of its execution.
- One objective-defined function **seeks** framework inputs from another.

All, or any combination, of these events may occur during a given run through a "chain" or "series" of interactions. Interaction may occur along the "lines of flow" of objective-defined functions, or it may occur in the form of "sequential interaction," i.e., at the start-and end-points of two objective-defined functions. The possible "points" of interaction may be complicated because of the following:

- —reservoirs are involved;
- —the sending and receiving points are not singular, due to the lack of non-successive repetition of the same flow function type, i.e., because of the intertwine complexity;
- —more than one other objective-defined function is interacted with, starting from one sending or receiving point.

Figures 11-11 a to e show how interactions may look when these complicating factors are involved.

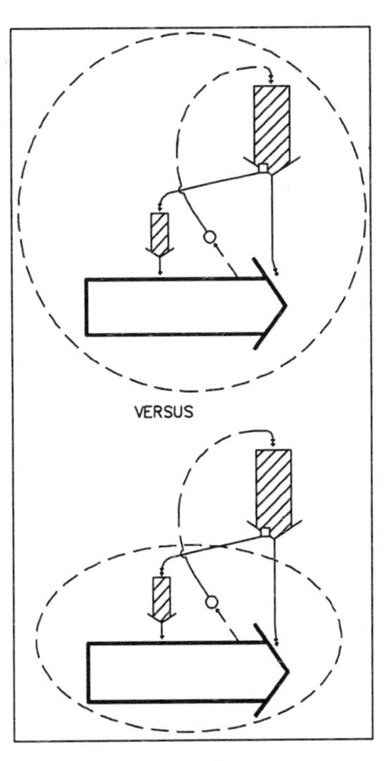

FIGURE 11-10: System boundary demarcations and externality of interaction. *This shows different perceptions of the same interaction as being internal or external, depending upon where the system boundary lines are drawn.*

Classification of Interaction Patterns

What follows is a scheme for classifying interactions based upon a number of independent dimensions. Before any of these are mentioned, it is worthwhile to observe that probably the most important factor used in organizing the discussion of interactions should be the order in which interactions appear in system flow, rather than any of the following dimensions.

- **Direct Versus Environment-Linked Dimension**: This dimension involves whether the interaction is direct, as opposed to environment-linked or reservoir-linked.
- **Degree of Distance From Operating Core Dimension**: This classifies interactions according to whether they are first-order, second-order, etc., based on distance from the operating core. Interactions between baseline and adaptation operations, and between basic framework and other adaptation operations, are classified as first-order.

- **One-Way Versus Two-Way Dimension**: This involves whether interaction is one-way or reciprocal, between an objective-defined function with itself, or between two different objective-defined functions of the system, or between an objective-defined function and an environmental entity or store.
- **Functional Significance Dimension**: This concerns the functional significance of the input/output involved in functional interaction, e.g., symbiotic, objective, competitive, offshoot, and feedback outputs form the basis for six classes of interaction. Also, the basis for classification may be the stimulus versus operand transactional input versus methodological framework input etc., distinctions.
- **Data Versus Physical Interchange Dimension**: This may simply be explained by an example. Thus, where personnel and machinery are involved in functional interaction, the interaction falls into the physical class.
- **Frequency Dimension**: The observer, looking at all the interactions going on in a system over a period of time, may see many different interactions in the same space of time. In fact, the same objective-defined functional sequence may undergo multiple, perhaps overlapping, runs during the observation period, just like a computer program with the "serially reusable" or "re-enterable" characteristic.

The frequency of interaction of any objective-defined function, or of any chain of objective-defined functions, varies with conditions inside and/or outside the system. Interaction may be regularized and/or variable in frequency, in the absolute sense of frequency and/or in the sense of relative frequencies of different interactions in the system. Where a chain of interaction is involved, the functional elements constituting the chain may each be part of it on a regularized or a variable basis. All these factors are important in describing the frequency of interaction.

Both baseline and adaptation interaction frequencies vary with conditions in the system environment. Also, these frequencies are related to one another. For example, a promotional campaign to sell the products of a business by the procurative or framework operations results in a growth in customer orders and a increase in baseline activity. On the other side of the picture, the higher the frequency of baseline interaction, the greater will also be the frequency of adaptation interactions, since increased baseline volume will, directly or indirectly, cause more problems of baseline flow to be referred for resolution to the adaptation operations; moreover, it may result in an increased rate of weardown of depreciable frameworks used in both baseline and adaptation operations.

FIGURE 11-11:
Factors which make interaction patterns more complex.
a) reservoir inserted into interaction;
b) multiple sending points because of intertwine of flow functions;
c) multiple receiving points because of intertwine of flow functions;
d) multiple destinations;
e) multiple originating points.

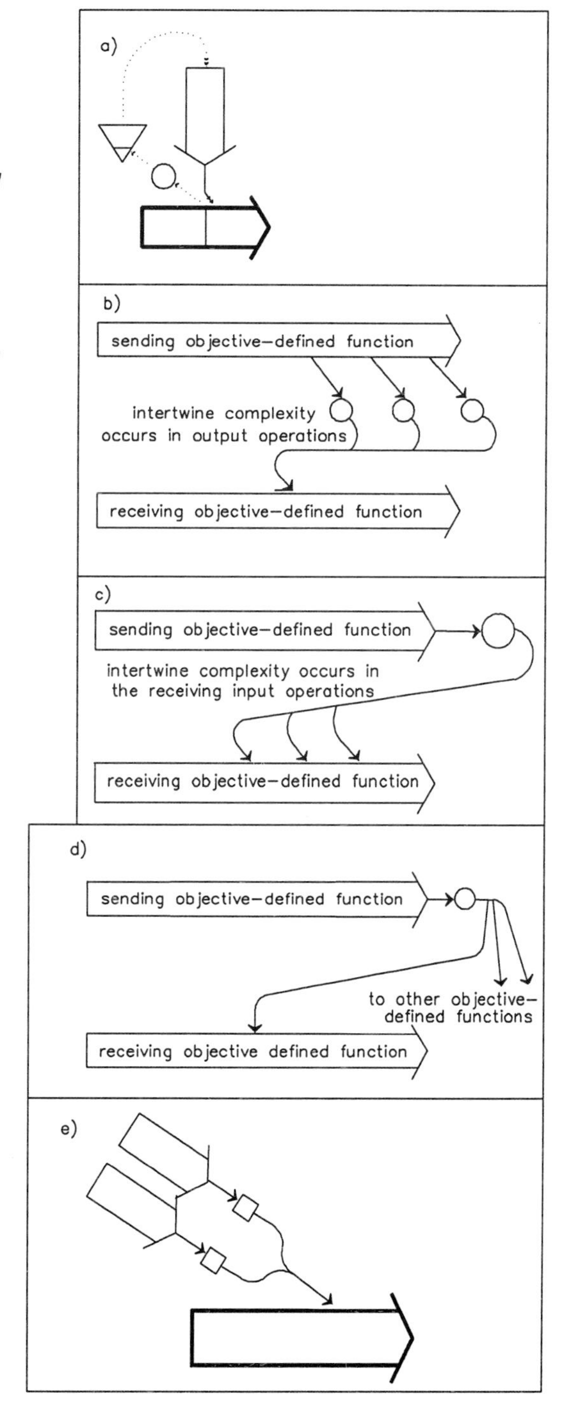

REFERENCES

Ahituv, N. and Neumann, S. Principles of Information Systems For Management. Dubuque, Iowa: W. C. Brown, 1982.

Ajenstat, J. Université du Québec, Montreal, Quebec, Canada.

Anthony, R. A. Planning and Control Systems: A Framework for Analysis. Boston, Mass.: Division of Research, Harvard Business School, Harvard University, 1965.

Ashby, W. R. Design for a Brain. 2nd edition. London: Chapman and Hall, 1960.

Baylin, E. "Logical System Structure," Journal of Systems Management. Cleveland, Ohio: Association for Systems Management, 31(8), August, 1980.

Baylin, E. "Functional Modeling of the Business Organization," Cybernetics and Systems: An International Journal. Washington, D.C.: Hemisphere Publishing Corp., 15(3—4), 1984.

Baylin, E. "A Clarification of Some Basic Application File Classification Concepts," Interface: The Computer Education Quarterly. Santa Cruz, California: Mitchell Publishing, 7(1), 1985.

Baylin, E. "A Comparative Review of System Charting Methods," Auerbach Information Management Series—System Development Management. Auerbach Publishers, #33-09-20, 1985.

Baylin, E. "Computer Systems Analyst's View of the Business Organization," Cybernetics and Systems: An International Journal. Washington, D.C.: Hemisphere Publishing Corp., 16:305—323, 1985.

Baylin, E. "System Diagramming Methods: Which Works Best?," Auerbach Information Management Series—Computer Programming Management. Auerbach Publishers, #14-01-30, 1986.

Baylin, E. "Identifying System Functions," International Journal of General Systems. Gordon and Breach Publishers, 12(1), 1986.

Baylin, E. "A Scheme for Handling Problems of Functional Class/ Time Orientation," working paper, 1985.

Baylin, E. "Analysis Versus Design: Similarities and Prototyping," working paper, 1985.

Baylin, E. "A Re-Explanation of the Cohesion Techniques in the Computer Literature," working paper, 1985.

Baylin, E. "Basic Principles of Identifying Subsystems," working paper, 1985.

Baylin, E. "The Structure-Flow Charting Method," Cybernetics and Systems: An International Journal. Washington, D.C.: Hemisphere Publishing Corp., 18:113—146, 1987.

Beer, S. Decision and Control. New York: John Wiley & Sons, 1967.

Blumenthal, S. C. Management Information Systems: A Framework for Planning and Development. Englewood Cliffs, N.J.: Prentice-Hall, 1969.

Boulding, K. E. "General Systems Theory: A Skeleton of Science," in P. P. Schoderbek (editor), Management Systems, 2nd edition. New York: John Wiley & Sons Inc., 1971.

Brown, D. B. and Herbanek, J. A. Systems Analysis for Applications Software Design. Oakland, California: Holden-Day, 1984.

Churchman, C. The Systems Approach. New York: Selacorte Press, 1968.

Database Design, Inc. "Action Diagrammer," Ann Arbor, Michigan.

Davis, G. B. Management Information Systems: Conceptual Foundations, Structure, and Development. New York: McGraw-Hill, 1974.

Davis, W. S. Systems Analysis and Design: A Structured Approach. Reading, Mass.: Addison-Wesley, 1983.

De Greene, K. B. Socio-Technical Systems: Factors in Design, Analysis and Management. Englewood Cliffs, N.J.: Prentice-Hall, 1969.

De Marco, T. Structured Analysis and System Specification. Englewood Cliffs, N.J.: Prentice-Hall, 1978 and 1979.

Deltak. Structured Design Series, 1981.

Fitzgerald, J., Fitzgerald, A.F., and Stallings, W.D. Jr. Fundamentals of Systems Analysis, 2nd edition. John Wiley & Sons, 1981.

Ein-Dor, P. and Segev, E. Managing Management Information Systems. Lexington, Mass.: D.C. Heath, 1978).

Fayol, H. General and Industrial Management, trans. C. Storrs. London: Sir Isaac Pitman & Sons, 1949.

Forrester, J. W. Industrial Dynamics. Cambridge, Mass.: M.I.T. Press, 1961.

Gane, C. and Sarson, T. Structured Systems Analysis: Tools and Techniques. Englewood Cliffs, N.J.: Prentice-Hall, 1979.

Gorry, G. A. and Scott Morton, M. S. "A Framework for Management Information Systems," Sloan Management Review, 13(1):55—70, 1971.

Hamilton, M. and Zeldin, S. Integrated Software Development System/Higher Order Software Conceptual Description. TR-3, Higher Order Software, Inc., 1976.

IBM. "HIPO: A Design Aid and Documentation Technique (GC20-185D)." White Plains, N.Y.: IBM Corp., 1974.

Martin, J. Program Design Which is Provably Correct. Carnforth, Lancashire, England: Savant Research Studies, 1983.

Martin, J. An Information Systems Manifesto. Englewood Cliffs, N.J.: Prentice-Hall, 1984.

Martin, J. and McClure, C. Structured Techniques for Computing. Englewood Cliffs, N.J.: Prentice-Hall, 1985.

Martin, J. and McClure, C. Charting Techniques for Analysts and Programmers. Englewood Cliffs, N.J.: Prentice-Hall, 1985.

Mintzberg, H. The Structuring of Organizations. Englewood Cliffs, N.J.: Prentice-Hall, 1979.

Nadler, G. Work Design: A Systems Concept, 2nd ed. Homewood, Illinois: G. Nadler©, Irwin Publishers, 1970.

Nadler, G. and The Planning, Design and Improvement Methods Group, Inc. The Planning and Design Approach. New York: John Wiley & Sons, 1981.

Open Systems Group. Systems Behaviour, 3rd ed. London: Harper and Row, 1981.

Orr, K. T. Structured Systems Development. New York: Yourdon Press, 1977.

Rakich, J. S., Longest, B. B., Jr., and Darr, K. Managing Health Service Organizations, 2nd edition. Philadelphia, PA: W. B. Saunders, 1985.

Riley, M. J. Management Information Systems (section of book written by W. M. Zani). San Francisco, California: Holden-Day, 1981.

Ross, D. T. "Structured Analysis for Requirements Definition," Proceedings of the Second International Conference on Software Engineering. New York: Association for Computing Machinery, 1976.

Schoderbek, C. G., Schoderbek, P. P. and Kefalas, A. G. Management Systems: Conceptual Considerations, 2nd. ed. Dallas, Texas: Business Publications Inc., 1980.

Schoderbek, P. P. Management Systems, 2nd ed. New York: John Wiley & Sons, 1971.

Senn, J. A. Information Systems in Management, 2nd ed. Belmont, California: Wadsworth, 1982.

Senn, J. A. Analysis and Design of Information Systems. New York: McGraw-Hill, 1984.

Shestowsky, B. J. "Hospital Management by Functional Areas: An Idea Whose Time Has Come," Dimensions in Health Service, Vol. xx, No. xx, Nov. 1988.

Simon, H. A. The New Science of Management Decisions. New York: Harper & Row, 1960.

Stevens, W., Myers, G. and Constantine, L. "Structured Design." IBM Systems Journal, 13(2):115—139, 1974.

Taylor, F. W. The Principles of Scientific Management. New York, N.Y.: W.W. Norton Co. Inc., 1911.

Thierauf, R. J. Effective Management Information Systems. Columbus, Ohio: Charles E. Merrill, 1984.

Thompson, J. D. Organizations in Action. New York: McGraw-Hill, 1967.

Warnier, J. D. Logical Construction of Programs. New York: Van Nostrand Reinhold Co., 1976.

Yourdon, E. and Constantine, L. Structured Design: Fundamentals of a Discipline of Computer Program and System Design. Englewood Cliffs, N.J.: Prentice-Hall, 1979.

INDEX

A

B

M

N

O

T

U

W

Y

Z